Applied Mathematical Sciences
Volume 180

For further volumes:
http://www.springer.com/series/34

Zeev Schuss

Nonlinear Filtering
and Optimal Phase Tracking

 Springer

Zeev Schuss
Department of Computer Sciences
School of Mathematical Science
Tel Aviv University
Tel Aviv
Israel

ISSN 0066-5452
ISBN 978-1-4614-0486-6   e-ISBN 978-1-4614-0487-3
DOI 10.1007/978-1-4614-0487-3
Springer New York Dordrecht Heidelberg London

Library of Congress Control Number: 2011939569

Mathematics Subject Classification (2010): 60G35, 93E10, 93E11, 94A05, 94A12, 94A14

Printed on acid-free paper

Springer is part of Springer Science+Business Media (www.springer.com)

Preface

Filtering theory is concerned with the extraction of information from noisy measurements of a random signal. The description of signals and their measurements as random processes started with Wiener's "yellow peril", his yellow-covered WWII monograph on the theory of random signals, in the first attempt to design a radar-guided antiaircraft gun. Wiener's theory, based on spectral characterization of continuous-time stochastic processes, set the tone of control and communications theory for a generation, and is still the dominant language of communications engineers. The state-space formulation of the filtering problem, which is a major departure from Wiener's spectral formulation, came about with the advent of Kalman's theory [75], [76], [77]. Both the signal and its measurement processes are described in this formulation in terms of stochastic differential equations rather than in terms of their spectral properties. Although originally formulated for linear problems [75], the state-space formulation led to the development of nonlinear filtering theory in terms of nonlinear stochastic differential equations [95], [164].

The probability density function of the random trajectories of stochastic differential equations is known to satisfy the Fokker–Planck and Kolmogorov forward and backward partial differential equations, respectively. The state-space formulation of nonlinear filtering theory in terms of nonlinear stochastic differential equations necessitated the derivation of analogous partial differential equations for the a posteriori probability density function of the signal, given its noisy measurements. A nonlinear stochastic integral and partial differential equation for the a posteriori density was derived by Kushner in 1964. Soon thereafter (1967), Zakai [164] derived a linear stochastic partial differential equation for an unnormalized version of the a posteriori density. Zakai's derivation relies heavily on measure theory in function space and in particular on Girsanov's theorem on change of measures induced by stochastic differential equations (see also [149]). Both Kushner's and Zakai's equations pose formidable mathematical challenges to analysts to extract from them useful information.

Many books, reviews, and research articles have been published on filtering; some are listed below in descending chronological order, but a Web search reveals many more.

- Books on filtering [11], [119], [151], [135], [71], [26], [56].
- Books on filtering and other topics [107], [8], [105], [15], [40], [133], [136].
- Books on filtering through Zakai's equation [6], [161], [74].
- Mathematical books on filtering [31], [55], [1].

The present book differs from all the above-mentioned texts in offering an analytical rather than measure-theoretical approach to the derivation and solution of the partial differential equations of nonlinear filtering theory. The basis for this approach is the discrete numerical scheme used in computer Monte–Carlo simulations of stochastic differential equations and Wiener's associated path integral representation of the transition probability density. The derivations rely on the Feynman–Kac formula and on the convergence of the discrete process to its continuous limit. Measure theory plays practically no role in this approach.

The analysis of the equations of diffusion and filtering theory, beginning with Einstein, through Fokker, Planck, Kolmogorov, Dynkin, Andronov, Vitt, Pontryagin, and Zakai, are equations of classical mathematical physics. They were studied by generations of chemists, physicists, engineers, economists, and mathematicians. Many of the mathematical tools of continuum theory (e.g., fluid dynamics, elasticity theory) and quantum mechanics, such as Feynman's path integrals, were adopted in this book for the analysis of these equations. The analysis of dynamical systems perturbed by small noise is central in this theory, in particular Kramers' problem of activated escape from an attractor. This is a typical singular perturbation problem for boundary value problems in partial differential equations. Boundary layer theory, the WKB (Wentzel–Kramers–Brillouin) method, and matched asymptotics, originally developed for aerodynamics and the turning point problem in quantum mechanics, turned out to be useful for studying stochastic differential equations. Their power is demonstrated in the asymptotic solution of the notorious cubic sensor problem in Section 4.2.

The problem of optimal phase tracking arises in many applications, beginning with the old problem of filtering phase- and frequency-modulated signals and up to modern radar, GPS, and cellular telephony. The problem is formulated in state space by highly nonlinear stochastic differential equations, most often driven by weak noise. The prevailing optimality criterion in filtering theory is minimum mean square estimation error (MMSEE), which results in the optimal estimator, which is the conditional average of the signal, given the measurements. This is the average of the signal with respect to the a posteriori probability density function, which is the solution of Kushner's nonlinear equation and can also be obtained from the solution of Zakai's equation. The prevailing methods of solution of these equations are the method of linearization, which produces various versions of the Kalman filter, simulations of the solution of Zakai's equation by means of various particle filters, large deviations theory, and singular perturbation analysis. All these methods result in phase trackers, such as phase-locked loops (PLL), which exhibit noise-induced losses of lock (phase slips), whose rate is one of the most important performance criteria for trackers. As the signal-to-noise ratio (SNR) decreases, the rate of phase slips rises sharply, causing a sharp performance threshold, beyond which the filtering quality deteriorates, rendering the tracker useless. The struggle to move the threshold farther away is as old as FM radio. There has been no significant improvement in this respect in the last sixty years.

The present book offers a new approach to the phase tracking problem. First, it proposes the maximal mean time to lose lock (MTLL) as the optimality criterion, rather than the mean square estimation error. The stochastic differential equations of the mathematical model now have a prescribed behavior at the boundaries of the lock domain. Thus optimizing the MTLL, a criterion often used in optimal stochastic control and differential games theory, leads to Zakai's equation with boundary conditions. These are derived here from path integrals, much as in modern diffusion theory and quantum mechanics (see, e.g., [137]), including new Zakai-type equations of nonlinear smoothing. Solutions are constructed by the classical and modern singular perturbation methods of applied mathematics, as mentioned above. Optimizing the MTLL produces the maximum a posteriori probability estimator and the Bellman–Mortensen minimum noise energy estimator. Numerical analysis of the optimization algorithm shows that the computational complexity of the optimization increases indefinitely as optimality is approached. Calculation of the MTLL in the benchmark first- and second-order models shows the incredible threshold improvement of 12 and 10.25 dB, respectively.

The benchmark examples worked out in this book are not aimed at introducing the reader to modern technology, but are rather well-known traditional, often outdated devices that illustrate the mathematical difficulties in analyzing nonlinear systems. They share these difficulties with modern systems, both continuous and discrete. Even as technology changes, many of the mathematical difficulties remain the same. Mastering the mathematical methodology offered in this book puts a powerful theoretical tool in the hands of the design engineer.

This book is based on lecture notes from a one-semester special topics course on stochastic processes and their applications that I taught many times to graduate students of mathematics, applied mathematics, physics, chemistry, computer science, electrical engineering, and other disciplines. The book contains exercises and worked-out examples aimed at illustrating the methods of mathematical modeling and performance analysis of phase trackers. The hands-on training in stochastic processes and nonlinear filtering, as my long teaching experience shows, consists in solving the exercises, without which understanding is only illusory. Students of nonlinear filtering and optimal tracking should have adequate training in the theory of stochastic processes and in the methods of applied mathematics (e.g., [137]).

The book is targeted at graduate and upper-level undergraduate students of the mathematical sciences and engineering (EE, ME, CS, physics, chemistry and CE, applied mathematics, mathematical finance, and so on). The presentation is based mostly on classical undergraduate probability theory, calculus, ordinary and partial differential equations of mathematical physics, and the asymptotic methods of applied mathematics [12]. A filtering course for students who studied chapters 1–6 and 10 of [137] can start at Chapter 3 of this book, whereas students with a less comprehensive background in stochastic processes should study Chapters 1 and 2 in this book or review Chapters 5, 6, and 10 of [137].

Acknowledgements Much of the material presented in this book is based on my collaboration with professors B.Z. Bobrovsky, Y. Steinberg, and R. Liptser and our graduate students R. Katzur, D. Ezri, E. Fishler, S. Landis, and many others. The scientific environment provided by Tel-Aviv University, my home institution, was conducive to interdisciplinary cooperation.

Zeev Schuss

Contents

List of Figures

List of Symbols

We use interchangeably $\langle \cdot \rangle$ and $\mathbb{E}(\cdot)$ to denote expectation (average) of a random variable, and $\mathbb{E}(\cdot \mid \cdot)$ and $\Pr\{\cdot \mid \cdot\}$ to denote conditional expectation and conditional probability, respectively.

$x,\ f(x)$	scalars - lowercase letters
$\boldsymbol{x}, \boldsymbol{f}(\boldsymbol{x})$	column vectors - bold lowercase letters
$\boldsymbol{A}$	matrices - bold uppercase letters
x_i	the ith element of the vector $\boldsymbol{x}$
$\boldsymbol{x}(\cdot)$	trajectory or function in function space
$J[\boldsymbol{x}(\cdot)]$	functional of the trajectory $\boldsymbol{x}(\cdot)$
$\hat{\boldsymbol{x}}(t)$	the estimator of $\boldsymbol{x}(t)$
$\boldsymbol{e}(t)$	the estimation error process: $\hat{\boldsymbol{x}}(t) - \boldsymbol{x}(t)$
$\boldsymbol{A}^T$	the transpose of $\boldsymbol{A}$
$\boldsymbol{A}^{-1}$	the inverse of $\boldsymbol{A}$
$\lvert \boldsymbol{x} \rvert$	the L_2 norm of $\boldsymbol{x}$: $\lvert \boldsymbol{x} \rvert = \sqrt{\boldsymbol{x}^T \boldsymbol{x}}$
V_x	the partial derivative of V with respect to x : $V_x = \dfrac{\partial V}{\partial x}$
$\mathfrak{f} = 2\pi\omega$	frequency
ω	angular frequency
$\Pr\{event\}$	the probability of $event$
$p(\boldsymbol{x})$	the probability density function of the vector $\boldsymbol{x}$
$\mathbb{E}(x)$	the expected value (expectation) of x
$\mathrm{Var}(x)$	the variance of x
$\boldsymbol{w}(t),\ \boldsymbol{v}(t),\ \tilde{\boldsymbol{v}}(t)$	vectors of independent Brownian motions
$\mathfrak{w},\ \mathfrak{w}(t)$	a continuous path
$\mathbb{R},\ \mathbb{R}^n$	the real line, n-dimensional Euclidean space
$L^2[a,b]$	square integrable functions on the interval $[a,b]$
$\mathbb{M}_{n,m}$	$n \times m$ real matrices

List of Acronyms

AGC	Automatic gain control
AM	Amplitude modulation
ARMA	Autoregressive moving average
BKE	Backward Kolmogorov equation
CDMA	Code division multiple access
CKE	Chapman–Kolmogorov equation
CMSEE	Conditional mean square estimation error
CNR	Carrier-to-noise ratio
DLL	Delay-locked loop
FM	Frequency modulation
FPE	Fokker–Planck equation
FPT	First passage time
GPS	Global positioning system
HJB	Hamilton–Jacobi–Bellman
i.i.d.	Independent identically distributed
MAP	Maximum a posteriori probability
MBM	Mathematical Brownian motion
MFPT	Mean first passage time
MNE	Minimum noise energy
MMSEE	Minimum mean square estimation error
MTLL	Mean time to lose lock
MSEE	Mean square estimation error
ODE	Ordinary differential equation
PDE	Partial differential equation
pdf	Probability density function
PDF	Probability distribution function
PLL	Phase-locked loop
PM	Phase modulation
radar	Radio detection and ranging
RMS	Root mean square
SAR	Synthetic aperture radar

SNR   Signal-to-noise ratio
SDE   Stochastic differential equation
DDR-SDRAM   Double data rate–synchronous dynamic random access memory
WKB   Wentzel–Kramers–Brillouin

Chapter 1
Diffusion and Stochastic Differential Equations

1.1 Classical Theory of Gaussian Noise

In the classical theory [124] one-dimensional zero-mean Gaussian noise $n(t)$ is a one-parameter family of real-valued Gaussian random variables such that for every sequence $0 = t_0 < t_1 < t_2 < \cdots < t_k$, the vector

$$n = \begin{pmatrix} n(t_1) \\ n(t_2) \\ \vdots \\ n(t_k) \end{pmatrix}$$

is zero-mean Gaussian with covariance matrix σ, given by

$$\sigma^{i,j} = \mathbb{E}n(t_i)n(t_j) \qquad (i, j = 0, 1, 2, \ldots, k). \tag{1.1}$$

The joint probability density function (pdf) of n is

$$\Pr\{n = x\} = p_n(x_1, t_1; x_2, t_2; \ldots, x_k, t_k)$$

$$= \frac{1}{(2\pi \det \sigma)^{k/2}} \exp\left\{-\frac{1}{2}x^T \sigma^{-1} x\right\}, \tag{1.2}$$

where $x^T = (x_1, x_2, \ldots, x_k)$. If the noise is uncorrelated, that is, if $n(t_i)$ is independent of $n(t_j)$ for $i \neq j$, then σ is a diagonal matrix with $\sigma^{i,i} = \mathrm{Var}[n(t_i)]$. This case is easy to simulate on a computer, because $n(t_i) \sim \mathcal{N}(0, \sigma^{i,i})$, that is, in a Monte–Carlo simulation we sample the random component $n(t_i)$ of the vector n, independently of all others, from the normal distribution $\mathcal{N}(0, \sigma^{i,i})$.

The Gaussian process $n(t)$ is *stationary* if its mean $m(t)$ is constant (e.g., $m(t) = 0$) and its covariance matrix $\sigma^{i,j}$ has the form $\sigma^{i,j} = \mathbb{E}n(t_i)n(t_j) =$

Z. Schuss, *Nonlinear Filtering and Optimal Phase Tracking*, Applied
Mathematical Sciences 180, DOI 10.1007/978-1-4614-0487-3_1,
© Springer Science+Business Media, LLC 2012

$R_n(t_j - t_i)$, where the *autocorrelation* $R_n(\tau)$ is a function or a generalized function [103], [73]. In this case,

$$p_{\boldsymbol{n}}(x_1, t_1; x_2, t_2; \ldots, x_k, t_k) = p_{\boldsymbol{n}}(x_1, 0; x_2, t_2 - t_1; \ldots, x_k, t_k - t_1).$$

The Fourier transform of the autocorrelation function,

$$S_n(\omega) = \int_{-\infty}^{\infty} R_n(\tau) e^{-i\omega\tau} \, d\tau, \tag{1.3}$$

is the *power-spectral density function* of the stationary process $n(t)$. The Fourier inversion formula gives

$$R_n(\tau) = \frac{1}{2\pi} \int_{-\infty}^{\infty} S_n(\omega) e^{i\omega\tau} \, d\omega. \tag{1.4}$$

For example, if $S_n(\omega) = 1$ (white spectrum), then $R_n(\tau) = \delta(\tau)$ (Dirac's function). Noise with white spectral density is called *white noise*.

1.1.1 Classical White Noise and Langevin's Equation

A modern construction of a classical Gaussian noise with a given autocorrelation function begins with a model of a particle in a potential field (e.g., an electron in a resistor), coupled linearly to a bath of harmonic oscillators (see [61] for references). The oscillators may represent, for example, the vibrations of the atoms in a resistor or a crystal. Consider, for example, the Hamiltonian

$$\mathcal{H} = \frac{p^2}{2M} + V(x) + \mathcal{H}_{\text{bath}}(x, \boldsymbol{q}, \boldsymbol{p}),$$

where

$$\mathcal{H}_{\text{bath}} = \frac{1}{2} \sum_{i=1}^{N} m_i \left[\dot{q}_i^2 + \omega_i^2 \left(q_i + \frac{C_i}{m_i \omega_i^2} x \right)^2 \right].$$

Here $\boldsymbol{q} = (q, q_2, \ldots, q_N)^T$ is the vector of displacements of the oscillators, $\boldsymbol{p} = (p_1, p_2, \ldots, p_N)^T$ is the vector of their momenta, m_i are their masses, and ω_i are their frequencies. The displacement of the particle is x, its momentum is p, its mass is M, and C_i are coupling constants. Although each oscillator may perturb the particle only weakly, the combined effect of all the bath modes on the particle motion may be significant. The coupling to the bath can cause strong dissipation and strong fluctuations of the particle's trajectory.

The equations of motion are given by

$$\dot{x} = \frac{\partial \mathcal{H}}{\partial p} = \frac{p}{M}, \quad \dot{p} = -\frac{\partial \mathcal{H}}{\partial x} = -V'(x) - \sum_{i=1}^{N} C_i \left(q_i + \frac{C_i}{m_i \omega_i^2} x \right),$$

$$m_i \dot{q}_i = \frac{\partial \mathcal{H}}{\partial p_i} = p_i, \quad \dot{p}_i = -\frac{\partial \mathcal{H}}{\partial q_i} = -m_i \omega_i^2 \left(q_i + \frac{C_i}{m_i \omega_i^2} x \right),$$

and their solutions for the motion of the forced harmonic oscillators give the generalized Langevin equation

$$\ddot{x} + \frac{\Gamma}{M} \int_0^t \varphi_N(t-s) \dot{x}(s)\, ds + \frac{V'(x)}{M} = \frac{\Xi_N(t)}{M}, \tag{1.5}$$

where

$$\Gamma \varphi_N(t) = \sum_{i=1}^{N} \frac{C_i^2}{m_i \omega_i^2} \cos \omega_i t,$$

$$\Xi_N(t) = -\sum_{i=1}^{N} C_i \left[\left(q_i(0) + \frac{C_i}{m_i \omega_i^2} x(0) \right) \cos \omega_i t + \frac{\dot{q}_i(0)}{\omega_i^2} \sin \omega_i t \right]. \tag{1.6}$$

Setting

$$U(x) = \frac{V(x)}{M}, \quad \gamma = \frac{\Gamma}{M}, \quad \xi_N(t) = \frac{\Xi_N(t)}{M}, \tag{1.7}$$

we can rewrite (1.8) as the *generalized Langevin equation*

$$\ddot{x} + \gamma \int_0^t \varphi_N(t-s) \dot{x}(s)\, ds + U'(x) = \xi_N(t). \tag{1.8}$$

If we assume that at time $t = 0$ the bath is in thermal equilibrium, such that the initial bath distribution in phase space is given as

$$\Pr\{q(0) = q, \dot{q}(0) = \dot{q}\} = C \exp\left\{ -\frac{\mathcal{H}_{\text{bath}}(x, q, p)}{k_B T} \right\}, \tag{1.9}$$

where C is a normalization constant, then $\xi_N(t)$ is a zero-mean stationary Gaussian process with autocorrelation function

$$\mathbb{E} \xi_N(t_1) \xi_N(t_2) = \frac{\gamma k_B T}{M} \varphi_N(|t_1 - t_2|), \tag{1.10}$$

which is called Einstein's *generalized fluctuation-dissipation* principle. In Einstein's original fluctuation-dissipation principle, $\varphi(\tau) = \delta(\tau)$. The spectral density of the noise $\xi_N(t)$ is given by

$$S_N(\omega) = \frac{\pi}{2} \sum_{i=1}^{N} \frac{C_i x(0)}{m_i \omega_i^2} \delta(\omega - \omega_i),$$

so the memory kernel can be represented as

$$\varphi_N(t) = \frac{2}{\pi} \int_{-\infty}^{\infty} \frac{S_N(\omega)}{\omega} \cos \omega t \, d\omega$$

with the Laplace transform

$$\hat{\varphi}_N(s) = \int_0^{\infty} e^{-st} \varphi_N(t) \, dt = \frac{2}{\pi} \int_{-\infty}^{\infty} \frac{S_N(\omega)}{\omega} \frac{s}{s^2 + \omega^2} \, d\omega.$$

Assuming that the frequencies ω_i form a dense set in $\mathbb{R}_+$ and that the coefficients of the random initial conditions are chosen in an appropriate way, the noise $\xi_N(t)$ can be made to converge to any stationary Gaussian process with sufficiently "nice" power spectral density function $S(\omega)$ as N is increased to infinity. In Langevin's original equation $\varphi(t) = \delta(t)$, that is, (1.8) is reduced to *Langevin's equation*

$$\ddot{x} + \gamma \dot{x} + U'(x) = \xi(t) \tag{1.11}$$

with

$$R_\xi(\tau) = \frac{k_B T}{M} \delta(\tau), \quad S_\xi(\omega) = \frac{k_B T}{M}, \quad \gamma = \frac{6\pi a \eta}{M}. \tag{1.12}$$

The friction coefficient γ has dimension of frequency, the radius of the heavy particle is a, the dynamical viscosity is η, and the factor $6\pi a \eta$ is Stokes's formula for the hydrodynamical drag coefficient on a spherical particle moving slowly in a viscous fluid. The noise $\xi(t)$ is Gaussian white noise.

1.1.2 Classical Theory of Brownian Motion

In his theory of Brownian motion, Einstein postulated in 1905 [39] that the disordered motion of a microscopic particle immersed in fluid is the manifestation of random collisions of the particle with the molecules of the surrounding medium (gas, liquid, solid). He derived the diffusion equation for the density $\rho(x, t)$ of the particles,

$$\frac{\partial \rho(x,t)}{\partial t} = D \frac{\partial^2 \rho(x,t)}{\partial x^2}, \tag{1.13}$$

and also derived the expression

$$D = \frac{k_B T}{6\pi a \eta} \tag{1.14}$$

for the diffusion coefficient D in terms of the absolute temperature T (in Kelvin's scale), Boltzmann's constant k_B, the particle's radius a, and the dynamical viscosity coefficient η [137, Section 1.1]. He interpreted the normalized solution of (1.13),

$$p(x,t) = \frac{\rho(x,t)}{\displaystyle\int_{\mathbb{R}} \rho(x,t)\,dx} = \frac{1}{\sqrt{4\pi Dt}} \exp\left\{-\frac{x^2}{4Dt}\right\}, \tag{1.15}$$

as the transition probability density function of a Brownian particle from the point $x = 0$ at time 0 to the point x at time t. Specifically, if $x(t)$ is the displacement of the particle at time t, then for any interval $A \subset \mathbb{R}$,

$$\Pr\{x(t) \in A\} = \int_A p(x,t)\,dx. \tag{1.16}$$

It follows that the mean value

$$\mathbb{E}x(t) = \int_{-\infty}^{\infty} x p(x,t)\,dx$$

and the variance of the displacement are respectively

$$\mathbb{E}x(t) = 0, \quad \mathbb{E}x^2(t) = 2Dt. \tag{1.17}$$

Obviously, if the particle starts at $x(0) = x_0$, then

$$\mathbb{E}[x(t) \mid x(0) = x_0] = x_0, \tag{1.18}$$

$$\mathrm{Var}[x(t) \mid x(0) = x_0] = \mathbb{E}[(x(t) - x_0)^2 \mid x(0) = x_0] = 2Dt.$$

Now, using (1.14) in (1.18), the mean square displacement of a Brownian particle along the x-axis is found as

$$\sigma = \sqrt{t}\sqrt{\frac{k_B T}{3\pi a \eta}}. \tag{1.19}$$

1.1.3 The Velocity Process and Colored Noise

Equation (1.19) indicates that the mean square displacement of a Brownian particle at times t not too short (compared with the mean free time between collisions of the Brownian particle with the molecules of the surrounding medium) is proportional to the square root of time. According to the Waterston–Maxwell equipartition theorem (see Wikipedia), the root mean square (RMS) velocity $\bar{v} = \sqrt{\langle v^2 \rangle}$ of a suspended particle should be determined by the equation

$$\frac{M}{2}\bar{v}^2 = \frac{3k_B T}{2}. \tag{1.20}$$

Each component of the velocity vector has the same variance, so that

$$\frac{M}{2}\bar{v}^2_{x,y,z} = \frac{k_B T}{2}, \tag{1.21}$$

which is the one-dimensional version of (1.20).

In 1908, Langevin [99] offered an alternative approach to Einstein's model of the Brownian motion. He assumed that the dynamics of a free Brownian particle is (1.19) with $U'(x) = 0$, that is, that the particle's motion is governed by the frictional force $-\gamma \dot{x}(t)$ and by the fluctuational force $\xi(t)$ described in Section 1.1.1 (i.e., with (1.12)). Setting $v = \dot{x}$ and multiplying (1.11) by x, he obtained

$$\frac{1}{2}\frac{d^2}{dt^2}x^2 - v^2 = -\frac{1}{2}\gamma\frac{d}{dt}x^2 + \xi x, \tag{1.22}$$

where γ is the damping coefficient (1.12). Averaging under the assumption that the fluctuational force $\xi(t)$ and the displacement of the particle $x(t)$ are mutually independent, he obtained, using (1.21), that

$$\frac{1}{2}\frac{d^2}{dt^2}\langle x^2 \rangle + \frac{1}{2}\gamma\frac{d}{dt}\langle x^2 \rangle = \frac{k_B T}{M}. \tag{1.23}$$

The solution is given by $d\langle x^2 \rangle/dt = 2k_B T/\gamma M + Ce^{-\gamma t}$, where C is a constant. The time constant in the exponent is about 10^{-8} sec, so the mean square speed decays on a time scale much shorter than that of observations. He concluded that $\langle x^2 \rangle - \langle x_0^2 \rangle = (2k_B T/\gamma M)t$. This, in turn (see (1.18)), he argued, implies that the diffusion coefficient is given by $D = k_B T/\gamma M$, as in Einstein's equation (1.14).

The conditional probability distribution function (PDF) of the velocity process of a Brownian particle, given that it started with velocity v_0 at time $t = 0$, is defined as $P(v, t \mid v_0) = \Pr\{v(t) < v \mid v_0\}$, and the conditional probability density function is defined by

$$p(v, t \mid v_0) = \frac{\partial P(v, t \mid v_0)}{\partial v}.$$

The conditioning implies that the initial condition for the pdf is $p\,(v, t \,|\, v_0) \to \delta(v - v_0)$ as $t \to 0$.

The solution of the Langevin equation (1.11) for a free Brownian particle is given by

$$v(t) = v_0 e^{-\gamma t} + \int_0^t e^{-\gamma(t-s)} \xi(s)\, ds. \tag{1.24}$$

To make sense of the stochastic integral in (1.24), we make a short mathematical digression on the definition of integrals of the type $\int_0^t g(s)\xi(s)\, ds$, where $g(s)$ is a deterministic square-integrable function. Such an integral is defined as the limit of finite Riemann sums of the form

$$\int_0^t g(s)\xi(s)\, ds = \lim_{\Delta s_i \to 0} \sum_i g(s_i)\xi(s_i)\, \Delta s_i, \tag{1.25}$$

where $0 = s_0 < s_1 < \cdots < s_N = t$ is a partition of the interval $[0, t]$. According to the assumptions about the noise $\xi(t)$, if we choose $\Delta s_i = \Delta t = t/N$ for all i, the increments $\Delta b_i = \xi(s_i)\,\Delta s_i$ are independent identically distributed (i.i.d.) random variables. Einstein's observation that the RMS velocity on time intervals of length Δt are inversely proportional to $\sqrt{\Delta t}$ implies that the normally distributed increments $\xi(s_i)\,\Delta s_i$ have zero-mean and their covariance matrix is $\langle \Delta b_i \Delta b_j \rangle = 2\gamma k_B T \delta_{ij}/M$. We therefore write

$$\Delta b_i \sim \mathcal{N}\left(0, \frac{2\gamma k_B T}{M} \Delta t\right).$$

It follows that

$$g(s_i)\xi(s_i)\,\Delta s_i \sim \mathcal{N}\left(0, |g(s_i)|^2 \frac{2\gamma k_B T}{M} \Delta t\right),$$

so that

$$\sum_i g(s_i)\xi(s_i)\,\Delta s_i \sim \mathcal{N}(0, \sigma_N^2),$$

where

$$\sigma_N^2 = \sum_i |g(s_i)|^2 \frac{2\gamma k_B T}{M} \Delta s_i.$$

As $\Delta t \to 0$, we obtain

$$\lim_{\Delta t \to 0} \sigma_N^2 = \frac{2\gamma k_B T}{M} \int_0^t g^2(s)\, ds$$

and $\int_0^t g(s)\xi(s)\,ds \frown \mathcal{N}(0,\sigma^2)$, where

$$\sigma^2 = \frac{2\gamma k_B T}{M} \int_0^t g^2(s)\,ds. \tag{1.26}$$

By considering Riemann sums of the form (1.25), we find that the cross-correlation between the integrals of two deterministic functions with respect to the white noise $\xi(t)$ is the expectation (average) of the Gaussian variables

$$\mathbb{E} \int_0^{t_1} f(s_1)\xi(s_1)\,ds_1 \int_0^{t_2} g(s_2)\xi(s_2)\,ds_2 = \frac{2\gamma k_B T}{M} \int_0^{t_1 \wedge t_2} f(s)g(s)\,ds, \tag{1.27}$$

where $t_1 \wedge t_2 = \min\{t_1, t_2\}$. We note that for the Heaviside function

$$H(t) = \begin{cases} 0 & \text{for } t < 0, \\ 1 & \text{for } t \geq 0, \end{cases}$$

the following identities hold:

$$\frac{\partial t_1 \wedge t_2}{\partial t_1} = H(t_2 - t_1), \qquad \frac{\partial^2 t_1 \wedge t_2}{\partial t_2 \partial t_1} = \delta(t_2 - t_1),$$

$$\frac{\partial t_1 \wedge t_2}{\partial t_1}\frac{\partial t_1 \wedge t_2}{\partial t_2} = H(t_2 - t_1)H(t_1 - t_2) = 0.$$

Therefore (1.27) means that

$$\langle \xi(s_1)\,ds_1 \xi(s_2)\,ds_2 \rangle = \frac{2\gamma k_B T}{M}\delta(s_1 - s_2)\,ds_1\,ds_2. \tag{1.28}$$

To interpret (1.24), we use (1.26) with $g(s) = e^{-\gamma(t-s)}$ and obtain

$$\sigma^2 = \frac{k_B T}{M}\left(1 - e^{-2\gamma t}\right). \tag{1.29}$$

Returning to the velocity $v(t)$, we obtain from the above considerations

$$v(t) - v_0 e^{-\gamma t} \frown \mathcal{N}\left(0, \sigma^2\right) \tag{1.30}$$

with σ^2 given by (1.29). The velocity process $v(t)$ is called *colored noise* or the *Ornstein–Uhlenbeck process* [137, Section 1.4].

We use (1.24) to calculate the autocorrelation function and spectral density of the velocity process $v(t)$ of a free Brownian particle. Equation (1.24) gives

$$R_v(\tau) = \lim_{t \to \infty} R_v(t, t + \tau) = \lim_{t \to \infty} \mathbb{E}v(t)v(t + \tau)$$

$$= \lim_{t \to \infty} \mathbb{E} \int_0^t e^{-\gamma(t-s)}\xi(s)\,ds \int_0^{t+\tau} e^{-\gamma(t+\tau-s)}\xi(s)\,ds, \tag{1.31}$$

so assuming $\tau > 0$ and setting $t_1 = t, t_2 = t + \tau,\ f(s_1) = e^{-\gamma(t_1-s_1)}, g(s_2) = e^{-\gamma(t_2-s_2)}$ in (1.27), we obtain

$$R_v(\tau) = \frac{k_B T}{M} e^{-\gamma\tau}.$$

For $\tau < 0$ we obtain

$$R_v(\tau) = \frac{k_B T}{M} e^{\gamma\tau},$$

so that the autocorrelation function of the stationary colored noise is

$$R_v(\tau) = \frac{k_B T}{M} e^{-\gamma|\tau|}. \tag{1.32}$$

Using the identity

$$\lim_{\gamma\to\infty} \int_0^\infty \phi(t_2)\gamma e^{-\gamma|t_2-t_1|}\, dt_2 = 2\phi(t_1) \text{ for } t_1 > 0 \tag{1.33}$$

for all test functions $\phi(t)$ in $\mathbb{R}^+$, we can approximate $R_v(\tau)$ for large γ by

$$R_v(\tau) = \frac{k_B T}{\gamma M}\delta(\tau). \tag{1.34}$$

The *correlation time* of the process is the time $R_v(\tau)$ decays by a factor of e. For colored noise the decay time is $\tau_{\text{decay}} = 1/\gamma$.

The spectral density of colored noise is given by

$$S_v(\omega) = \frac{k_B T}{M} \int_{-\infty}^\infty e^{-i\omega\tau} e^{-\gamma|\tau|}\, d\tau = \frac{k_B T}{M} \frac{2\gamma}{\gamma^2 + \omega^2}, \tag{1.35}$$

which is called the *Lorentzian power spectrum*. The *bandwidth* ω_B of colored noise is defined as the frequency at which the power-spectral density is reduced to half its maximal height, that is, $\omega_B = \gamma = 1/\tau_{\text{decay}}$. For large γ and $\omega \ll \gamma$ the power spectral density of the colored noise can be approximated by

$$S_v(\omega) \approx \frac{2k_B T}{\gamma M}, \tag{1.36}$$

that is, for short correlation time, the spectral density is practically constant in a wide range of frequencies, which gives colored noise the name *wideband noise*.

1.1.4 The Origin of Thermal Noise in Resistors

In 1928 Johnson [72] measured the random fluctuating voltage across a resistor and found that the power-spectral density function of the random electromotive force produced by the resistor was white with spectral height proportional to resistance and temperature. A theoretical derivation of this result was presented by Nyquist [121] in the same issue of *Physical Reviews*. Here, we derive Nyquist's result for an ionic solution, where the ions are assumed identical independent Brownian particles in a uniform electrostatic field.

The Ramo–Shockley theorem [140], [130] relates the microscopic motion of mobile charges in a domain (open connected set) D to the electric current measured at any given electrode. For a single charge q, moving with velocity v at location x, the instantaneous current at the jth electrode is given by

$$I_j = q v \cdot \nabla u_j(x), \tag{1.37}$$

where u_j is the solution of the Laplace equation

$$\nabla \cdot \left[\varepsilon(x) \nabla u_j \right] = 0 \ \text{ for } x \in D \tag{1.38}$$

with the boundary conditions

$$u_j \Big|_{\partial D_j} = 1 \quad u_j \Big|_{\partial D_i} = 0, \quad (i \neq j), \tag{1.39}$$

where ∂D_j is the boundary of the jth electrode. In addition, the normal component of the field is continuous at dielectric interfaces [69],

$$\varepsilon_1 \frac{\partial u_j}{\partial n} - \varepsilon_2 \frac{\partial u_j}{\partial n} = 0,$$

where derivatives are taken in the normal direction to the interface, and ε_1 and ε_2 are the dielectric coefficients on the two sides. In the case of many particles, due to superposition, the total current recorded at the jth electrode is given by $I_j = \sum_i q_i v_i \cdot \nabla u_j(x_i)$. Consider, for example, an infinite conducting parallel-plate capacitor, shorted through an ammeter. The separation between the plates is L, and a point charge q is moving with instantaneous velocity $v(t)$ in a direction perpendicular to the electrodes. The solution of (1.38), (1.39) is $u(x) = x/L$ for $0 \leq x \leq L$. Therefore, according to (1.37), the current on the ammeter is

$$I = \frac{q v}{L}. \tag{1.40}$$

Now an electrostatically neutral resistor (e.g., electrolytic solution of concentration ρ) is placed between the plates of the capacitor, and a voltage V is maintained across the plates. We assume, for simplicity, that the positive charges in the resistor have charge q and the negative ones have charge $-q$ and that they have the same constant diffusion coefficient. Thus, we do not distinguish between positive and negative charges, because they make the same contribution to the current and to the noise. Under these conditions the electrostatic field E on the resistor is uniform.

The average motion $\bar{x}(t) = \mathbb{E}x(t)$ of a charged particle in the resistor is described by

$$\ddot{\bar{x}}(t) + \gamma \dot{\bar{x}}(t) = \frac{qE}{M}. \tag{1.41}$$

In the steady state, the velocity is given by $\lim_{t \to \infty} \dot{\bar{x}}(t) = qE/\gamma M$, so that the steady-state average current per particle is given by $\bar{I}_p = q\dot{\bar{x}}(t)/L = q^2 E/\gamma ML$. The voltage across the capacitor is $V = EL$, so that $\bar{I}_p$ can be written as $\bar{I}_p = q^2 V/\gamma ML^2$. If N identical charges are uniformly distributed between the plates of the capacitor with density (per unit length) $\rho = N/L$, the average current is given by $\bar{I} = Nq^2 V/\gamma ML^2 = q^2 \rho V/\gamma ML$, so that Ohm's law gives the resistance

$$R = \frac{V}{\bar{I}} = \frac{\gamma ML}{q^2 \rho}. \tag{1.42}$$

Thus the resistance of the one-dimensional ionic solution is proportional to the friction coefficient, to the mass of the moving charge, and to the length of the resistor, and inversely proportional to the density of the charges and to the square of the particle's charge.

Setting $\Delta x(t) = x(t) - \bar{x}(t)$, (1.11) takes the form $\Delta \ddot{x}(t) + \gamma \Delta \dot{x}(t) = \xi$. Writing the noisy current per particle as $I_p(t) = \bar{I}_p + \Delta I_p(t)$, we find from (1.40)

$$I_p = \frac{d}{dt}Q(t) = -\frac{q}{L}\frac{d}{dt}\Delta x(t) = \frac{qv}{L}, \tag{1.43}$$

that $\Delta I_p(t) = q\Delta \dot{x}(t)/L$.

Thus, according to Section 1.1.3, the autocorrelation function $R_{I_p}(\tau)$ of current fluctuations per particle is given by

$$R_{I_p}(\tau) = \frac{q^2}{L^2}\frac{k_B T}{M}e^{-\gamma|\tau|} \tag{1.44}$$

(see (1.32)). If γ is large, as is the case in liquids and solids, we obtain from (1.34) that

$$R_{I_p}(\tau) \approx \frac{2q^2}{L^2}\frac{k_B T}{\gamma M}\delta(\tau). \tag{1.45}$$

For N identical noninteracting particles eqs. (1.45) and (1.42) give

$$R_I(\tau) = NR_{I_p}(\tau) = \frac{2k_BT}{R}\delta(\tau). \tag{1.46}$$

It follows that the power spectrum of the current fluctuations is given by $S_I(\omega) = 2k_BT/R$. The power spectrum of the voltage fluctuations is given by $S_V(\omega) = R^2S_I(\omega) = 2k_BTR$, which is Nyquist's formula for the random electromotive force of a resistor [121], [97].

1.1.5 White Noise and the Wiener Process

We consider now a lattice $t_i = i\,\Delta t$ for a fixed Δt and $-\infty < i < \infty$. We define for $t_i \leq t < t_{i+1}$,

$$n_{\Delta t}(t) = \frac{1}{\sqrt{\Delta t}}n_i,$$

where $n_i \frown \mathcal{N}(0,1)$ are independent identically distributed (i.i.d.) standard normal (Gaussian) variables. Setting

$$\mathbf{1}_{[a,b]}(x) = \begin{cases} 1 \text{ if } a \leq x \leq b, \\ 0 \text{ otherwise,} \end{cases}$$

the autocorrelation of this process for $t_i \leq t < t_{i+1}$ on the lattice can be written as

$$R_{n_{\Delta t}}(t+\tau,t) = \mathbb{E}n_{\Delta t}(t+\tau)n_{\Delta t}(t) = \frac{1}{\Delta t}\mathbf{1}_{[-\Delta t/2+t-t_i,\Delta t/2+t-t_i]}(\tau).$$

It follows that

$$\int_{-\infty}^{\infty} R_{n_{\Delta t}}(t+\tau,t)\,d\tau = 1,$$

so that

$$R_{n_{\Delta t}}(t+\tau,t) \to \delta(\tau) \text{ as } \Delta t \to 0.$$

According to the definition (1.3), the power-spectral density of $n_{\Delta t}(t)$ is

$$S_{n_{\Delta t}}(t,\omega) = \frac{\sin\omega\Delta t/2}{\omega\Delta t/2}e^{-i\omega(t-t_i)}$$

$$= \frac{\sin\omega\Delta t/2}{\omega\Delta t/2}(1 + O(\omega\Delta t)) \to 1 \text{ as } \Delta t \to 0.$$

Thus, when $n_{\Delta t}(t)$ converges in some sense to a limit $n(t)$ as $\Delta t \to 0$, the noise $n(t)$ is called *a δ-correlated Gaussian white noise*, which is stationary. Obviously, $n(t)$ cannot be a function, because it should be infinite everywhere. However, its discrete integral $w_{\Delta t}(t)$, defined on the lattice by the scheme

$$w_{\Delta t}(t_{i+1}) = w_{\Delta t}(t_i) + \Delta t n(t_{i+1}), \quad w_{\Delta t}(0) = 0, \tag{1.47}$$

can converge to a limit function $w(t)$ as $\Delta t \to 0$. If we interpolate the values of $w_{\Delta t}(t_i)$ linearly between the lattice points, the limit function $w(t)$ can even be continuous. The function $w(t)$ is called the *Wiener process* or *Brownian motion*. Obviously, its increments

$$\Delta w_{\Delta t}(t_i) = w_{\Delta t}(t_i) - w_{\Delta t}(t_{i-1}) = \Delta t n_i \sim \mathcal{N}(0, \Delta t) \tag{1.48}$$

are independent zero-mean Gaussian variables with variance Δt. It follows that on the lattice,

$$\mathbb{E} w_{\Delta t}(t_i) = 0, \quad \mathbb{E} w_{\Delta t}^2(t_i) = t_i, \tag{1.49}$$

so that the Wiener process is not stationary.

There are several theoretical issues that have to be clarified in order to give probabilistic meaning to the limits $n(t)$ and $w(t)$. While there is no difficulty in calculating the probability density function of a trajectory of $w_{\Delta t}(t)$ or of $n_{\Delta t}(t)$, for example, on the lattice, the concept of a probability density function of an entire continuous trajectory is not as simple. For example, the probability density of sampling the trajectory $n_{\Delta t}(t_i) = x_i$ for $i = 1, 2, \ldots, k$ on the lattice is

$$\Pr\{n_{\Delta t}(t_1) = x_1, \ldots, n_{\Delta t}(t_k) = x_k\} = \left(\frac{\Delta t}{2\pi}\right)^{k/2} \exp\left\{-\frac{\Delta t}{2} \sum_{i=1}^{k} x_i^2\right\}. \tag{1.50}$$

It follows that the probability that the trajectory is contained in a finite strip $a_i \leq n_{\Delta t}(t_i) < b_i$ $(i = 1, 2, \ldots, k)$ is

$$\Pr\{a_1 \leq n_{\Delta t}(t_1) < b_1, a_2 \leq n_{\Delta t}(t_2) < b_2, \ldots, a_k \leq n_{\Delta t}(t_k) < b_k\}$$

$$= \left(\frac{\Delta t}{2\pi}\right)^{k/2} \int_{a_1}^{b_1} \exp\left\{-\frac{\Delta t x_1^2}{2}\right\} dx_1 \cdots \int_{a_k}^{b_k} \exp\left\{-\frac{\Delta t x_k^2}{2}\right\} dx_k$$

$$= \left(\frac{1}{2\pi}\right)^{k/2} \prod_{i=1}^{k} \int_{a_i \sqrt{\Delta t}}^{b_i \sqrt{\Delta t}} \exp\left\{-\frac{z^2}{2}\right\} dz \to 0 \text{ as } \Delta t \to 0. \tag{1.51}$$

Thus practically no trajectories of the white noise $n(t)$ are contained in any finite strip.

In contrast, the probability density of a trajectory of $w_{\Delta t}(t)$ on the lattice is given by

$$\Pr\{w_{\Delta t}(t_1) = x_1, w_{\Delta t}(t_2) = x_2, \ldots, w_{\Delta t}(t_k) = x_k\}$$

$$= \Pr\{\Delta w_{\Delta t}(t_1) = x_1, \Delta w_{\Delta t}(t_2) = x_2 - x_1, \ldots, \Delta w_{\Delta t}(t_k) = x_k - x_{k-1}\}$$

$$= \prod_{i=1}^{k} \frac{1}{\sqrt{2\pi(t_i - t_{i-1})}} \exp\left\{-\frac{(x_i - x_{i-1})^2}{2(t_i - t_{i-1})}\right\},$$

because $w_{\Delta t}(0) = 0$. It follows that the probability that a trajectory is contained in a finite strip $a_i \leq w_{\Delta t}(t_i) < b_i$ $(i = 1, 2, \ldots, k)$ is

$$\Pr\{a_1 \leq w_{\Delta t}(t_1) < b_1, a_2 \leq w_{\Delta t}(t_2) < b_2, \ldots, a_k \leq w_{\Delta t}(t_k) < b_k\}$$

$$= \int_{a_1}^{b_1} \int_{a_2}^{b_2} \cdots \int_{a_k}^{b_k} \prod_{i=1}^{k} \frac{dx_i}{\sqrt{2\pi(t_i - t_{i-1})}} \exp\left\{-\frac{(x_i - x_{i-1})^2}{2(t_i - t_{i-1})}\right\}. \qquad (1.52)$$

Keeping the points $0 = t_0 < t_1 < t_2 < \cdots < t_k$ fixed on a sequence of lattices such that $\Delta t \to 0$, equation (1.52) shows that the probability of the trajectories of the limit process $w(t)$ in the strip is independent of Δt, so that there is a finite probability of sampling trajectories in the strip even in the limit $\Delta t \to 0$, in contrast to (1.51).

The above examples of white noise and the Wiener process indicate that continuous-time models of random signals and their noisy measurements require the specification of relevant sample spaces of events and appropriate probability measures. The relevant mathematical formalism that can extract useful information from such models is diffusion theory and stochastic differential equations. The different stochastic differential equations assign different probabilities to random trajectories, which are the same continuous functions generated by different models. The driver of randomness in stochastic dynamics is classical white noise and its integral, Brownian motion. The objective of this chapter is to present a short summary of the basic notions, constructions, simulations, and methods of analysis of random trajectories and their probabilities (see [137, Chapter 1] for a more thorough treatment).

1.2 Mathematical Brownian Motion

To assign probabilities to random continuous trajectories, probability theory calls for a sample space, elementary events, events, and a probability measure on these events. The most natural choice of elementary events for the definition and construction of Brownian motion are its random paths.

1.2.1 The Space of Brownian Trajectories

A continuous-time *random process* (or *stochastic process*) $x(t, \mathfrak{w})$: $\mathbb{R}^+ \times \Omega \to \mathbb{R}$ is a function of two variables: a real variable t, usually interpreted as time, and a variable $\mathfrak{w}$ that varies in a *probability space* (or *sample space*) Ω, in which *events* are defined. More generally, the random process $x(t, \mathfrak{w})$ can take values in a set X, called *state space*, such as the real line $\mathbb{R}$, or the Euclidean space $\mathbb{R}^d$, or any other set. For each $\mathfrak{w} \in \Omega$ the stochastic process is a function of t, called a *trajectory*.

Throughout this book the state space of a stochastic process $x(t, \mathfrak{w})$ is X= $\mathbb{R}^d$ ($d \geq 1$), and for fixed $\mathfrak{w} \in \Omega$ the trajectory $x(t, \mathfrak{w})$ is assumed a continuous curve in $\mathbb{R}^d$. We define events in Ω in terms of these continuous trajectories. All continuous functions are possible paths of the Brownian motion. An elementary event can represent the time-dependent voltage on a capacitor charged by cosmic radiation, that is, integrated white noise, or the path of a microscopic particle immersed in solution. Thus an elementary event in the probability space Ω represents the outcome of the experiment of continuous recording of the random voltage or the path of a particle diffusing without jumps. Outcomes of path-recording experiments that have jumps require a different probability space, depending on the properties of the paths, for example, as is the case for the paths of the Poisson jump process [124, p. 290], [80, p. 22, Example 2].

In many cases we consider sets of elementary events, called *Brownian events*, or *events* for short. A typical Brownian event that corresponds to an experiment consists of (uncountably many) elementary events. Thus, the experimental tracing of the trajectory of a cellular protein labeled by fusing it with the green-fluorescent protein, obtained from the jellyfish *Aequorea victoria* consists in sampling the trajectory at discrete times and with finite resolution of the recording apparatus. Therefore the trajectory may have made practically any excursion between the sampling times. Thus the experiment actually samples all possible Brownian paths that are found in given spatial intervals at sampling times, for example, in a microscope window.

The mathematical formulation of such Brownian elementary events, and events in general, is given in the following definition.

Definition 1.2.1 (The space of elementary events in $\mathbb{R}$). *The space Ω of elementary events for one-dimensional Brownian motion is the set of all continuous real-valued functions of $t \geq t_0$.*

For example,

$$\Omega = \{\mathfrak{w}(\cdot) \mid \mathbb{R}_+ \mapsto \mathbb{R}\},$$

where $\mathfrak{w}(t)$ is continuous. To define Brownian events that consist of uncountably many Brownian trajectories, we define first events called "cylinders."

Definition 1.2.2 (Cylinder sets in $\mathbb{R}$). *A cylinder set of Brownian trajectories is defined by times $t_0 \leq t_1 < t_2 < \cdots < t_n$ and real intervals $I_k = (a_k, b_k)$ ($k = 1, 2, \ldots, n$) as*

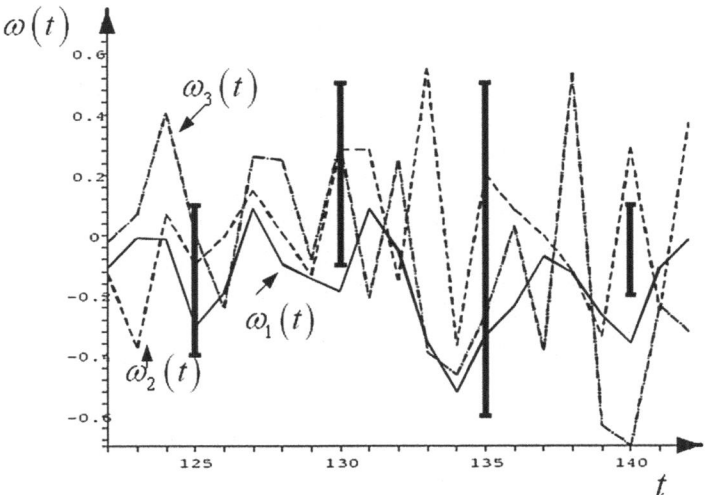

Fig. 1.1 Three Brownian trajectories sampled at discrete times. The cylinder $C(125, 135;$ $[-0.4, 0.1], [-0.6, 0.5])$ contains all three trajectories, $C(130, [-0.1, 0.5])$ contains the dotted and dashed lines, whereas $C(140, [-0.2, 0.10])$ contains none of them. The trajectories were sampled according to the scheme (1.60).

$$C\left(t_1, \ldots, t_n; I_1, \ldots, I_n\right) = \left\{\mathfrak{w}(\cdot) \in \Omega \mid \mathfrak{w}(t_k) \in I_k, \ k = 1, \ldots, n\right\}. \qquad (1.53)$$

For example, the strip whose probability is calculated in (1.52) is a cylinder set. Obviously, for any $t_0 \leq t_1 < t$ and any interval I_1,

$$C\left(t; \mathbb{R}\right) = \Omega, \quad C\left(t_1, t; I_1, \mathbb{R}\right) = C\left(t_1; I_1\right). \qquad (1.54)$$

For a Brownian trajectory $\mathfrak{w}(t)$ *not* to belong to the cylinder

$$C = C\left(t_1, t_2, \ldots, t_n; I_1, I_2, \ldots, I_n\right),$$

it suffices that for at least one of the times t_k, the value of $\mathfrak{w}(t_k)$ be not in the interval I_k, for example, the trajectory $\mathfrak{w}_1(t)$ (solid line) in Figure 1.1 does not belong to the cylinder $C(130; [-0.1, 0.5])$. For the ray $I^x = (-\infty, x]$ $(x \in \mathbb{R})$ the cylinder $C\left(t; I^x\right)$ is the set of all continuous functions $\mathfrak{w}(\cdot)$ such that $\mathfrak{w}(t) \leq x$.

For example, the trajectory $\mathfrak{w}_1(t)$ in Figure 1.1 (solid line) belongs to $C(125;$ $[-0.4, 0.1])$ and to $C(125, 135; [-0.4, 0.1], [-0.6, 0.5])$, but not to the cylinder $C(130; [-0.1, 0.5])$. The cylinder $C(140; [-0.2, -0.1])$ is empty. The cylinder $C(130, 135; [-0.1, 0.5], [-0.6, 0.5])$ contains $\mathfrak{w}_2(t)$ (dashed line) and $\mathfrak{w}_3(t)$ (dotted line), but not $\mathfrak{w}_1(t)$. The trajectories were sampled at discrete points according to the scheme (1.60) below.

Exercise 1.1 (Structure of cylinders). Show that the intersection of two cylinders is a cylinder. Is the union of two cylinders a cylinder? Is $\Omega - C(t; I^x)$ a cylinder? $\square$

Definition 1.2.3 (Brownian events). *Brownian events are all sets of Brownian trajectories that can be obtained from cylinders by the operations of countable union, intersection, and complement.*

We denote the set of all Brownian events by $\mathcal{F}$. The pair $(\Omega, \mathcal{F})$ is called the *probability space* of Brownian motion.

Definition 1.2.4 (Random variables in $(\Omega, \mathcal{F})$). *A random variable $X(\mathfrak{w})$ in $(\Omega, \mathcal{F})$ is a real function $X(\cdot) : \Omega \to \mathbb{R}$ such that $\{\mathfrak{w} \in \Omega \mid X(\mathfrak{w}) \leq x\} \in \mathcal{F}$ for all $x \in \mathbb{R}$.*

That is, the set of continuous functions $\{\mathfrak{w} \in \Omega \mid X(\mathfrak{w}) \leq x\}$ can be constructed by a finite or infinite sequence of operations of union, intersection, and complement of cylinders [68, Section 1.4], [137, Chapter 2].

Example 1.1 (Random functions). For each $t \geq 0$ consider the random variable $X_t(\mathfrak{w}) = \mathfrak{w}(t)$ in Ω. This random variable is the outcome of the experiment of sampling the position of a Brownian particle (trajectory) at a fixed time t. Thus $X_t(\mathfrak{w})$ takes different values on different trajectories. Obviously, $\{\mathfrak{w} \in \Omega \mid X_t(\mathfrak{w}) \leq x\} = \{\mathfrak{w} \in \Omega \mid \mathfrak{w}(t) \leq x\} = C(t; I^x) \in \mathcal{F}$, so that $X_t(\mathfrak{w})$ is a random variable in $(\Omega, \mathcal{F})$. $\square$

Example 1.2 (Average velocity). Although the trajectories of the Brownian motion are nondifferentiable [68, Section 1.4, Problem 7], [137, Section 2.4], the *average velocity process* of a Brownian trajectory $\mathfrak{w}$ in the time interval $[t, t + \Delta t]$ can be defined as $\bar{V}_t(\mathfrak{w}) = [\mathfrak{w}(t + \Delta t) - \mathfrak{w}(t)]/\Delta t$. The time averaging here is not expectation, because it is defined separately on each trajectory. Therefore $\bar{V}_t(\mathfrak{w})$ is a random variable, which takes different values on different trajectories [137, Example 2.2]. $\square$

Example 1.3 (Integrals of random functions). It can be shown [137, Example 2.3] that $X(\mathfrak{w}) = \int_0^T \mathfrak{w}(t)\, dt$ is a random variable in Ω. $\square$

Example 1.4 (First passage times). Consider all $\mathfrak{w} \in \Omega$ such that $\mathfrak{w}(0) < y$ for a given $y \in \mathbb{R}$ and define $\tau_y(\mathfrak{w}) = \inf\{t \geq 0 \mid \mathfrak{w}(t) \geq y\}$, that is, $\tau_y(\mathfrak{w})$ is the *first passage time* (FPT) of a Brownian trajectory $\mathfrak{w}(t)$ through the value y. It can be shown that the FPT is a random variable [137, Example 2.4]. $\square$

Example 1.5 (Indicators). For any set $A \in \Omega$ the *indicator* function of A is defined by

$$\mathbf{1}_A(\mathfrak{w}) = \begin{cases} 1 & \text{if } \mathfrak{w} \in A, \\ 0 & \text{otherwise.} \end{cases} \tag{1.55}$$

For all $A \in \mathcal{F}$ the function $\mathbf{1}_A(\mathfrak{w})$ is a random variable in $(\Omega, \mathcal{F})$ [137, Example 2.6]. $\square$

Exercise 1.2 (Positive random variables). For a random variable $X(\mathfrak{w})$ define the functions $X^+(\mathfrak{w}) = \max\{X(\mathfrak{w}), 0\}$ and $X^-(\mathfrak{w}) = \min\{X(\mathfrak{w}), 0\}$. Show that $X^+(\mathfrak{w})$ and $X^-(\mathfrak{w})$ are random variables (that is, show that $\{\mathfrak{w} \in \Omega \mid X^{\pm}(\mathfrak{w}) \leq x\} \in \mathcal{F}$ for all $x \in \mathbb{R}$). $\square$

Definition 1.2.5 (Stochastic processes in $(\Omega, \mathcal{F})$). *A function $x(t, \mathfrak{w}) : \mathbb{R}_+ \times \Omega \mapsto \mathbb{R}$ is called a* stochastic process *with continuous trajectories in $(\Omega, \mathcal{F})$ if*

(i) $x(t, \mathfrak{w})$ is a continuous function of t for every $\mathfrak{w} \in \Omega$,
(ii) for every fixed $t \geq 0$ the function $x(t, \mathfrak{w}) : \Omega \mapsto \mathbb{R}$ is a random variable in Ω.

The variable $\mathfrak{w}$ of a stochastic process $x(t, \mathfrak{w})$ is the Brownian path (elementary event) to which $x(t, \mathfrak{w})$ assigns a value. Point (ii) of the definition means that the sets $\{\mathfrak{w} \in \Omega \mid x(t, \mathfrak{w}) \leq x\}$ are Brownian events for each $t \geq 0$ and $x \in \mathbb{R}$, that is, they belong to $\mathcal{F}$. When they do, we say that the process $x(t, \mathfrak{w})$ is *measurable* with respect to $\mathcal{F}$ or simply $\mathcal{F}$-measurable.

Definition 1.2.6 (Adapted processes). *The process $x(t, \mathfrak{w})$ is said to be* adapted *to the Brownian motion if for every $t \geq 0$ and $x \in \mathbb{R}$ the event $\{\mathfrak{w} \in \Omega \mid x(t, \mathfrak{w}) \leq x\}$ is generated by cylinders defined by times $t_i \leq t$.*

Thus an adapted process at each moment of time t is independent of future displacements of Brownian trajectories, but depends on their past displacements.

Exercise 1.3 (A random time). Consider a stochastic process $x(t, \mathfrak{w})$ and an open set $O \subset \mathbb{R}$. What does the integral $\tau(\mathfrak{w}) = \int_0^T \mathbf{1}_O(x(t, \mathfrak{w})) \, dt$ represent? Is it a random variable in $(\Omega, \mathcal{F})$? $\square$

1.2.2 Probability in $(\Omega, \mathcal{F})$

Definition 1.2.7 (Probability measure in $(\Omega, \mathcal{F})$). *A nonnegative function* $\Pr :$ $\mathcal{F} \mapsto \mathbb{R}_+$, *such that* $\Pr\{\Omega\} = 1$ *and*

$$\Pr\left\{\bigcup_{i=1}^{\infty} A_i\right\} = \sum_{i=1}^{\infty} \Pr\{A_i\}$$

for any sequence of pairwise disjoint events $A_i \in \mathcal{F}$ is called a probability measure *in $(\Omega, \mathcal{F})$.*

Thus the probability of an event is a number between 0 and 1. An event that always occurs is called a *sure event*; thus Ω is a sure event. An event whose probability is 1 is called an *almost sure event* (see discussion in *http://en.wikipedia.org/wiki/Almost_surely*), or we say that the event occurs almost surely (a.s.), or that the event occurs with probability 1 (w.p. 1). Obviously, $\Pr\{\emptyset\} = 0$, where $\emptyset$ is the empty event. There are many ways for assigning probabilities to events, depending on the degree of uncertainty we have about a given event; different persons may assign different probabilities to the same events. We may think of probability as a mathematical model of our degree of uncertainty concerning events [33], but in effect, it is a measure-theoretical model of statistics.

Definition 1.2.8 (Integration with respect to a probability measure). *The probability* $\Pr\{A\}$ *defines an integral of a random variable* $X(\mathfrak{w})$ *by*

$$\int_{\Omega} X(\mathfrak{w}) \, d\Pr\{\mathfrak{w}\} = \lim_{h \to 0} \lim_{M,N \to \infty} \sum_{n=-M}^{N} nh \Pr\{\mathfrak{w} : nh \leq X(\mathfrak{w}) \leq (n+1)h\},$$

(1.56)

whenever the limit exists. In this case, we say that $X(\mathfrak{w})$ *is an integrable random variable.*

For any set $A \in \mathcal{F}$, the indicator function $\mathbf{1}_A(\mathfrak{w})$ (see Example 1.5) is integrable and $\int_{\Omega} \mathbf{1}_A(\mathfrak{w}) \, d\Pr\{\mathfrak{w}\} = \Pr\{A\}$. We define an integral over an event A by

$$\int_A X(\mathfrak{w}) \, d\Pr\{\mathfrak{w}\} = \int_{\Omega} \mathbf{1}_A(\mathfrak{w}) X(\mathfrak{w}) \, d\Pr\{\mathfrak{w}\}.$$

If $\int_A X(\mathfrak{w}) \, d\Pr\{\mathfrak{w}\}$ exists, we say that $X(\mathfrak{w})$ is *integrable in* A. In that case $X(\mathfrak{w})$ is integrable in every subevent of A. All random variables $X(\mathfrak{w})$ are integrable in all events A such that $\Pr\{A\} = 0$, and $\int_A X(\mathfrak{w}) \, d\Pr\{\mathfrak{w}\} = 0$.

Definition 1.2.9 (PDF and pdf). *For an integrable random variable* $X(\mathfrak{w})$ *the function*

$$F_X(x) = \Pr\{\mathfrak{w} \mid X(\mathfrak{w}) \leq x\}$$

is called the probability distribution function (PDF) *of* $X(\mathfrak{w})$. *The function (or generalized function* [103], [73]*)*

$$f_X(x) = \frac{d}{dx} F_X(x)$$

is called the probability density function *(pdf) of* $X(\mathfrak{w})$. *The expectation* $\mathbb{E}X(\mathfrak{w})$ *is defined as*

$$\mathbb{E}X(\mathfrak{w}) = \int_{-\infty}^{\infty} x \, dF_X(x) = \int_{-\infty}^{\infty} x f_X(x) \, dx.$$

(1.57)

If the PDF is not a differentiable function, then the density cannot be a function. It can be defined, however, in the sense of *distributions* [103], [73]. We assume henceforward that every random variable has a pdf in this sense.

1.2.3   The Wiener Measure of Brownian Trajectories

Having constructed the set $\mathcal{F}$ of events for the Brownian trajectories, we proceed
to construct a probability measure of these events. The probability measure will be
used to construct a mathematical theory of the Brownian motion that can describe
experiments.

A probability measure $\mathrm{Pr}\{\cdot\}$ can be defined on Ω (that is, on the events $\mathcal{F}$ in
Ω) to conform with the Einstein–Langevin description of Brownian motion. It is
enough to define the probability measure $\mathrm{Pr}\{\cdot\}$ on cylinder sets and then to extend it
to all events in $\mathcal{F}$ by the elementary properties of a probability measure [68]. The
following probability measure in $\mathcal{F}$ is called *Wiener's measure* [155]. Consider the
cylinder $C(t; I)$, where $t \geq 0$ and $I = (a, b)$, and set

$$\mathrm{Pr}\left\{C(t; I)\right\} = \frac{1}{\sqrt{2\pi t}} \int_a^b e^{-x^2/2t}\, dx. \tag{1.58}$$

If $0 = t_0 < t_1 < t_2 < \cdots < t_n$ and I_k $(k = 1, 2, \ldots, n)$ are real intervals, set

$$\mathrm{Pr}\left\{C\left(t_1, t_2, \ldots, t_n; I_1, I_2, \ldots, I_n\right)\right\}$$

$$= \int_{I_1} \int_{I_2} \cdots \int_{I_n} \prod_{k=1}^n \frac{dx_k}{\sqrt{2\pi(t_k - t_{k-1})}} \exp\left\{-\frac{(x_k - x_{k-1})^2}{2(t_k - t_{k-1})}\right\}, \tag{1.59}$$

where $x_0 = 0$ (the extension of the Wiener probability measure from cylinders to $\mathcal{F}$
is described in [68], [132]). This definition is consistent with (1.52). The obvious
features of the Wiener probability measure that follow from (1.54) and (1.59) are

$$\mathrm{Pr}\{\Omega\} = \frac{1}{\sqrt{2\pi t}} \int_{-\infty}^\infty e^{-x^2/2t}\, dx = 1,$$

and for $t_1 < t$,

$$\mathrm{Pr}\left\{C\left(t_1, t; I_1, \mathbb{R}\right)\right\} = \int_{I_1} \int_{-\infty}^\infty \frac{dx\, dx_1}{2\pi\sqrt{(t - t_1)t_1}} \exp\left\{-\frac{(x - x_1)^2}{2(t - t_1)}\right\} \exp\left\{-\frac{x_1^2}{2t_1}\right\}$$

$$= \frac{1}{\sqrt{2\pi t_1}} \int_{I_1} \exp\left\{-\frac{x_1^2}{2t_1}\right\}\, dx_1 = \mathrm{Pr}\left\{C(t_1; I_1)\right\}.$$

The Wiener probability measure (1.59) of a cylinder is the probability of
sampling points of a trajectory in the cylinder by the Monte–Carlo simulation

$$x(t_k) = x(t_{k-1}) + \Delta w(t_k), \quad k = 1, \ldots, n, \tag{1.60}$$

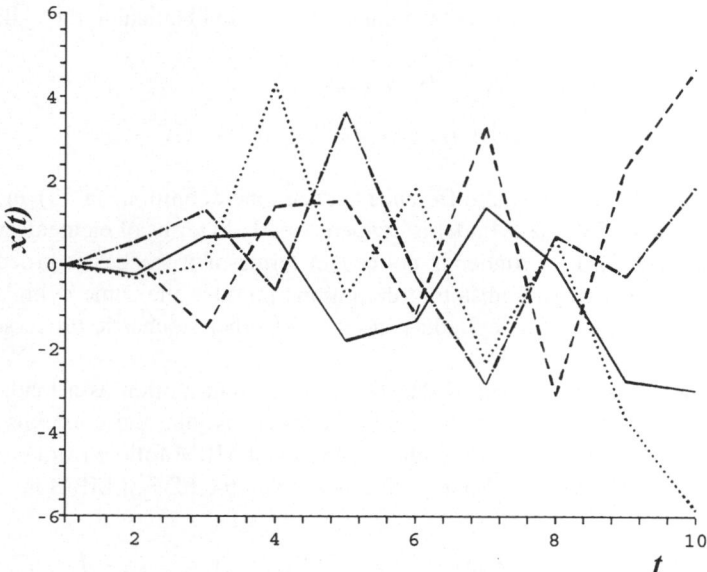

Fig. 1.2 Four Brownian trajectories sampled at discrete points according to the Wiener probability measure $\mathrm{Pr}_0\{\cdot\}$ by the scheme (1.60).

where t_k are ordered as above and $\Delta w(t_k) \backsim \mathcal{N}(0, t_k - t_{k-1})$ are independent normal variables. It is the same as (1.47). The vertices of the trajectories in Figures 1.1 and 1.2 were sampled according to (1.60) and interpolated linearly. Those of Figure 1.2 were sampled according to the Wiener probability measure $\mathrm{Pr}_0\{\cdot\}$. Skorokhod's theorem (Theorem 2.1.1) asserts that the scheme (1.60) converges (in some sense) as $\max_k (t_k - t_{k-1}) \to 0$.

1.2.4 Definition of Mathematical Brownian Motion

The axiomatic definition of Brownian motion, consistent with the formal properties of the simulation (1.47), is given as follows.

Definition 1.2.10 (MBM). *A real-valued stochastic process $w(t, \mathfrak{w})$ defined on $\mathbb{R}_+ \times \Omega$ is a mathematical Brownian motion if*

1. *$w(0, \mathfrak{w}) = 0$ w.p. 1;*
2. *$w(t, \mathfrak{w})$ is almost surely a continuous function of t;*
3. *for every $t, s \geq 0$, the increment $\Delta w(s, \mathfrak{w}) = w(t+s, \mathfrak{w}) - w(t, \mathfrak{w})$ is independent of $w(\tau, \mathfrak{w})$ for all $\tau \leq t$, and is a zero-mean Gaussian random variable with variance*

$$\mathbb{E} |\Delta w(s, \mathfrak{w})|^2 = s. \tag{1.61}$$

According to this definition, the cylinders (1.53) are identical to the cylinders

$$C\left(t_1, t_2, \ldots, t_n; I_1, I_2, \ldots, I_n\right)$$
$$= \left\{\mathfrak{w}(t) \in \Omega \mid w(t_k, \mathfrak{w}) \in I_k, \text{ for all } k = 1, 2, \ldots, n\right\}. \tag{1.62}$$

To understand the conceptual difference between the definitions (1.53) and (1.62), we note that in (1.53), the cylinder is defined directly in terms of elementary events, whereas in (1.62), the cylinder is defined in terms of a stochastic process. It is coincidental that two such different definitions produce the same cylinder. Later in the book we will define cylinders in terms of other stochastic processes, as in (1.62).

Properties (1)–(3) are axioms that define Brownian motion as a mathematical entity. It can be shown that a stochastic process satisfying these axioms actually exists [137, Section 2.3]. Some of the properties of MBM follow from the axioms in a straightforward manner. First, (1.58) means that the PDF of MBM is

$$F_w(x, t) = \Pr\left\{\mathfrak{w} \in \Omega \mid w(t, \mathfrak{w}) \le x\right\} = \Pr\left\{C(t, I^x)\right\} = \frac{1}{\sqrt{2\pi t}} \int_{-\infty}^{x} e^{-y^2/2t} \, dy,$$

and the pdf is

$$f_w(x, t) = \frac{\partial}{\partial x} F_w(x, t) = \frac{1}{\sqrt{2\pi t}} e^{-x^2/2t}. \tag{1.63}$$

It is well known (and easily verified) that $f_w(x, t)$ is the solution of the initial value problem for the diffusion equation

$$\frac{\partial f_w(x, t)}{\partial t} = \frac{1}{2} \frac{\partial^2 f_w(x, t)}{\partial x^2}, \quad \lim_{t \downarrow 0} f_w(x, t) = \delta(x). \tag{1.64}$$

Second, we note that (1) and (2) are not contradictory, despite the fact that not all continuous functions vanish at time $t = 0$. Property (1) asserts that all trajectories of Brownian motion that do not start at the origin are assigned probability 0. In view of the above, $x_0 = 0$ in the definition (1.59) of the Wiener probability measure of a cylinder means that the Brownian paths are those continuous functions that take the value 0 at time 0. That is, the Brownian paths are conditioned on starting at time $t = 0$ at the point $x_0 = w(0, \mathfrak{w}) = 0$. To emphasize this point, we modify the notation of the Wiener probability measure to $\Pr_0\{\cdot\}$. If this condition is replaced with $x_0 = x$ in (1.59), then $\Pr_x\left\{w(0, \mathfrak{w}) = x\right\} = 1$ under the modified Wiener probability measure, now denoted by $\Pr_x\{\cdot\}$ [137, Section 2.2].

Thus conditioning reassigns probabilities to the Brownian paths; the set of trajectories $\left\{\mathfrak{w} \in \Omega \mid w(0, \mathfrak{w}) = x\right\}$, which was assigned the probability 0 under the measure $\Pr_0\{\cdot\}$, is now assigned the probability 1 under the measure $\Pr_x\{\cdot\}$. Similarly, replacing the condition $t_0 = 0$ with $t_0 = s$ and setting $x_0 = x$ in (1.59) shifts the Wiener measure, now denoted by $\mathrm{Pr}_{x,s}$, so that

$$\mathrm{Pr}_{x,s}\{C(t;[a,b])\} = \mathrm{Pr}_0\{C(t-s;[a-x,b-x])\}. \tag{1.65}$$

This means that for all positive t, the increment $\Delta w(s,\mathfrak{w}) = w(t+s,\mathfrak{w}) - w(t,\mathfrak{w})$, as a function of s, is an MBM; so that the probabilities of any Brownian event of $\Delta w(s,\mathfrak{w})$ are independent of t, that is, the increments of the MBM are stationary. Accordingly, the first two moments of the MBM are

$$\mathbb{E}w(t,\mathfrak{w}) = \int_{-\infty}^{\infty} \frac{x}{\sqrt{2\pi t}} e^{-x^2/2t}\,dx = 0,$$

$$\mathbb{E}w^2(t,\mathfrak{w}) = \frac{1}{\sqrt{2\pi t}} \int_{-\infty}^{\infty} x^2 e^{-x^2/2t}\,dx = t. \tag{1.66}$$

Note that (1.61) follows from (1.66) and the independence of the increments of the MBM.

We recall that the autocorrelation function of a stochastic process $x(t,\mathfrak{w})$ is defined as the expectation $R_x(t,s) = \mathbb{E}x(t,\mathfrak{w})x(s,\mathfrak{w})$. Using the notation $t \wedge s = \min\{t,s\}$, we have the following theorem

Theorem 1.2.1 (Property (5)). *The autocorrelation function of $w(t,\mathfrak{w})$ is*

$$\mathbb{E}w(t,\mathfrak{w})w(s,\mathfrak{w}) = t \wedge s. \tag{1.67}$$

Proof. Assuming that $t \geq s \geq 0$ and using property (3), we find that

$$\mathbb{E}w(t,\mathfrak{w})w(s,\mathfrak{w}) = \mathbb{E}\Big[w(t,\mathfrak{w})-w(s,\mathfrak{w})\Big]\Big[w(s,\mathfrak{w})-w(0,\mathfrak{w})\Big] + \mathbb{E}w(s,\mathfrak{w})w(s,\mathfrak{w})$$

$$= s = t \wedge s.$$

$\square$

1.2.5   MBM in $\mathbb{R}^d$

If $w_1(t,\mathfrak{w}_1), w_2,(t,\mathfrak{w}_2), \ldots, w_d(t,\mathfrak{w}_d)$ are independent Brownian motions, the vector process

$$\boldsymbol{w}(t,\mathfrak{w}) = \begin{pmatrix} w_1(t,\mathfrak{w}_1) \\ w_2(t,\mathfrak{w}_2) \\ \vdots \\ w_d(t,\mathfrak{w}_d) \end{pmatrix}$$

is defined as a *d-dimensional Brownian motion*. The probability space $\boldsymbol{\Omega}$ for n-dimensional Brownian motion consists of all $\mathbb{R}^d$-valued continuous functions of t. The elementary events $\mathfrak{w}$ are trajectories

$$\mathfrak{w}(t) = \begin{pmatrix} \mathfrak{w}_1(t) \\ \mathfrak{w}_2(t) \\ \vdots \\ \mathfrak{w}_d(t) \end{pmatrix},$$

where $\mathfrak{w}_j(t) \in \Omega$. Cylinder sets are defined as follows.

Definition 1.2.11 (Cylinder sets in $\mathbb{R}^d$). *A cylinder set of d-dimensional Brownian trajectories is defined by times $0 \le t_1 < t_2 < \cdots < t_k$ and open sets I_k ($k = 1, 2, \ldots, k$) as*

$$C\left(t_1, t_2, \ldots, t_k; I_1, I_2, \ldots, I_k\right) = \left\{ \mathfrak{w}(t) \in \Omega : \mathfrak{w}(t_j) \in I_j, \ j = 1, 2, \ldots, k \right\}.$$

The open sets I_j can be, for example, open boxes or balls in $\mathbb{R}^d$. In particular, we write $I_x = \{\mathfrak{w} \le x\} = \{\mathfrak{w}_1 \le x_1, \ldots, \mathfrak{w}_d \le x_d\}$.

Definition 1.2.12 (The Wiener measure for d-dimensional MBM). *The d-dimensional Wiener probability measure of a cylinder is defined as*

$$\Pr\{C\left(t_1, t_2, \ldots, t_k; I_1, I_2, \ldots, I_k\right)\}$$

$$= \int_{I_1} \int_{I_2} \cdots \int_{I_k} \prod_{j=1}^{k} \frac{dx_j}{[2\pi(t_j - t_{j-1})]^{n/2}} \exp\left\{-\frac{|x_j - x_{j-1}|^2}{2(t_j - t_{j-1})}\right\}. \tag{1.68}$$

The PDF of the d-dimensional MBM is

$$F_{\mathbf{w}}(x, t) = \Pr\{\mathfrak{w} \in \Omega \mid w(t, \mathfrak{w}) \le x\}$$

$$= \frac{1}{(2\pi t)^{n/2}} \int_{-\infty}^{x_1} \cdots \int_{-\infty}^{x_d} e^{-|y|^2/2t} \, dy_1 \cdots dy_d, \tag{1.69}$$

and the pdf is

$$f_{\mathbf{w}}(x, t) = \frac{\partial^d F_{\mathbf{w}}(x, t)}{\partial x_1 \partial x_2 \cdots \partial x_d} = \frac{1}{(2\pi t)^{n/2}} e^{-|x|^2/2t}. \tag{1.70}$$

Equations (1.64) imply that $f_{\mathbf{w}}(x, t)$ satisfies the d-dimensional diffusion equation and the initial condition

$$\frac{\partial f_{\mathbf{w}}(x, t)}{\partial t} = \frac{1}{2} \Delta f_{\mathbf{w}}(x, t), \quad \lim_{t \downarrow 0} f_{\mathbf{w}}(x, t) = \delta(x). \tag{1.71}$$

It can be seen from (1.68) that any rotation of d-dimensional Brownian motion is d-dimensional Brownian motion. Higher-dimensional stochastic processes are defined as follows.

Definition 1.2.13 (Vector-valued processes). *A vector-valued function* $x(t, \mathfrak{w})$:
$\mathbb{R}_+ \times \Omega \mapsto \mathbb{R}^d$ *is called a* stochastic process *in* $(\Omega, \mathcal{F})$ *with continuous trajectories if*

(i) $x(t, \mathfrak{w})$ *is a continuous function of t for every* $\mathfrak{w} \in \Omega$,
(ii) *for every* $t \geq 0$ *and* $x \in \mathbb{R}^d$, *the sets* $\{\mathfrak{w} \in \Omega : x(t, \mathfrak{w}) \leq x\}$ *are Brownian events, that is, if* $\{\mathfrak{w} \in \Omega : x(t, \mathfrak{w}) \leq x\} \in \mathcal{F}$.

Note that the dimension of the process, n, and the dimension of the space in which the trajectories move, d, are not necessarily the same. The PDF of $x(t, \mathfrak{w})$ is defined as

$$F_x(y, t) = \Pr\{\mathfrak{w} \in \Omega : x(t, \mathfrak{w}) \leq y\}, \tag{1.72}$$

and the pdf is defined as

$$f_x(y, t) = \frac{\partial^d F_x(y, t)}{\partial y^1 \partial y^2 \cdots \partial y^d}. \tag{1.73}$$

The expectation of a matrix-valued function $g(x)$ of a vector-valued process $x(t, \mathfrak{w})$ is the matrix

$$\mathbb{E}g(x(t, \mathfrak{w})) = \int_{\mathbb{R}^d} g(y) f_x(y, t) \, dy. \tag{1.74}$$

Definition 1.2.14 (Autocorrelation and autocovariance). *The autocorrelation matrix of* $x(t, \mathfrak{w})$ *is defined as the* $d \times d$ *matrix*

$$R_x(t, s) = \mathbb{E}x(t)x^T(s), \tag{1.75}$$

and the autocovariance matrix *is defined as*

$$Cov_x(t, s) = \mathbb{E}[x(t) - \mathbb{E}x(t)][x - Ex(s)]^T. \tag{1.76}$$

The autocovariance matrix of the d*-dimensional Brownian motion is found from* (1.67) *as*

$$Cov_w(t, s) = I(t \wedge s), \tag{1.77}$$

where I *is the identity matrix.*

Exercise 1.4 (Transformations preserving MBM). Show, by verifying properties (1)–(3), that the following processes are Brownian motions:

(i) $w_1(t) = w(t + s) - w(s)$
(ii) $w_2(t) = cw(t/c^2)$, where c is any positive constant
(iii) $w_3(t) = tw(1/t)$. $\qquad\qquad\qquad\qquad\qquad\qquad\qquad\qquad\qquad\qquad\qquad\square$

Exercise 1.5 (Changing scale). Give necessary and sufficient conditions on the functions $f(t)$ and $g(t)$ such that the process $w_4(t) = f(t)w(g(t))$ is an MBM.   □

Exercise 1.6 (The joint pdf of the increments). Define

$$\Delta w = \begin{pmatrix} \Delta w(t_1) \\ \Delta w(t_2) \\ \vdots \\ \Delta w(t_n) \end{pmatrix}.$$

Find the joint pdf of Δw.   □

Exercise 1.7 (Radial MBM). Define *radial MBM* by $y(t) = |w(t)|$, where $w(t)$ is d-dimensional MBM. Find the pdf of $y(t)$, the partial differential equation, and the initial condition it satisfies.   □

1.2.6 Constructions of MBM

Consider a sequence of standard Gaussian i.i.d. random variables $\{Y_k\}$, for $k = 0, 1, \ldots$, defined in a probability space $\tilde{\Omega}$. We denote by $\mathfrak{w}$ any realization of the infinite sequence $\{Y_k\}$ and construct a continuous path corresponding to this realization. We consider a sequence of binary partitions of the unit interval,

$$T_1 = \{0, 1\}, \quad T_2 = \left\{ 0, \frac{1}{2}, 1 \right\}, \quad T_3 = \left\{ 0, \frac{1}{4}, \frac{1}{2}, \frac{3}{4}, 1 \right\} \ldots,$$

$$T_{n+1} = \left\{ \frac{k}{2^n}, \, k = 0, 1, \ldots, 2^n \right\}.$$

The set $T_0 = \bigcup_{n=1}^{\infty} T_n$ contains all the binary numbers in the unit interval. The binary numbers are dense in the unit interval in the sense that for every $0 \le x \le 1$ there is a sequence of binary numbers $x_j = k_j 2^{n_j}$ with $0 \le k_j \le 2^{n_j}$ such that $x_j \to x$ as $j \to \infty$. Figure 1.3 shows the graphs of $X_1(t)$ (dots), its first refinement $X_2(t)$ (dash dot), and second refinement $X_3(t)$ (dash). A Brownian trajectory sampled at 1024 points is shown in Figure 1.4. Define $X_1(\mathfrak{w}) = tY_1(\mathfrak{w})$ for $0 \le t \le 1$. Keeping in mind that $T_2 = \{0, \frac{1}{2}, 1\}$ and $T_1 \setminus T_1 = \{\frac{1}{2}\}$, we refine by keeping the "old" points, that is, by setting $X_2(t, \mathfrak{w}) = X_1(t, \mathfrak{w})$ for $t \in T_1$, and in the "new" point, $T_2 \setminus T_1 = \{\frac{1}{2}\}$, we set $X_2\left(\frac{1}{2}, \mathfrak{w}\right) = \frac{1}{2}[X_1(0, \mathfrak{w}) + X_1(1, \mathfrak{w})] + \frac{1}{2}Y_2(\mathfrak{w})$. The process $X_2(t, \mathfrak{w})$ is defined in the interval by linear interpolation between the points of T_2.

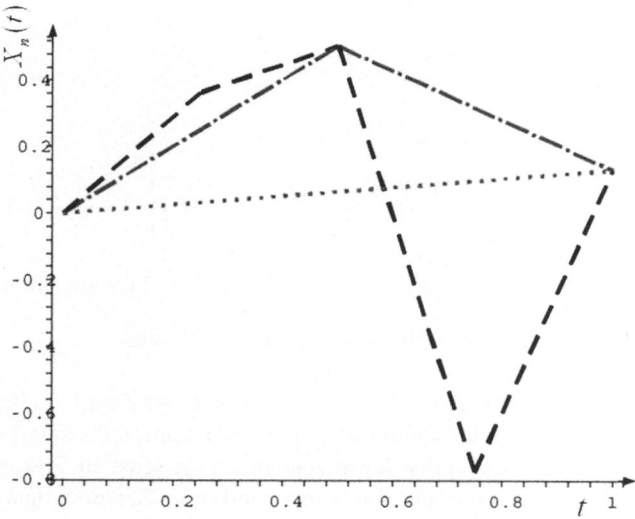

Fig. 1.3 The graphs of $X_1(t)$ (dotted line), its first refinement $X_2(t)$ (dash dot), and second refinement $X_3(t)$ (dash).

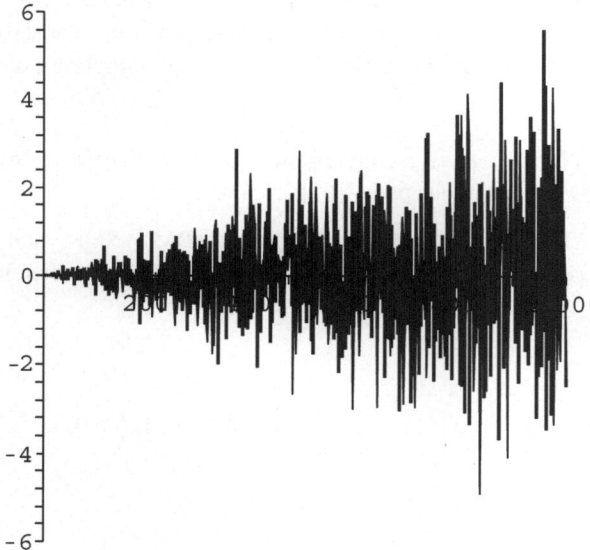

Fig. 1.4 A Brownian trajectory sampled at 1024 points.

We proceed by induction:

$$X_{n+1}(t, \mathfrak{w})$$

$$= \begin{cases} X_n(t, \mathfrak{w}) \text{ for } t \in T_n \text{ (old points)} \\ \frac{1}{2}\left\{X_n\left(t + \frac{1}{2^n}, \mathfrak{w}\right) + X_n\left(t - \frac{1}{2^n}, \mathfrak{w}\right)\right\} + \frac{1}{2^{\frac{n+1}{2}}} Y_k(\mathfrak{w}) \\ \text{for } t \in T_{n+1} \setminus T_n, \ k = 2^{n-1} + \frac{1}{2}(2^n t - 1) \text{ (new points)} \\ \text{connect linearly between consecutive points.} \end{cases}$$

Thus $X_{n+1}(t)$ is a refinement of $X_n(t)$. Old points stay put! So far, for every realization $\mathfrak{w}$, we constructed an infinite sequence of continuous functions. It can be shown [137, Theorem 2.3.2] that for almost all (in the sense of $\tilde{\Omega}$) realizations $\mathfrak{w}$, the sequence $X_n(t)$ converges uniformly to a continuous function, thus establishing a correspondence between $\mathfrak{w}$ and a continuous function. Obviously, the correspondence can be reversed in this construction.

Exercise 1.8 (MBM at binary points). Show that at binary points, $t_{k,n} = k2^{-n}$, $0 \le k \le 2^n$, the process $X_n(t, \mathfrak{w})$ has the properties of the Brownian motion $w(t)$. $\qquad\square$

Exercise 1.9 (Refinements). If a Brownian trajectory is sampled at points $0 = t_0 < t_1 < \cdots < t_n = T$ according to the scheme (1.60) or otherwise, how should the sampling be refined by introducing an additional sampling point $\tilde{t}_i$ such that $t_i < \tilde{t}_i < t_{i+1}$? $\qquad\square$

Exercise 1.10 ($L^2([0, 1] \times \Omega)$ convergence*). Show that $X_n(t, \mathfrak{w}) \xrightarrow{L^2} X(t, \mathfrak{w})$, where $X(t, \mathfrak{w})$ has continuous paths [65]. $\qquad\square$

Exercise 1.11 (Lévy's construction gives an MBM). Show that if $X_1(t)$ and $X_2(t)$ are independent Brownian motions on the interval $[0, 1]$, then the process

$$X(t) = \begin{cases} X_1(t) & \text{for } 0 \le t \le 1, \\ \\ X_1(1) + tX_2\left(\frac{1}{t}\right) - X_2(1) \text{ for } t > 1, \end{cases}$$

is a Brownian motion on $\mathbb{R}^+$. $\qquad\square$

1.2.7 Analytical and Statistical Properties of MBM

The Wiener probability measure assigns probability 0 to several important classes of Brownian paths. These classes include all differentiable paths, all paths that satisfy the Lipschitz condition at some point, all continuous paths with bounded variation on some interval, and so on. Brownian paths have many interesting properties [68], [65], [132]; here we list only a few of the most prominent features of Brownian paths.

Although continuous, Brownian paths are nondifferentiable at any given point with probability 1 [123], [68, Section 1.4, Problem 7], [137, Section 2.4]. This means that the Wiener probability measure assigns probability 0 to all differentiable paths. This fact implies that the white noise process $\dot{w}(t)$ does not exist, so that strictly speaking, none of the calculations carried out under the assumption that $\dot{w}(t)$ exists are valid. This means that the velocity process of MBM (white noise) should be interpreted as the overdamped limit of the Brownian velocity process described in [137, Section 1.2]. The *level-crossing property* of MBM is that for any level a the times t such that $w(t) = a$ form a perfect set (i.e., every point of this set is a limit of points in this set). Thus, when a Brownian path reaches a given level at time t, it recrosses it infinitely many times in every interval $[t, t + \Delta t]$.

Exercise 1.12 (Level crossing). Use the scheme (1.60) with step size $\Delta t = 0.5$ to sample a Brownian path in the interval $0 \leq t \leq 1$ and refine it several times at binary points. Count the number of given level crossings as the trajectory is refined. □

Definition 1.2.15 (Markov process). *A stochastic process $\zeta(t)$ on $[0, T]$ is called a Markov process if for any sequences $0 \leq t_0 < \cdots < t_n \leq T$ and $x_0, x_1, \ldots, x_n$, its transition probability distribution function has the property*

$$\Pr\left\{\zeta(t_n) < x_n \mid \zeta(t_{n-1}) < x_{n-1}, \zeta(t_{n-2}) < x_{n-2}, \ldots, \zeta(t_0) < x_0\right\}$$

$$= \Pr\left\{\zeta(t_n) < x_n \mid \zeta(t_{n-1}) < x_{n-1}\right\}. \tag{1.78}$$

The transition probability density function, defined by

$$p\left(x_n, t_n \mid x_{n-1}, t_{n-1}, \ldots, x_1, t_1\right)$$

$$= \frac{\partial}{\partial x_n} \Pr\left\{\zeta(t_n) < x_n \mid \zeta(t_{n-1}) = x_{n-1}, \zeta(t_{n-2}) = x_{n-2}, \ldots, \zeta(t_0) = x_0\right\},$$

then satisfies

$$p\left(x_n, t_n \mid x_{n-1}, t_{n-1}, \ldots, x_1, t_1\right) = p\left(x_n, t_n \mid x_{n-1}, t_{n-1}\right). \tag{1.79}$$

The Markov property (1.78) means that the process "forgets" the past in the sense that if the process is observed at times $t_0, t_1, \ldots, t_{n-1}$ such that $0 \leq t_0 < \cdots < t_{n-1} \leq T$, its "future" evolution (at times $t > t_{n-1}$) depends only on the "latest" observation (at time t_{n-1}).

Theorem 1.2.2 (The Chapman–Kolmogorov equation [28], [93]). *The transition probability density function of a Markov process satisfies the Chapman–Kolmogorov equation*

$$p(y, t \mid x, s) = \int_{-\infty}^{\infty} p(y, t \mid z, \tau) p(z, \tau \mid x, s)\, dz. \tag{1.80}$$

Proof. For any three times $t < \tau < s$ and any points x, y, z, the identities

$$p(y, t, z, \tau \mid x, s) = p(y, t \mid z, \tau, x, s) p(z, \tau \mid x, s)$$
$$= p(y, t \mid z, \tau) p(z, \tau \mid x, s) \tag{1.81}$$

are consequences of the Markov property. Using these identities and writing $p(y, t \mid x, s)$ as a marginal density of $p(y, t, z, \tau \mid x, s)$, we obtain equation (1.80). $\square$

Theorem 1.2.3. *A MBM is a Markov process.*

Proof. To determine the Markov property of Brownian motion, consider any sequences $0 = t_0 < t_1 < \cdots < t_n$ and $x_0 = 0, x_1, \ldots, x_n$. The joint pdf of the vector

$$w = \begin{pmatrix} w(t_1) \\ w(t_2) \\ \vdots \\ w(t_n) \end{pmatrix} \tag{1.82}$$

is given by (see (1.59))

$$p\left(x_1, t_1; x_2, t_2; \ldots; x_n, t_n\right) = \Pr\left\{w(t_1) = x_1, w(t_2) = x_2, \ldots, w(t_n) = x_n\right\}$$
$$= \prod_{k=1}^{n} \left[\{2\pi(t_k - t_{k-1})\}^{-1/2} \exp\left\{ -\frac{(x_k - x_{k-1})^2}{2(t_k - t_{k-1})} \right\} \right], \tag{1.83}$$

where $\Pr\{w(t_1) = x_1, w(t_2) = x_2, \ldots, w(t_n) = x_n\}$, with some abuse of notation, is meant here as probability density, not probability (see below). Hence for $0 = t_0 < t_1 < \cdots < t_n < t = t_{n+1}$ and $0 = x_0, x_1, \ldots, x_n, x = x_{n+1}$,

$$\Pr\left\{w(t) = x \mid w(t_n) = x_n, \ldots, w(t_1) = x_1\right\}$$
$$= \frac{\Pr\left\{w(t_{n+1}) = x_{n+1}, w(t_n) = x_n, \ldots, w(t_1) = x_1\right\}}{\Pr\left\{w(t_n) = x_n, \ldots, w(t_1) = x_1\right\}}$$

$$
\prod_{k=1}^{n+1} \left[\{2\pi(t_k - t_{k-1})\}^{-1/2} \exp\left\{ -\frac{(x_k - x_{k-1})^2}{2(t_k - t_{k-1})} \right\} \right]
$$
$$
= \frac{\displaystyle\prod_{k=1}^{d} \left[\{2\pi(t_k - t_{k-1})\}^{-1/2} \exp\left\{ -\frac{(x_k - x_{k-1})^2}{2(t_k - t_{k-1})} \right\} \right]}{}
$$

$$
= \frac{1}{\sqrt{2\pi(t - t_n)}} \exp\left\{ -\frac{(x_{n+1} - x_n)^2}{2(t_{n+1} - t_n)} \right\} = \Pr\left\{ w(t) = x \mid w(t_n) = x_n \right\},
$$

that is, Brownian motion is a Markov process.   □

It follows that it suffices to know the two-point transition pdf of Brownian motion, $p(y, t \mid x, s)\, dy = \Pr\{w(t) \in y + dt \mid w(s) = x\}$ for $t > s$, to calculate the joint and conditional probability densities of the vector (1.82), that is, $p(x_1, t_1; x_2, t_2; \ldots; x_n, t_n) = \prod_{k=1}^{n} p(x_k, t_k \mid x_{k-1}, t_{k-1})$.

Exercise 1.13 (The velocity process). Consider the velocity process of the physical Brownian motion (see [137, Section 1.2]) $y(t) = w(t) - \int_0^t e^{-(t-s)} w(s)\, ds$, and define the displacement process $x(t) = \int_0^t y(s)\, ds$.

 (i) Prove that $y(t)$ is a Markov process.
 (ii) Prove that $x(t)$ is not a Markov process.
 (iii) Prove that the two-dimensional process $z(t) = (x(t), y(t))$ is a Markov process.

                                                                                           □

Exercise 1.14 (The integrated MBM). Consider the integrated MBM, $y(t) = \int_0^t w(s)\, ds$.

 (i) Prove that $y(t)$ is not a Markov process.
 (ii) Prove that the two-dimensional process $z(t) = (w(t), y(t))$ is a Markov process.   □

1.3   Integration With Respect to MBM

1.3.1   The Itô Integral

The intuitive meaning of Definition 1.2.6 is that a stochastic process $f(t, \mathfrak{w})$ is *adapted* to the Brownian motion $w(t, \mathfrak{w})$ if it is independent of the increments of the Brownian motion $w(t, \mathfrak{w})$ "in the future," that is, $f(t, \mathfrak{w})$ is independent of $w(t + s, \mathfrak{w}) - w(t, \mathfrak{w})$ for all $s > 0$. For example, if $f(x)$ is an integrable deterministic function, then the functions $f(w(t, \mathfrak{w}))$ and $\int_0^t f(w(s, \mathfrak{w}))\, ds$ are adapted. We denote by $H_2[0, T]$ the class of adapted stochastic processes $f(t, \mathfrak{w})$ on an interval $[0, T]$ such that $\int_0^T \mathbb{E} f^2(s, \mathfrak{w}) \, ds < \infty$. Integration with respect to white noise is defined in this class of stochastic processes. The Itô integral of a function $f(t, \mathfrak{w}) \in H_2[0, T]$ is defined by the sums over partitions $0 \le t_0 < t_1 < \cdots < t_n = t \le T$ in the form

$$\sigma_n(t, \mathfrak{w}) = \sum_{i=1}^n f(t_{i-1}, \mathfrak{w}) \left[w(t_i, \mathfrak{w}) - w(t_{i-1}, \mathfrak{w}) \right]. \tag{1.84}$$

Note that the increment $\Delta_i w = w(t_i, \mathfrak{w}) - w(t_{i-1}, \mathfrak{w})$ is independent of $f(t_{i-1}, \mathfrak{w})$, because $f(t, \mathfrak{w})$ is adapted. It can be shown (see [113, Chapter 2], [137, Section 3.3]) that for any sequence of partitions of the interval, such that $\max_i (t_i - t_{i-1}) \to 0$, the sequence $\{\sigma_n(t, \mathfrak{w})\}$ converges in probability to the same limit, denoted by

$$(I) \int_0^t f(s, \mathfrak{w}) \, dw(s, \mathfrak{w}) \overset{\text{Pr}}{=} \lim_{\max_i (t_i - t_{i-1}) \to 0} \sigma_n(t, \mathfrak{w}), \tag{1.85}$$

and called the *Itô integral of* $f(t, \mathfrak{w})$. It can also be shown that the convergence in (1.85) is uniform in t with probability one, that is, on almost every trajectory $w(t, \mathfrak{w})$ of Brownian motion. The Itô integral is also an adapted stochastic process in Ω. It takes different values on different realizations $\mathfrak{w}$ of Brownian trajectories. If $f(t)$ is an integrable deterministic function, then

$$\int_0^t f(s) \, dw(s) \sim \mathcal{N} \left(0, \int_0^t f^2(s) \, ds \right).$$

For $f(t) \in H_2[0, T]$, and any $0 \le \tau \le t \le T$,

$$\mathbb{E} \int_0^t f(s) \, dw(s) = 0,$$

$$\mathbb{E} \left[\int_0^t f(s) \, dw(s) \,\middle|\, \int_0^\tau f(s) \, dw(s) = x \right] = xm \tag{1.86}$$

$$\mathbb{E} \left[\int_0^T f(s) \, dw(s) \right]^2 = \int_0^T \mathbb{E} f^2(s) \, ds, \tag{1.87}$$

and for $f(t), g(t) \in H_2[0, T]$,

$$\mathbb{E} \left[\int_0^T f(s) \, dw(s) \int_0^T g(s) \, dw(s) \right] = \int_0^T \mathbb{E} \left[f(s) g(s) \right] \, ds. \tag{1.88}$$

Property (1.86) follows from the construction of the Itô integral, and the independence of $f(t)$ from the increments $w(t'') - w(t')$ for all $t \le t' \le t''$. It is easy to see that properties (1.87) and (1.88) are equivalent.

Exercise 1.15 (Integral of $w(t, \mathfrak{w})$). Show that

$$(I) \int_a^b w(s) \, dw(s) = \frac{1}{2}[w^2(b) - w^2(a)] - \frac{1}{2}(b - a) \tag{1.89}$$

and derive an equation analogous to (1.86) for the conditional expectation

$$\mathbb{E}\left[\left(\int_0^t f(s) \, dw(s)\right)^2 \bigg| \int_0^\tau f(s) \, dw(s) = x\right].$$

$\square$

Exercise 1.16 (Conditional variance). Derive an equation analogous to (1.88) for

(i) the conditional expectations

$$\mathbb{E}\left[\int_0^T f(s) \, dw(s) \int_0^T g(s) \, dw(s) \bigg| \int_0^{\tau_1} f(s) \, dw(s) = x, \int_0^{\tau_2} g(s) \, dw(s) = y\right],$$

where $0 \le \tau_1 \le \tau_2 < T$,

(ii) the conditional expectations

$$\mathbb{E}\left[\int_0^{T_1} f(s) \, dw(s) \int_0^{T_2} g(s) \, dw(s) \bigg| \int_0^{\tau_1} f(s) \, dw(s) = x, \int_0^{\tau_2} g(s) \, dw(s) = y\right],$$

where $0 \le \tau_1 < T_1$, $0 \le \tau_2 < T_2$. Assume $\tau_1 < \tau_2 < T_2$ and consider the different possibilities for T_1. $\square$

1.3.2   The Stratonovich Integral

The Stratonovich integral for $f(t, \mathfrak{w}) \in H_2[0, T]$ is defined by the sums

$$\sigma_n(t, \mathfrak{w}) = \frac{1}{2} \sum_{i=1}^n [f(t_i, \mathfrak{w}) + f(t_{i-1}, \mathfrak{w})] [w(t_i, \mathfrak{w}) - w(t_{i-1}, \mathfrak{w})] \tag{1.90}$$

as

$$\lim_{n \to \infty} \sigma_n(t, \mathfrak{w}) = \int_0^t f(s, \mathfrak{w}) \, d_S w(s, \mathfrak{w}). \tag{1.91}$$

According to Theorem 1.3.1 below, the Stratonovich integral (1.90) exists for functions $f(t, \mathfrak{w}) \in H_2[0, T]$ that are independent of the Brownian motion $w(\cdot, \mathfrak{w})$.

If $f(t, \mathfrak{w}) = f(w(t, \mathfrak{w}), t) \in H_2[0, T]$, then we can replace in the integral sum
(1.84) the left endpoint t_{i-1} of the integrand with the midpoint $t_{\frac{1}{2}i} = \frac{1}{2}(t_i + t_{i-1})$ to
obtain

$$\tilde{\sigma}_n(t, \mathfrak{w}) = \sum_{i=1}^{n} f\left(w\left(t_{\frac{1}{2}i}, \mathfrak{w}\right), t_{i-1}\right) \Delta_i w, \qquad (1.92)$$

and get the limit

$$\int_a^b f(w(t, \mathfrak{w}), t) \, d_S w(t) \overset{\mathrm{Pr}}{=} \lim_{\max_i (t_i - t_{i-1}) \to 0} \tilde{\sigma}_n(t, \mathfrak{w}). \qquad (1.93)$$

Exercise 1.17 (Another Stratonovich sum). Show that

(i) If $f(x, t)$ is continuously differentiable in $\mathbb{R}$ and $f(w(t, \mathfrak{w}), t) \in H_2[0, T]$, then
    the sums (1.90), (1.93), and

$$\tilde{\tilde{\sigma}}_n(t, \mathfrak{w}) = \sum_{i=1}^{n} f\left(\frac{w(t_i, \mathfrak{w}) + w(t_{i-1}, \mathfrak{w})}{2}, t_{i-1}\right) \Delta_i w \qquad (1.94)$$

define the same Stratonovich integral (use the calculations of [137, Section
3.1]).
(ii) Show that

$$\int_a^b w(s, \mathfrak{w}) \, d_S w(s, \mathfrak{w}) = \frac{1}{2}[w^2(b, \mathfrak{w}) - w^2(a, \mathfrak{w})]. \qquad (1.95)$$

$\square$

The Stratonovich and Itô integrals are related by the following theorem.

Theorem 1.3.1 (The Wong–Zakai correction [156]). *If $f(x, t)$ has a continuous
derivative of second-order such that $|f_{xx}(x, t)| < A(t)e^{a(t)|x|}$ for some positive
continuous functions $\alpha(t)$ and $A(t)$ for all $a \le t \le b$, then*

$$\int_a^b f(w(t), t) \, d_S w(t) = \int_a^b f(w(t), t) \, dw(t) + \frac{1}{2} \int_a^b \frac{\partial}{\partial x} f(w(t), t) \, dt \qquad (1.96)$$

*in the sense that the left-hand side of (1.96) exists if and only if the right-hand side
exists and they are equal.*

Exercise 1.18. (The backward integral). The backward integral is defined by the
integral sums

$$\int_a^b f(t)\, d_B w(t) \stackrel{\text{Pr}}{=} \lim_{\max_i (t_i - t_{i-1}) \to 0} \sum_{i=1}^n f(t_i)\, [w(t_i) - w(t_{i-1})]. \tag{1.97}$$

(i) Show that the Wong–Zakai formula is now [137, Section 3.2.3]

$$\int_a^b f(w(t), t)\, d_B w(t) = \int_a^b f(w(t), t)\, dw(t) + \int_a^b \frac{\partial}{\partial x} f(w(t), t)\, dt. \tag{1.98}$$

(ii) Show that

$$\int_a^b w(s)\, d_B w(s) = \frac{1}{2}[w^2(b) - w^2(a)] + \frac{1}{2}(b - a). \tag{1.99}$$

(iii) Use the Wong–Zakai correction (1.96) and (1.98) to derive the relationship between the Stratonovich and the backward integrals. $\qquad\square$

1.3.3 Itô and Stratonovich Differentials

Consider two processes, $a(t), b(t)$, of class $H_2[0, T]$ and define the stochastic process

$$x(t) = x_0 + \int_0^t a(s)\, ds + \int_0^t b(s)\, dw(s), \tag{1.100}$$

where x_0 is a random variable independent of $w(t)$ for all $t > 0$. Then, for $0 \le t_1 < t_2 \le T$,

$$x(t_2) - x(t_1) = \int_{t_1}^{t_2} a(s)\, ds + \int_{t_1}^{t_2} b(s)\, dw(s). \tag{1.101}$$

We abbreviate this notation as

$$dx(t) = a(t)\, dt + b(t)\, dw(t). \tag{1.102}$$

If the Itô integral in (1.100) is replaced with the Stratonovich integral (1.91), then (1.102) is written as

$$d_S x(t) = a(t)\, dt + b(t)\, d_S w(t). \tag{1.103}$$

Example 1.6 (The Itô differential of $w^2(t)$). Equation (1.89) gives $w^2(t_2) - w^2(t_1) = 2\int_{t_1}^{t_2} w(t)\, dw(t) + \int_{t_1}^{t_2} 1\, dt$ for the process $x(t) = w^2(t)$. According to eqs.(1.101) and (1.102), this can be written as $dw^2(t) = 1\,dt + 2w(t)\,dw(t)$, that is, $a(t) = 1$ and $b(t) = 2w(t)$. If, however, the Itô integral in the definition (1.101) is replaced with the Stratonovich integral, then (1.95) gives $d_S w^2(t) = 2w(t)\,d_S w(t)$. Thus the Itô differential (1.102) does not satisfy the usual rule $dx^2 = 2x\,dx$.   □

Example 1.7 (The Itô differential of $f(t)w(t)$). If $f(t)$ is a smooth deterministic function, then integration by parts is possible so that

$$\int_{t_1}^{t_2} f(t)\,dw(t) = f(t_2)w(t_2) - f(t_1)w(t_1) - \int_{t_1}^{t_2} f'(t)w(t)\,dt.$$

Thus, setting $x(t) = f(t)w(t)$, we obtain

$$dx(t) = f'(t)w(t)\,dt + f(t)dw(t) = w(t)\,df(t) + f(t)\,dw(t),$$

as in the classical calculus. In this case, $a(t) = f'(t)w(t)$ and $b(t) = f(t)$. Note that the same conclusion holds if $f(t) = f(t, w)$ is a smooth function in $H_2[0, t]$.
□

1.3.4 The Chain Rule for Stochastic Differentials

The essence of the differentiation rules is captured in the chain rule for differentiating composite functions. Consider n Itô differentiable processes $dx^i = a^i\,dt + \sum_{j=1}^{m} b^{ij}\,dw^j$ for $i = 1, 2, \ldots, n$, where $a^i, b^{ij} \in H_2[0, T]$ for $i = 1, 2, \ldots, n$, $j = 1, 2, \ldots, m$, and w^j are independent Brownian motions, and a function $f(x^1, x^2, \ldots, x^n, t)$ that has continuous partial derivatives of second-order in $x^1, x^2, \ldots, x^n$ and a continuous partial derivative with respect to t. For an n-dimensional process $x(t)$ that is differentiable in the ordinary sense, the classical chain rule is

$$df(x(t), t) = \frac{\partial f\,(x(t), t)}{\partial t}\,dt + \nabla_x f\,(x(t), t) \cdot dx(t) \tag{1.104}$$

$$= \left(\frac{\partial f(x(t), t)}{\partial t} + \sum_{i=1}^{n} a^i(x(t), t) \frac{\partial f\,(x(t), t)}{\partial x^i} \right) dt$$

$$+ \sum_{i=1}^{n} \sum_{j=1}^{m} b^{ij}\,(x(t), t) \frac{\partial f\,(x(t), t)}{\partial x^i}\,dw^j.$$

For processes differentiable in the Itô sense, but not in the ordinary sense, (1.104) does not hold. Rather, we have the following theorem.

Theorem 1.3.2 (Itô's formula).

$$df\,(x(t),t) = \left[\frac{\partial f\,(x(t),t)}{\partial t} + \mathcal{L}_x^* f\,(x,t)\right] dt$$

$$+ \sum_{i=1}^{n}\sum_{j=1}^{m} b^{ij}\,(x(t),t)\,\frac{\partial f\,(x(t),t)}{\partial x^i}\,dw^j, \tag{1.105}$$

where

$$\mathcal{L}_x^* f\,(x,t) = \sum_{i=1}^{n}\sum_{j=1}^{n}\sigma^{ij}\,(x,t)\,\frac{\partial^2 f\,(x,t)}{\partial x^i\,\partial x^j} + \sum_{i=1}^{n} a^i\,(x,t)\,\frac{\partial f\,(x,t)}{\partial x^i} \tag{1.106}$$

and

$$\sigma^{ij}\,(x,t) = \frac{1}{2}\sum_{k=1}^{m} b^{ik}\,(x,t)\,b^{jk}\,(x,t). \tag{1.107}$$

The $n \times n$ matrix $\left\{\sigma^{ij}\,(x,t)\right\}$ is called the *diffusion matrix*. In matrix notation,

$$B\,(x,t) = \left\{b^{ij}\,(x,t)\right\}_{n\times m} \tag{1.108}$$

is the *noise matrix*, and the diffusion matrix $\sigma\,(x,t)$ is given by

$$\sigma\,(x,t) = \frac{1}{2}B\,(x,t)\,B^T\,(x,t).$$

The operator $\mathcal{L}_x^*$ in (1.106) is called the *backward Kolmogorov operator* (see [137, Section 3.4]).

Exercise 1.19 (Itô's formula in 1-D). Specialize Itô's formula (1.105) to the one-dimensional case: for a process $x(t)$ with differential $dx = a(t)\,dt + b(t)\,dw$, where $a(t), b(t) \in H_2[0,T]$, and a twice continuously differentiable function $f(x,t)$,

$$df(x(t),t) = \left[\frac{\partial f(x(t),t)}{\partial t} + a(t)\frac{\partial f(x(t),t)}{\partial x} + \frac{1}{2}b^2(t)\frac{\partial^2 f(x(t),t)}{\partial x^2}\right] dt$$

$$+ b(t)\frac{\partial f(x(t),t)}{\partial x}\,dw(t). \qquad \square$$

Exercise 1.20 (Itô's formula as the chain rule).

(i) Apply Itô's formula (1.105) to the function $f(x^1, x^2) = x^1 x^2$ and obtain the rule for differentiating a product.

(ii) Apply Itô's one-dimensional formula of Exercise 1.19 to the function $f(x) = e^x$. Obtain a differential equation for the function $y(t) = e^{\alpha w(t)}$.

(iii) Use the transformation $y = \log x$ to solve the linear stochastic differential equation

$$dx(t) = ax(t)\,dt + bx(t)\,dw(t), \quad x(0) = x_0. \tag{1.109}$$

Show that the solution cannot change sign.   □

Exercise 1.21 (Applications to moments).

(i) Use the one-dimensional Itô formula to prove

$$\mathbb{E}e^{w(t)} = 1 + \frac{1}{2}\int_0^t \mathbb{E}e^{w(s)}\,ds = e^{t/2}.$$

(ii) Calculate the first and the second moments of $e^{aw(t)}, e^{iw(t)}, \sin aw(t)$, and $\cos aw(t)$, where a is a real constant.   □

Exercise 1.22 (Rotation of white noise). If $w_1(t), w_2(t)$ are independent Brownian motions and $x(t)$ is a process in $H_2[0, T]$, the processes $u_1(t), u_2(t)$ can be defined by their differentials

$$du_1(t) = -\sin x(t)\,dw_1(t) + \cos x(t)\,dw_2(t),$$
$$du_2(t) = \cos x(t)\,dw_1(t) + \sin x(t)\,dw_2(t).$$

Show that $u_1(t)$ and $u_2(t)$ are independent Brownian motions.   □

Theorem 1.3.3 (The Stratonovich chain rule). *The chain rule for Stratonovich differentials is the usual rule* (1.104), *that is,*

$$d_S f(x, t) = \frac{\partial f}{\partial x}\,d_S x + \frac{\partial f}{\partial t}\,dt = \left[\frac{\partial f}{\partial t} + a\frac{\partial f}{\partial x}\right]dt + b\frac{\partial f}{\partial x}\,d_S w. \tag{1.110}$$

Proof. First, we convert the Stratonovich equation (1.103) to Itô's form by introducing the Wong–Zakai correction,

$$dx(t) = \left(a(t) + \frac{1}{2}\frac{\partial b}{\partial w}\right)dt + b(t)\,dw(t). \tag{1.111}$$

If the dependence of $b(t)$ on $w(t)$ is expressed as

$$b(t) = B\left(x(t), t\right),$$

where $B(x,t)$ is a differentiable function in both variables, then the Wong–Zakai correction is found as follows:

$$\frac{\Delta b(t)}{\Delta w(t)} = \frac{\Delta B(x(t),t)}{\Delta x(t)} \frac{\Delta x(t)}{\Delta w(t)} = \frac{\Delta B(x(t),t)}{\Delta x(t)} \frac{a(t)\,\Delta t + b(t)\,\Delta w(t) + o(\Delta t)}{\Delta w(t)}$$

$$= \frac{\partial B(x(t),t)}{\partial x} \left[b(t) + a(t) O\left(\frac{\Delta t}{\Delta w(t)} \right) \right].$$

Note that

$$\Pr\left\{ \left| \frac{\Delta t}{\Delta w} \right| > \varepsilon \right\} = \Pr\left\{ |\Delta w| < \frac{\Delta t}{\varepsilon} \right\} = \frac{1}{\sqrt{2\pi \Delta t}} \int_{-\Delta t/\varepsilon}^{\Delta t/\varepsilon} e^{-x^2/2\Delta t}\, dx$$

$$= \frac{1}{\sqrt{2\pi}} \int_{-\sqrt{\Delta t}/\varepsilon}^{\sqrt{\Delta t}/\varepsilon} e^{-z^2/2}\, dz \to 0 \quad \text{as} \quad \Delta t \to 0,$$

so that $\lim_{\Delta t \to 0} \Delta t / \Delta w(t) \overset{\Pr}{=} 0$. It follows that in this case the Wong–Zakai correction is

$$\frac{1}{2} \frac{\partial b}{\partial w} = \frac{1}{2} B(x,t) \frac{\partial B(x,t)}{\partial x}. \tag{1.112}$$

Next, from Itô's formula and (1.111), we have

$$f(x(t),t) = f(x(t_0),t_0) + \int_{t_0}^{t} \left\{ \frac{\partial f(x(s),s)}{\partial t} \right.$$

$$+ \left[a(s) + \frac{1}{2} \frac{\partial b(s)}{\partial w(s)} \right] \frac{\partial f(x(s),s)}{\partial x} + \frac{1}{2} b^2(s) \frac{\partial^2 f(x(s),s)}{\partial x^2} \right\}\, ds$$

$$+ \int_{t_0}^{t} b(s) \frac{\partial f(x(s),s)}{\partial x}\, dw(s).$$

Now we convert the Itô integral into a Stratonovich integral using the Wong–Zakai correction:

$$\int_{t_0}^{t} b(s) \frac{\partial f(x(s),s)}{\partial x}\, dw(s) = \int_{t_0}^{t} b(s) \frac{\partial f(x(s),s)}{\partial x}\, dw_S(s)$$

$$- \frac{1}{2} \int_{t_0}^{t} \frac{\partial}{\partial w(s)} \left[b(s) \frac{\partial f(x(s),s)}{\partial x} \right]\, ds. \tag{1.113}$$

Using the differentiation rule (1.112), we find that

$$\frac{\partial}{\partial w(t)} \left[b(t) \frac{\partial f(x(t),t)}{\partial x} \right] = \frac{\partial b(t)}{\partial w(t)} \frac{\partial f(x(t),t)}{\partial x} + \frac{\partial^2 f(x(t),t)}{\partial x^2} b^2(t), \tag{1.114}$$

so (1.113) gives

$$
\begin{aligned}
f(x(t), t) = f(x(t_0), t_0) + \int_{t_0}^{t} \Bigg\{ & \frac{\partial f(x(s), s)}{\partial t} \\
& + \left[a(s) + \frac{1}{2}\frac{\partial b(s)}{\partial w(s)} \right] \frac{\partial f(x(s), s)}{\partial x} + \frac{1}{2}b^2(s)\frac{\partial^2 f(x(s), s)}{\partial x^2} \Bigg\} \, ds \\
& + \int_{t_0}^{t} b(s)\frac{\partial f(x(s), s)}{\partial x} \, dw_S(s) \\
& - \frac{1}{2}\int_{t_0}^{t} \left[\frac{\partial b(t)}{\partial w(t)}\frac{\partial f(x(t), t)}{\partial x} + \frac{\partial^2 f(x(t), t)}{\partial x^2}b^2(t) \right] ds \\
= f(x(t_0), t_0) + & \int_{t_0}^{t} \left[f_t(x(s), s) + a(s)f_x(x(s), s) \right] ds \\
& + \int_{t_0}^{t} b(s)f_x(x(s), s) \, dw_S(s),
\end{aligned}
$$

as asserted. In differential form this is identical to (1.110).   □

Thus the differentials in Exercises 1.6 and 1.7 are $d_S w^2(t) = 2w(t)\,d_S dw(t)$ and $d_S f(t)w(t) = w(t)\,df(t) + f(t)\,d_S w(t)$.

Theorem 1.3.4 (The Stratonovich chain rule). *The chain rule for Stratonovich differentials is the usual rule* (1.104), *that is,*

$$
d_S f(x, t) = \frac{\partial f}{\partial x}\,d_S x + \frac{\partial f}{\partial t}\,dt = \left[\frac{\partial f}{\partial t} + a\frac{\partial f}{\partial x} \right] dt + b\frac{\partial f}{\partial x}\,d_S w.
$$

1.4 Itô and Stratonovich SDEs

Dynamics driven by white noise, often written as

$$
dx = a(x, t)\,dt + B(x, t)\,dw, \quad x(0) = x_0, \tag{1.115}
$$

is usually understood as the integral equation

$$
x(t) = x(0) + \int_0^t a(x(s), s)\,ds + \int_0^t B(x(s), s)\,dw(s), \tag{1.116}
$$

where $a(x, t)$ and $B(x, t)$ are random coefficients, which can be interpreted in several different ways, depending on the interpretation of the stochastic integral in (1.116) as Itô, Stratonovich, or otherwise. Different interpretations lead to very different solutions and to qualitative differences in the behavior of the solution. For example, a noisy dynamical system of the form (1.115) may be stable if the Itô integral is used in (1.116), but unstable if the Stratonovich or the backward integral (see Exercise 1.18) is used instead (see Exercise 1.30 below). Different interpretations lead to different numerical schemes for the computer simulation of the equation. A different approach, based on path integrals, is given in Chapter 2.

In modeling stochastic dynamics with equations of the form (1.116), a key question arises of which of the possible interpretations is the right one to use. This question is particularly relevant if the noise is state dependent, that is, if the coefficients $\boldsymbol{B}(\boldsymbol{x}, t)$ depend on $\boldsymbol{x}$. This situation is encountered in many different applications, for example, when the friction coefficient or the temperature in Langevin's equation is not constant. The answer to this question depends on the origin of the noise. The correlationless white noise (or the nondifferentiable MBM) is an idealization of a physical process that may have finite, though short, correlation time (or differentiable trajectories). The white noise approximation may originate in a model with discontinuous paths in the limit of small or large frequent jumps, and so on.

Thus, the choice of the integral in (1.116) is not arbitrary, but rather derives from the underlying more microscopic model and from the passage to the white noise limit. In certain situations this procedure leads to an Itô interpretation and in others to a Stratonovich interpretation. The limiting procedures are described in [137, Section 3.2]. In this chapter, we consider the Itô and Stratonovich interpretations and their interrelationship. The backward interpretation is left as an exercise.

1.4.1 Stochastic Differential Equations of Itô Type

First, we consider the one-dimensional version of equation (1.115) and interpret it in the Itô sense as the output of an Euler numerical scheme of the form

$$x_E(t + \Delta t, \mathfrak{w}) = x_E(t, \mathfrak{w}) + a(x_E(t, \mathfrak{w}), \mathfrak{w})\Delta t + b(x_E(t, \mathfrak{w}), \mathfrak{w})\Delta w(t, \mathfrak{w})$$
$$(1.117)$$

in the limit $\Delta t \to 0$. To each realization of the MBM $w(t, \mathfrak{w})$ constructed numerically, for example, by any of the methods of Section 1.2.6, equation (1.117) assigns a realization $x_E(t, \mathfrak{w})$ of the solution at grid points. Because $\Delta w(t, \mathfrak{w}) = w(t + \Delta t, \mathfrak{w}) - w(t, \mathfrak{w})$ is a Gaussian random variable, the right-hand side of (1.117) can assume any value in $\mathbb{R}$, so that $x_E(t, \mathfrak{w})$ can assume any value at every time t. This implies that $a(x, t, \mathfrak{w})$ and $b(x, t, \mathfrak{w})$ have to be defined for all $x \in \mathbb{R}$. If for each $x \in \mathbb{R}$ the random coefficients $a(x, t, \mathfrak{w})$ and $b(x, t, \mathfrak{w})$ are adapted processes, say of class $H_2[0, T]$ for all $T > 0$, the output process $x_E(t, \mathfrak{w})$ is also an adapted process.

The output process at grid times $t_j = j\Delta t$, given by

$$x_E(t_j, \mathfrak{w}) = x_0 + \sum_{k=0}^{j-1} [a(x_E(t_k), t_k, \mathfrak{w})\Delta t + b(x_E(t_k), t_k, \mathfrak{w})\Delta w(t_k, \mathfrak{w})], \quad (1.118)$$

has the form of two integral sums. The first one is for the Riemann integral $\int_0^t a(x(s, \mathfrak{w}), s, \mathfrak{w})\, ds$ and the other is for the stochastic Itô integral $\int_0^t b(x(s, \mathfrak{w}),$ $s, \mathfrak{w})\, dw(s, \mathfrak{w})$, where $x(t, \mathfrak{w}) = \lim_{\Delta t \to 0} x(t_j, \mathfrak{w})$ for $t_j \to t$ if the limit exists in some sense.

If the coefficients $a(x, t, \mathfrak{w})$ and $b(x, t, \mathfrak{w})$ are adapted processes (of class $H_2[0, T]$ for all $T > 0$), equation (1.115) is written in the Itô form

$$dx = a(x, t, \mathfrak{w})\, dt + b(x, t, \mathfrak{w})\, dw(t, \mathfrak{w}), \quad x(0, \mathfrak{w}) = x_0, \quad (1.119)$$

or as an equivalent integral equation

$$x(t, \mathfrak{w}) = x_0 + \int_0^t a(x(s, \mathfrak{w}), s, \mathfrak{w})\, ds + \int_0^t b(x(s, \mathfrak{w}), s, \mathfrak{w})\, dw(s, \mathfrak{w}). \quad (1.120)$$

The initial condition x_0 is assumed independent of $w(t)$.

There are several different definitions of a solution to the stochastic differential equation (1.119), including strong, weak, a solution to the martingale problem, path integral interpretation (see Chapter 2), and so on. Similarly, there are several different notions of uniqueness, including uniqueness in the strong sense, pathwise uniqueness, and uniqueness in probability law. For the definitions and relationships between the different definitions, (see [105], [79]). We consider here only strong solutions (abbreviated as *solutions*) of (1.119).

Definition 1.4.1 (Solution of an SDE). *A stochastic process $x(t, \mathfrak{w})$ is a solution of the initial value problem* (1.119) *in the Itô sense if*

I. $x(t, \mathfrak{w}) \in H_2[0, T]$ for all $T > 0$
II. Equation (1.120) *holds for almost all $\mathfrak{w} \in \Omega$.*

We assume that the coefficients $a(x, t, \mathfrak{w})$ and $b(x, t, \mathfrak{w})$ satisfy the *uniform Lipschitz condition*, that is, there exists a constant K such that

$$|a(x, t, \mathfrak{w}) - a(y, t, \mathfrak{w})| + |b(x, t, \mathfrak{w}) - b(y, t, \mathfrak{w})| \leq K|x - y| \quad (1.121)$$

for all $x, y \in \mathbb{R}$, $t \geq 0$, and $\mathfrak{w} \in \Omega$.

Theorem 1.4.1 (Existence and uniqueness). *If $a(x, t, \mathfrak{w})$ and $b(x, t, \mathfrak{w})$ satisfy the Lipschitz condition* (1.121), *uniformly for all x, t, and for almost all $\mathfrak{w} \in \Omega$, then there exists a unique solution to the initial value problem* (1.119). *Its trajectories are continuous with probability* 1.

So far the solution of an SDE has been defined on the entire line $\mathbb{R}$. However, if $D \subset \mathbb{R}^n$ is a domain such that $x_0 \in D$ and the coefficients $a(x, t, \mathfrak{w})$ and $b(x, t, \mathfrak{w})$

are not defined for all x, but only for $x \in D$, the definition of the solution has to be modified. To this end we denote the first exit time from D by

$$\tau_\Omega(\mathfrak{w}) = \inf\{t > 0 \mid x(t, \mathfrak{w}) \notin D\}.$$

First, we need the following theorem [60].

Theorem 1.4.2 (Localization principle). *Assume that $a_i(x, t, \mathfrak{w})$ and $b_i(x, t, \mathfrak{w})$ $(i = 1, 2)$ satisfy the Lipschitz condition uniformly for all $x \in \mathbb{R}, t \geq 0$, and*

$$a_1(x, t, \mathfrak{w}) = a_2(x, t, \mathfrak{w}), \quad b_1(x, t, \mathfrak{w}) = b_2(x, t, \mathfrak{w}),$$

for all $x \in D$, $\mathfrak{w} \in \Omega$, $t \geq 0$, and that $x_0 \in D$. Let $x_1(t, \mathfrak{w})$ and $x_2(t, \mathfrak{w})$ be the solutions of

$$dx_i(t, \mathfrak{w}) = a_i(x, t, \mathfrak{w})\, dt + b_i(x, t, \mathfrak{w})\, dw(t, \mathfrak{w}), \quad x_i(0, \mathfrak{w}) = x_0, \quad i = 1, 2,$$

respectively, and let $\tau_1(\mathfrak{w})$, $\tau_2(\mathfrak{w})$ be their first exit times from D. Then $\tau_1(\mathfrak{w}) = \tau_2(\mathfrak{w})$ with probability 1, and $x_1(t, \mathfrak{w}) = x_2(t, \mathfrak{w})$ for all $t < \tau_1(\mathfrak{w})$ and almost all $\mathfrak{w} \in \Omega$.

The localization theorem can be used to define solutions to Itô equations in finite domains. Assume that $a(x, t, \mathfrak{w})$ and $b(x, t, \mathfrak{w})$ are defined only for $x \in D$ and satisfy there the Lipschitz condition and can be extended to all $x \in \mathbb{R}$ as uniformly Lipschitz functions. Then solutions are well defined for the extended equations. The localization principle ensures that all solutions, corresponding to different extensions, are the same for all $t < \tau_\Omega(\mathfrak{w})$, that is, as long as the solutions do not leave D.

Exercise 1.23 (Proof of localization*). Prove the localization principle (see [60], [137, Theorem 4.1.2]).   □

Exercise 1.24 (Growth estimate). Use Itô's formula to show that if $\mathbb{E}x_0^{2m} < \infty$, then the solution of (1.120) satisfies the inequality $\mathbb{E}x^{2m}(t, \mathfrak{w}) \leq \mathbb{E}\left(1 + x_0^{2m}\right) e^{Ct}$, where C is a constant.   □

Exercise 1.25 (Modulus of continuity). Show that

$$\mathbb{E}|x(t, \mathfrak{w}) - x(0)|^{2m} \leq C_1 \mathbb{E}\left(1 + |x_0|^{2m}\right) e^{C_2 t} t^m, \tag{1.122}$$

where C_1 is another constant.   □

Exercise 1.26 (Test of uniqueness). For which values of α does the equation $dx = |x|^\alpha\, dw$ have a unique solution satisfying the initial condition $x(0) = 0$?   □

Exercise 1.27 (Example of nonuniqueness). For any $T \geq 0$, denote by $\tau_T(\mathfrak{w})$ the first passage time of the MBM to the origin after time T, that is, $\tau_T(\mathfrak{w}) = \inf\{s \geq T \mid w(s, \mathfrak{w}) = 0\}$. Show that the stochastic equation $dx = 3x^{1/3}\, dt + 3x^{2/3}\, dw$,

with the initial condition $x(0) = 0$, has infinitely (uncountably) many solutions of
the form

$$x_T(t, \mathfrak{w}) = \begin{cases} 0 & \text{for} \quad 0 \le t < \tau_T(\mathfrak{w}), \\ w^3(t, \mathfrak{w}) & \text{for} \quad t \ge \tau_T(\mathfrak{w}). \end{cases}$$

This example is due to Itô and Watanabe.   □

Next, we consider a system of Itô equations of the form

$$dx^i = a^i(x, t)\, dt + \sum_{j=1}^{m} b^{ij}(x, t)\, dw^j, \quad x^i(0) = x_0^i, \; i = 1, 2, \dots, n, \quad (1.123)$$

where $w^j(t)$ are independent MBMs and $x = (x^1, x^2, \dots, x^n)$. If the coefficients
satisfy a uniform Lipschitz condition, the proofs of the existence and uniqueness
theorem and of the localization principle are generalized in a straightforward
manner to include the case of systems of the form (1.123).

Exercise 1.28 (Existence and uniqueness for (1.123)). Generalize the above
existence and uniqueness theorem and the localization principle for the system
(1.123).   □

1.4.2 Change of Time Scale

In changing the independent variable in Itô equations, the Brownian scaling laws of
Exercise 1.4 have to be borne in mind. Thus, changing the time scale $t = \alpha s$, where
α is a constant, transforms the Brownian motion and its differential as follows:

$$w(t) = w(\alpha s) = \sqrt{\alpha} \left[\frac{1}{\sqrt{\alpha}} w(\alpha s) \right] = \sqrt{\alpha} w_\alpha(s), \quad (1.124)$$

where $w_\alpha(s)$ is a Brownian motion. The differential $dw(t)$ is expressed in terms of
the differential $dw_\alpha(s)$ as

$$d_t w(t) = d_t w(\alpha s) = \sqrt{\alpha}\, d_s w_\alpha(s). \quad (1.125)$$

Setting $x(t) = x_\alpha(s)$, the integral equation (1.120) becomes

$$x_\alpha(s) = x_0 + \alpha \int_0^s a(x_\alpha(u), \alpha u)\, du + \sqrt{\alpha} \int_0^s b(x_\alpha(u), \alpha u)\, dw_\alpha(u). \quad (1.126)$$

The Itô differential equation (1.119) is therefore transformed into

$$dx_\alpha(s) = \alpha\, a(x_\alpha(s), \alpha s)\, ds + \sqrt{\alpha}\, b(x_\alpha(s), \alpha s)\, dw_\alpha(s). \quad (1.127)$$

1.4.3   Conversion of SDEs Between Different Forms

The conversion of SDEs between the Itô and Stratonovich types uses Theorem 1.3.1. More specifically, the Itô equation equivalent to the Stratonovich SDE (1.103) is obtained by applying the identity (1.112) in the Wong–Zakai correction (1.96). The resulting Itô equation is given by

$$ dx = \left[a(x,t) + \frac{1}{2} b(x,t) \frac{\partial}{\partial x} b(x,t) \right] dt + b(x,t)\, dw. \tag{1.128} $$

To convert in the other direction, that is, from Itô form to Stratonovich form, the Wong–Zakai correction is subtracted. Thus, the Itô equation $dx = a(x,t)\, dt + b(x,t)\, dw$ is converted to the equivalent Stratonovich form $d_S x = [a(x,t) - \frac{1}{2} b(x,t) b_x(x,t)]\, dt + b(x,t)\, d_S w$.

In d dimensions the Stratonovich system

$$ \boldsymbol{x}(t) = \boldsymbol{x}(0) + \int_0^t \boldsymbol{a}(\boldsymbol{x}(s), s)\, ds + \int_0^t \boldsymbol{B}(\boldsymbol{x}(s), s)\, d\boldsymbol{w}_S(s) \tag{1.129} $$

is converted to Itô form by the Wong–Zakai correction

$$ dx^i(t) = \left[a^i(\boldsymbol{x}(t), t) + \frac{1}{2} \sum_{k=1}^{n} \sum_{j=1}^{m} b^{i,j}(\boldsymbol{x}(t), t) \frac{\partial}{\partial x^k} b^{ij}(\boldsymbol{x}(t), t) \right] dt $$

$$ + \sum_{j=1}^{m} b^{kj}(\boldsymbol{x}(t), t)\, dw^j. $$

Exercise 1.29 (The differential of exp{w(t)}). Set $x(t) = e^{w(t)}$. Show that $d_S x(t) = x(t)\, d_S w(t)$ and $dx(t) = x(t)\, d_I w(t) + \frac{1}{2} x(t)\, dt$. This can be done by power series expansion or using Itô's formula.   □

Exercise 1.30 (Stability of Itô and Stratonovich linear equations).

 (i) Convert the stochastic linear differential equation (1.109),

$$ dx(t) = ax(t)\, dt + bx(t)\, dw(t), \quad x(0) = x_0, \tag{1.130} $$

   from Itô to Stratonovich form.

(ii) Convert the Stratonovich linear equation

$$ d_S x(t) = ax(t)\, dt + bx(t)\, d_S w(t), \quad x(0) = x_0, \tag{1.131} $$

   to Itô form.

(iii) Evaluate $\mathbb{E}x(t)$ for the solutions of (1.130) and of (1.131)
(iv) Find values of a and b such that the origin is stable for (1.131) but unstable for (1.130) in the sense that $\mathbb{E}x(t)$ decays or diverges in time. Are there real values of a and b such that the stability of the origin is reversed between the two forms?
(v) Repeat the steps (i)–(iv) for the nonlinear equation

$$dx(t) = ax(t)\,dt + b\sqrt{1 + x^2(t)}\,dw(t), \quad x(0) = x_0, \qquad (1.132)$$

$\square$

1.4.4   The Markov Property

It was shown above that the Markov property (1.78) implies that the transition probability density function of a Markov process $p(y, t \mid x, s)$ can be expressed in terms of the transitions probabilities at intermediate times by the Chapman–Kolmogorov equation (1.80). The solution of the Itô SDE (1.119) also has the Markov property, that is, it is a Markov process. Indeed, for $t > s$,

$$x(t) = x(s) + \int_s^t a(x(u), u)\,du + \int_s^t b(x(u), u)\,dw(u), \qquad (1.133)$$

and the existence and uniqueness theorem asserts that the initial condition $x(s)$ determines the solution of the Itô integral equation (1.133) uniquely. Because $a, b \in H[0, T]$ and dw is a forward difference of Brownian motion, the solution in the interval $[s, t]$ depends only on $x(s)$ and on a, b, and the increments of w in this interval. It follows from (1.133) that for $t > s > s_1 > \cdots > s_n$,

$$\Pr\{x(t) < x \mid x(s) = x_0,\, x(s_1) = x_1,\, \ldots,\, x(s_n) = x_n,\}$$
$$= \Pr\{x(t) < x \mid x(s) = x_0\},$$

which means that $x(t)$ is a Markov process.

1.4.5   Diffusion Processes

Definition 1.4.2 (Diffusion process in $\mathbb{R}$). *A one-dimensional Markov process $x(t)$ is called a diffusion process with (deterministic) drift $a(x, t)$ and (deterministic) diffusion coefficient $b^2(x, t)$ if it has continuous trajectories,*

$$\lim_{\Delta t \to 0} \frac{1}{\Delta t} \mathbb{E}\left\{ x(t + \Delta t) - x(t) \mid x(t) = x \right\} = a(x, t), \qquad (1.134)$$

$$\lim_{\Delta t \to 0} \frac{1}{\Delta t} \mathbb{E}\left\{ [x(t + \Delta t) - x(t)]^2 \mid x(t) = x \right\} = b^2(x, t), \qquad (1.135)$$

and for some $\delta > 0$,

$$\lim_{\Delta t \to 0} \frac{1}{\Delta t} \mathbb{E}\left\{ [x(t + \Delta t) - x(t)]^{2+\delta} \mid x(t) = x \right\} = 0. \qquad (1.136)$$

Definition 1.4.3 (Diffusion process in $\mathbb{R}^d$). *A d-dimensional Markov process $x(t)$ is called a diffusion process with (deterministic) drift $a(x, t)$ and (deterministic) diffusion Matrix $\sigma(x, t)$ if it has continuous trajectories,*

$$\lim_{\Delta t \to 0} \frac{1}{\Delta t} \mathbb{E}\left\{ x(t + \Delta t) - x(t) \mid x(t) = x \right\} = a(x, t),$$

$$\lim_{\Delta t \to 0} \frac{1}{\Delta t} \mathbb{E}\left\{ [x^i(t + \Delta t) - x^i(t)] \left[x^j(t + \Delta t) - x^j(t) \right] \mid x(t) = x \right\} = \sigma^{ij}(x, t)$$
$$(1.137)$$

for $i, j = 1, 2, \ldots, d$, and for some $\delta > 0$,

$$\lim_{\Delta t \to 0} \frac{1}{\Delta t} \mathbb{E}\left\{ |x(t + \Delta t) - x(t)|^{2+\delta} \mid x(t) = x \right\} = 0.$$

Theorem 1.4.3 (SDEs and diffusions). *Solutions of the Itô SDE (1.119) are diffusion processes.*

See [137, Theorem 4.3.1] for a proof. Theorem 1.4.3 holds also for the solution of a system of Itô SDEs

$$dx(t) = a(x, t)\, dt + B(x, t)\, dw, \quad x(0) = x_0, \qquad (1.138)$$

where

$$x(t) = \left(x^1(t), \ldots, x^d(t) \right)^T,$$

$$a(x, t) = \left[a^1 \left(x^1(t), \ldots, x^d(t) \right), \ldots, a^d \left(x^1(t), \ldots, x^d(t) \right) \right]^T,$$

$$B(x, t) = \left\{ b^{ij} \left(x^1(t), \ldots, x^d(t) \right) \right\}_{i \leq d, j \leq m},$$

$$w(t) = \left(w^1(t), \ldots, w^m(t) \right)^T,$$

$w^i(t)$ are independent Brownian motions, $a^i, a^{ij} \in H[0, T]$, and x_0 is independent of $w(t)$ and a^i, b^{ij}. Also in this case the existence and uniqueness theorem implies that the solution is a d-dimensional Markov process with continuous trajectories and that it is a diffusion process with drift vector $a(x, t)$ and diffusion matrix

$$\sigma(x, t) = \frac{1}{2} B(x, t) B^T(x, t).$$

Also a partial converse is true: assume that $x(t)$ is a diffusion process with (deterministic) drift $a(x, t)$ and (deterministic) diffusion matrix $\sigma(x, t)$. If $a(x, t)$ is a uniformly Lipschitz continuous vector and $\sigma(x, t)$ is a uniformly Lipschitz continuous strictly positive definite matrix, then there exists a uniformly Lipschitz continuous matrix $B(x, t)$ and a Brownian motion $w(t)$ such that $x(t)$ is a solution of (1.138) (see, e.g., [79]).

1.5 SDEs and Partial Differential Equations

Many useful functionals of solutions of stochastic differential equations, such as the transition probability density function, conditional and weighted expectations, functionals of the first passage times, escape probabilities from a given domain, and others, can be found by solving *deterministic* partial differential equations. These include Kolmogorov's representation formulas, the Andronov–Vitt–Pontryagin equation for the expected first passage time [127], [126], the Feynman–Kac formula for the transition pdf when trajectories can be terminated at random times, and so on. These partial differential equations reflect the continuum macroscopic properties of the underlying stochastic dynamics of the individual trajectories.

Throughout this section, $x_{x,s}(t)$ with $t > s$ denotes the solution of the Itô system

$$dx(t) = a(x(t), t)\,dt + B(x(t), t)\,dw(t), \quad x(s) = x, \tag{1.139}$$

where $a(x, t) : \mathbb{R}^d \times [0, T] \mapsto \mathbb{R}^d$, $B(x, t) : \mathbb{R}^d \times [0, T] \mapsto \mathbb{M}_{n,m}$, and $w(t)$ is an m-dimensional Brownian motion. We assume that $a(x, t)$ and $B(x, t)$ satisfy the conditions of the existence and uniqueness theorem.

1.5.1 The Feynman–Kac Representation and Killing

The Feynman–Kac formula provides a representation of the solution to a backward parabolic terminal value problem of the form

$$\frac{\partial v(x, t)}{\partial t} + \mathcal{L}_x^* v(x, t) + g(x, t) v(x, t) = 0, \quad t < T, \tag{1.140}$$

$$\lim_{t \uparrow T} v(x, t) = f(x), \tag{1.141}$$

where $\mathcal{L}_x^*$ is the backward Kolmogorov operator (1.106), and $g(x,t)$ and $f(x)$ are given sufficiently smooth functions, as a conditional expectation of a certain functional of the solution to the Itô system (1.139).

Theorem 1.5.1 (The Feynman–Kac formula). *Assume that the initial value problem* (1.139) *and the terminal value problem* (1.140), (1.141) *have unique solutions. Then*

$$v(x,s) = \mathbb{E}\left[f(x(T))\exp\left\{ \int_s^T g(x(t),t)\,dt \right\} \,\Big|\, x(s) = x \right], \qquad (1.142)$$

where $x(t) = x_{x,s}(t)$ *is the solution of the Itô system* (1.139) *for* $t > s$ *with the initial condition* $x(s) = x$.

The proof is given in [137, Theorem 4.4.2] (see also Exercise 2.8 below). The Feynman–Kac formula can be interpreted as the expectation of $f(x(T))$, where $x(t)$ is a solution of the stochastic dynamics (1.139) whose trajectories can terminate at any point and at any time with a certain probability. Such dynamics are referred to as *stochastic dynamics with killing*. The *killing rate* $-g(x,t)$ is defined as follows. Assume that at each point x and time t there is a probability $-g(x,t)$ per unit time that the trajectory of the solution $x(t)$ terminates there and then, independently of the past. Partition the time interval $[t,T]$ into N small intervals of length Δt, $t = t_0 < t_1 < \cdots < T$. Then the probability at time t that the solution $x(t)$ survives by time T is the product of the probabilities that it survives each one of preceding N time intervals,

$$\mathrm{Pr}_N \{\text{killing time} > T\} = \prod_{i=1}^N [1 + g(x(t_i),t_i)\,\Delta t] + o(\Delta t). \qquad (1.143)$$

Under mild assumptions on the regularity of $g(x,t)$, the limit of the product (1.143) as $N \to \infty$,

$$\mathrm{Pr}\{\text{killing time} > T\} = \lim_{N \to \infty} \mathrm{Pr}_N \{\text{killing time} > T\}$$

$$= \exp\left\{ \int_t^T g(x(t'),t')\,dt' \right\}, \qquad (1.144)$$

is uniform for $t \le T \le T_0$ for every continuous trajectory $x(s)$ and $T_0 < \infty$. Hence,

$$\mathbb{E}\,[f(x(T)), \text{killing time} > T \mid x(t) = x]$$

$$= \mathbb{E}\left[f(x(T))\exp\left\{ \int_t^T g(x(s),s)\,ds \right\} \,\Big|\, x(t) = x \right],$$

which is (1.142).

Exercise 1.31 (Representation for an inhomogeneous problem). Use Itô's formula to derive the representation

$$v(x, t) = \mathbb{E} \left[\int_t^T f(x(s), s) \, ds \, \Big| \, x(t) = x \right]$$

for the solution of the terminal value problem

$$\frac{\partial v(x, t)}{\partial t} + \mathcal{L}_x^* v(x, t) + f(x, t) = 0 \; \text{for } t < T, \; x \in \mathbb{R}^d, \; \lim_{t \uparrow T} v(x, t) = 0.$$

$\square$

1.5.2 The Andronov–Vitt–Pontryagin Equation

The first passage time of the solution $x(t)$ of (1.139) to the boundary ∂D of a domain D is a random variable defined on Brownian trajectories, as in Exercise 1.4 (or on the trajectories of (1.139)), by

$$\tau_D(\mathfrak{w}) = \inf \{t > s \mid x(t, \mathfrak{w}) \notin D\}.$$

The mean first passage time (MFPT) from any point $x \in D$ to the boundary ∂D is defined as the conditional expectation $\mathbb{E}[\tau_D(\mathfrak{w}) \mid x(s, \mathfrak{w}) = x]$. If it is finite, then the following theorem holds [127], [126].

Theorem 1.5.2 (The Andronov–Vitt–Pontryagin formula). *Assume that the boundary value problem*

$$\frac{\partial u(x, s)}{\partial s} + \mathcal{L}_x^* u(x, s) = -1 \; \text{for } x \in D, \text{ for all } s \in \mathbb{R}, \qquad (1.145)$$

$$u(x, s) = 0 \; \text{for } x \in \partial D, \qquad (1.146)$$

where $\mathcal{L}_x^*$ is the backward Kolmogorov operator (1.106), has a unique bounded solution. Then the MFPT $\mathbb{E}[\tau_D(\mathfrak{w}) \mid x(s, \mathfrak{w}) = x]$ of the solution $x(t)$ of (1.139) from every point x in a bounded domain D to the boundary ∂D is finite and

$$\mathbb{E}[\tau_D(\mathfrak{w}) \mid x(s, \mathfrak{w}) = x] = s + u(x, s). \qquad (1.147)$$

The assumptions of the theorem are satisfied if the coefficients are continuously differentiable functions and $\sigma(x, t)$ is a uniformly positive definite matrix in the domain. If the coefficients a and B are independent of t, the solution of (1.145) is independent of s, so that the backward parabolic boundary value problem (1.145), (1.146) reduces to the elliptic boundary value problem of Andronov, Vitt, and Pontryagin:

$$\mathcal{L}_x^* u(x) = -1 \text{ for } x \in D, \quad u(x) = 0 \text{ for } x \in \partial D. \tag{1.148}$$

The representation formula (1.147) simplifies to $\mathbb{E}\left[\tau_D(w) \mid x(0, w) = x\right] = u(x)$. The proof is given in Section 1.5.4 below. It can be shown that if the boundary value problem (1.148) has a finite solution, the MFPT is finite [60].

Example 1.8 (The MFPT of the MBM). To find the mean exit time of Brownian motion from an interval $[a, b]$, given that it starts at a point x in the interval, we have to solve equation (1.148), $\frac{1}{2}u''(x) = -1$ for $a < x < b$, with the boundary conditions $u(a) = u(b) = 0$. The solution is given by $\mathbb{E}\left[\tau_{[a,b]} \mid w(0) = x\right] = u(x) = (b - x)(x - a)$. In particular, $\lim_{b \to \infty} u(x) = \infty$, that is, the mean time to exit a half-line $[0, \infty)$ is infinite. This means that in a simulation of random walks almost every trajectory will reach the endpoint of the half-line in a finite number of steps. However, if the number of steps to get there is averaged over a sample of N trajectories, the average will grow indefinitely as $N \to \infty$. $\qquad\square$

Example 1.9 (The MFPT of the Ornstein–Uhlenbeck process). To solve the same problem for the Ornstein–Uhlenbeck process, recall that it is defined by the SDE $dx = -\alpha x\, dt + \gamma\, dw$. Equation (1.148) is now $\frac{1}{2}\gamma^2 u''(x) - \alpha x u'(x) = -1$ for $a < x < b$, and the boundary conditions are $u(a) = u(b) = 0$. The solution is given by,

$$u(x) = C \int_a^x e^{-\alpha y^2/\gamma^2}\, dy - \frac{2}{\gamma^2} \int_a^x \int_a^y e^{-\alpha(y^2-z^2)/\gamma^2}\, dz\, dy,$$

where

$$C = 2 \int_a^b \int_a^y e^{-\alpha(y^2-z^2)/\gamma^2}\, dz\, dy \bigg/ \gamma^2 \int_a^b e^{-\alpha y^2/\gamma^2}\, dy.$$

Does $\lim_{b \to \infty} u(x) = \infty$ hold in this case as well? $\qquad\square$

Exercise 1.32 (Higher moments of the FPT). Derive boundary value problems similar to (1.148) for higher moments of the FPT (HINT: replace -1 on the right-hand side of the equation with an appropriate power of t). $\qquad\square$

1.5.3   The Exit Distribution

We consider again the system (1.139) in a domain D and assume that the solution exits the domain in finite time τ_D with probability 1.

Theorem 1.5.3 (Representation of the exit distribution). *The conditional probability density function of the exit points $x(\tau_D)$ $(\tau_D > s)$ of trajectories of (1.139) given $x(s) = x$, is Green's function for the boundary value problem*

$$\frac{\partial u(x,t)}{\partial t} + \mathcal{L}_x^* u(x,t) = 0 \ \text{ for } x \in D, \, t \geq s, \tag{1.149}$$

$$u(x,t) = f(x) \ \text{ for } x \in \partial D,$$

where $\mathcal{L}_x^*$ is the backward Kolmogorov operator (1.106), and

$$u(x,s) = \mathbb{E}\left[f(x(\tau_D)) \,|\, x(s) = x\right] \tag{1.150}$$

(see [137, Section 4.4.2]).

If $a(x,t)$ and $B(x,t)$ are independent of t, the boundary value problem (1.149) becomes the elliptic boundary value problem

$$\mathcal{L}_x^* u(x) = 0 \ \text{ for } x \in D, \, u(x) = f(x) \ \text{ for } x \in \partial D, \tag{1.151}$$

and Kolmogorov's formula (1.150) becomes

$$u(x) = \mathbb{E}\left[f(x(\tau_D)) \,|\, x(0) = x\right]. \tag{1.152}$$

Kolmogorov's equation indicates that the solution of the boundary value problem can be constructed by running trajectories of the SDE that start at x until they hit ∂D and averaging the boundary function at the points where the trajectories hit ∂D.

Equation (1.152) leads to an important interpretation of Green's function for the elliptic boundary value problem (1.151). By definition, Green's function, $G(x, y)$, is characterized by the relation

$$u(x) = \oint_{\partial D} f(y) G(x, y) \, dS_y, \tag{1.153}$$

where dS_y is a surface area element on ∂D. On the other hand, (1.152) can be written as

$$u(x) = \oint_{\partial D} f(y) p\left(x(\tau_D) = y \,|\, x(0) = x\right) dS_y. \tag{1.154}$$

Because eqs.(1.153) and (1.154) hold for all smooth functions $f(y)$ on ∂D, we must have

$$G(x, y) = \Pr\{x(\tau_D) = y \,|\, x(0) = x\},$$

that is, Green's function is the pdf of the exit points on ∂D of trajectories of (1.139) that start at x. In a simulation it counts the fraction of trajectories that, starting at x, hit the boundary at y.

Exercise 1.33 (Exit distribution of the MBM from a half-space—the Cauchy process*). Assume that the MBM $w(t) = (w_1(t), w_2(t), \ldots, w_d(t))^T$ in $\mathbb{R}^d$ starts in the upper half-space at $w(0) = (0, 0, \ldots, 0, z)^T$ with $z > 0$.

(i) Find the distribution of its exit points in the plane $z = 0$.
(ii) Let τ_z be the FPT to the line $z = 0$ in $\mathbb{R}^2$. Show that $x(z) = w_1(\tau_z)$ is the Cauchy process defined by the transition probability density function

$$p(y, z \mid x, 0) = \frac{z}{\pi} \frac{1}{(x - y)^2 + z^2}.$$

See [145] for more details. $\square$

1.5.4 The Distribution of the First Passage Time

We consider again the solution $x(t)$ of the Itô system (1.139) that starts at time s in a domain D. The PDF of the FPT τ_D to the boundary ∂D, conditioned on $x(s) = x \in D$, is the conditional probability $P(T \mid x, s) = \Pr\{\tau_D < T \mid x(s) = x\}$ for every $T > s$. Obviously, if the starting point is on the boundary, that is, if $x \in \partial D$, then $P(T \mid x, s) = 1$, because in this case the trajectories of $x(t)$ start out on the boundary so that surely $\tau_D = s < T$. Similarly, $P(T \mid x, T) = 0$ for all $x \in D$, because the trajectories of the solution $x(t)$ cannot be at the same time T both inside D and on its boundary ∂D.

Theorem 1.5.4 (A PDE for the PDF of the FPT).

$$\Pr\{\tau_D < T \mid x(s) = x\} = u(x, s, T), \tag{1.155}$$

where $u(x, t, T)$ is the solution of the backward parabolic terminal–boundary value problem

$$\frac{\partial u(x, t, T)}{\partial t} + \mathcal{L}_x^* u(x, t, T) = 0 \text{ for } x \in D, \, t < T, \tag{1.156}$$

$$u(x, t, T) = 1 \text{ for } x \in \partial D, \, t < T,$$

$$u(x, T, T) = 0 \text{ for } x \in D \tag{1.157}$$

(see [137, Section 4.4.3]).

In the autonomous case that the coefficients in the stochastic system (1.139) are independent of t, the solution of (1.156) is a function of the difference $T - t$, so that the change of the time variable $\tau = T - t$ and the substitution $v(x, \tau) = 1 - u(x, \tau)$ transform the terminal–boundary value problem (1.156)–(1.157) into the forward homogeneous initial–boundary value problem

$$\frac{\partial v(x, \tau)}{\partial \tau} = \mathcal{L}_x^* v(x, \tau) \text{ for } x \in D, \ \tau > 0,$$

$$v(x, \tau) = 0 \text{ for } x \in \partial D, \ \tau > 0, \ v(x, 0) = 1 \text{ for } x \in D. \qquad (1.158)$$

Because the above problem is invariant to time shifts, we may assume that $s = 0$ and then

$$v(x, \tau) = \Pr\{\tau_D > \tau \mid x(0) = x\}. \qquad (1.159)$$

1.6 The Fokker–Planck Equation

The transition probability density function of the solution $x_{x,s}(t)$ of the stochastic differential equation (1.139), denoted $p(y, t \mid x, s)$, satisfies two different partial differential equations, one with respect to the "forward variables" (y, t) and one with respect to the "backward variables" (x, s). The former is called the *Fokker–Planck equation* or the *forward Kolmogorov equation*, and is the subject of this section. The latter is called *the backward Kolmogorov equation* and is derived in Section 1.6.1.

Definition 1.6.1 (The Fokker–Planck operator). *The operator*

$$\mathcal{L}_y p = \sum_{i=1}^{d} \frac{\partial}{\partial y^i} \left\{ \sum_{j=1}^{d} \frac{\partial}{\partial y^j} \sigma^{ij}(y, t) p - a^i(y, t) p \right\} \qquad (1.160)$$

is called the Fokker–Planck operator, or the forward Kolmogorov operator.

Note that the forward operator $\mathcal{L}_y$ is the formal adjoint, with respect to the $L^2(\mathbb{R}^d)$ inner product $\langle \cdot, \cdot \rangle_{L^2}$, of the operator $\mathcal{L}_x^*$, defined by (1.106), that appears in Itô's formula (1.105), in the sense that for all sufficiently smooth functions $f(x), g(x)$ in $\mathbb{R}^d$ that vanish sufficiently fast at infinity,

$$\int_{\mathbb{R}^d} g(y) \mathcal{L}_y f(y) \, dy = \langle \mathcal{L}_y f, g \rangle_{L^2} = \langle f, \mathcal{L}_y^* g \rangle_{L^2} = \int_{\mathbb{R}^d} f(y) \mathcal{L}_y^* g(y) \, dy. \qquad (1.161)$$

Theorem 1.6.1 (The FPE). *The pdf $p(y, t \mid x, s)$ satisfies the initial value problem*

$$\frac{\partial p(y, t \mid x, s)}{\partial t} = \mathcal{L}_y \, p(y, t \mid x, s) \quad \text{for } x, y \in \mathbb{R}^d, \ t > s, \tag{1.162}$$

$$\lim_{t \to s} p(y, t \mid x, s) = \delta(x - y). \tag{1.163}$$

The proof is given in Section 2.2 below and in [137, Section 4.5]. A classical solution of (1.162) is a function that has all the derivatives that appear in the equation and such that the equation is satisfied at all points. It is known from the theory of parabolic partial differential equations [54] that under mild regularity assumptions, if $\sigma(y, \tau)$ is a strictly positive definite matrix, the initial value problem (1.162), (1.163) has a unique classical solution.

The one-dimensional Fokker–Planck equation has the form

$$\frac{\partial p(y, t \mid x, s)}{\partial t} = \frac{1}{2} \frac{\partial^2 \left[b^2(y, t) p(y, t \mid x, s) \right]}{\partial y^2} - \frac{\partial \left[a(y, t) p(y, t \mid x, s) \right]}{\partial y} \tag{1.164}$$

with the initial condition

$$\lim_{t \downarrow s} p(y, t \mid x, s) = \delta(y - x). \tag{1.165}$$

Exercise 1.34 (The solution of the FPE solves the CKE).

(i) Use the existence and uniqueness theorem for linear parabolic initial value problems to show that the solution $p(y, t \mid x, s)$ of (1.162), (1.163) satisfies the CKE (1.80).

(ii) Prove that if $a(x, t)$ and $\sigma(x, t)$ are sufficiently regular, then the solution $p(y, t \mid x, s)$ of (1.162), (1.163) satisfies (1.134)–(1.136). □

The simplest example of the Fokker–Planck equation corresponds to the case $a(x, t) = 0$ and $b(x, t) = 1$, that is, $x(t)$ is the Brownian motion $w(t)$. In this case, the Fokker–Planck equation (1.164) and the initial condition (1.165) reduce to the diffusion equation and the initial condition (1.64), moved from the origin to the point x.

The Fokker–Planck equation corresponding to the Ornstein–Uhlenbeck process (or colored noise), defined by the stochastic dynamics

$$dx(t) = -ax(t) \, dt + b \, dw(t), \quad x(s) = x, \tag{1.166}$$

is

$$\frac{\partial p(y, t \mid x, s)}{\partial t} = \frac{b^2}{2} \frac{\partial^2 p(y, t \mid x, s)}{\partial y^2} + a \frac{\partial y p(y, t \mid x, s)}{\partial y}, \tag{1.167}$$

$$p(y, t \mid x, s) \to \delta(y - x) \quad \text{as } t \downarrow s. \tag{1.168}$$

Exercise 1.35 (Explicit solution of the FPE (1.167), (1.168)). Use the explicit solution of (1.166) to find the explicit solution of the Fokker–Planck equation (1.167), (1.168). □

1.6.1 The Backward Kolmogorov Equation

Theorem 1.6.2 (The backward Kolmogorov equation). *The transition probability density function $p(y, t \mid x, s)$ of the solution $x_{x,s}(t)$ of the stochastic differential equation (1.139) satisfies, with respect to the backward variables (x, s), the backward Kolmogorov equation*

$$\frac{\partial p(y, t \mid x, s)}{\partial s} = -\sum_{i=1}^{d} a^i(x, s) \frac{\partial p(y, t \mid x, s)}{\partial x^i} - \sum_{i=1}^{d} \sum_{j=1}^{d} \sigma^{ij}(x, s) \frac{\partial^2 p(y, t \mid x, s)}{\partial x^i \partial x^j}$$

$$= -\mathcal{L}_x^* p(y, t \mid x, s) \tag{1.169}$$

with the terminal condition

$$\lim_{s \to t} p(y, t \mid x, s) = \delta(x - y) \tag{1.170}$$

(see [137, Section 4.6]).

Exercise 1.36 (The solution of the FPE solves the BKE). Use Exercise (1.34) and the Chapman–Kolmogorov equation (1.80) to prove that if $a(x, t)$ and $\sigma(x, t)$ are sufficiently regular, then the solution $p(y, t \mid x, s)$ of the FPE (1.162), (1.163) satisfies the BKE (1.169), (1.170). □

1.7 Diffusion Approximation to $1/\mathfrak{f}$ Noise

Not all Gaussian processes with continuous trajectories are necessarily diffusions, or even Markovian, so that the Fokker–Planck, Kolmogorov, Andronov–Vitt–Pontryagin partial differential equations cannot be used for the evaluation of their functionals. If, however, a stochastic process can be approximated by a diffusion process, its functionals can be approximated by solutions of these equations (see [137, Chapter 7] for a full discussion). We consider here the case of a Gaussian process whose power spectral density function is $1/\mathfrak{f}$ ($\mathfrak{f}$ denotes frequency). We construct an approximate Gaussian noise with $1/\mathfrak{f}$ spectrum by passing white Gaussian noise through a filter, whose transfer function in the Laplace domain is [83]

$$H(s) = \frac{1}{\sqrt{s}}. \tag{1.171}$$

This noise cannot be realized as a Markovian process in a straightforward fashion, and therefore the standard tools of Markov processes are not available for the study of the effects of $1/\mathfrak{f}$ noise in dynamical systems. We construct a sequence of rational approximations to (1.171) by truncating its continued fraction representation in a fashion similar to that used in [38] and references therein. Although (1.171) is not an analytic function near the origin, it is analytic at any nonzero s such as $s = \omega_0 = 2\pi\mathfrak{f}_0 > 0$, so it has the continued fraction representation

$$\frac{1}{\sqrt{\tilde{s} + \omega_0}} = \cfrac{1}{1 + \cfrac{1}{\cfrac{2\omega_0}{\tilde{s}} + \cfrac{1}{2 + \cfrac{1}{\cfrac{2\omega_0}{\tilde{s}} + \cfrac{1}{2 + \ddots}}}}}, \tag{1.172}$$

where $\tilde{s} = \omega_0(s - 1)$. Thus (1.171) is

$$H(s) = \cfrac{1}{1 + \cfrac{1}{\cfrac{2}{s-1} + \cfrac{1}{2 + \cfrac{1}{\cfrac{2}{s-1} + \cfrac{1}{2 + \ddots}}}}}, \tag{1.173}$$

which converges uniformly for $|s - \omega_0| < \omega_0$. Next, we define an approximate $1/\mathfrak{f}$ noise through the Laplace transform relation

$$\Phi_n(s) = H(s)V(s), \tag{1.174}$$

where the power spectral density of a white Gaussian process is given by

$$S_{vv}(f) = N_{ph}. \tag{1.175}$$

Truncating the continued fraction and using (1.173), (1.174), we obtain the system of $2N$ equations

$$V(s) = \Phi_n(s) + Y_1(s), \quad \Phi_n(s) = \frac{2}{s-1}Y_1(s) + Y_2(s), \tag{1.176}$$

$$Y_1(s) = 2Y_2(s) + Y_3(s), \quad Y_2(s) = \frac{2}{s-1}Y_3(s) + Y_4(s),$$

$$Y_3(s) = 2Y_4(s) + Y_5(s), \quad Y_4(s) = \frac{2}{s-1}Y_5(s) + Y_6(s),$$

$$\vdots$$

$$Y_{2N-3}(s) = 2Y_{2N-2}(s) + Y_{2N-1}(s), \quad Y_{2N-2}(s) = \frac{2}{s-1}Y_{2N-1}(s) + Y_{2N}(s),$$

$$2Y_{2N-1}(s) = 4Y_{2N}(s) + (s-1)Y_{2N},$$

where N denotes the order of approximation to the $1/\mathfrak{f}$ noise.

We note that all the state variables $Y_{2j+1}(s)$ in (1.176) can be eliminated by a linear transformation. To transform the system (1.176) into the time domain, we denote by $v(t)$ a standard Gaussian white noise and denote the state variables in the time domain by lowercase letters. Then (1.176) is transformed into the Itô system

$$\dot{y}_{2N}(t) = y_{2N}(t) - y_{2N}(0) + 2\left[v(t) - \phi_n(t) - 2\sum_{k=1}^{N} y_{2k}(t) \right],$$

$$\dot{y}_{2N-2}(t) = y_{2N-2}(t) - y_{2N-2}(0) - 4y_{2N}(t) + 4\left[v(t) - \phi_n(t) - 2\sum_{k=1}^{N-1} y_{2k}(t) \right],$$

$$\dot{y}_{2N-4}(t) = y_{2N-4}(t) - y_{2N-4}(0) - 4y_{2N}(t) - 8y_{2N-2}(t)$$
$$+ 6\left[v(t) - \phi_n(t) - 2\sum_{k=1}^{N-2} y_{2k}(t) \right], \tag{1.177}$$

$$\vdots$$

$$\dot{y}_2(t) = y_2(t) - y_2(0) - 4\sum_{m=1}^{N-1} m y_{2(N-m+1)}(t) + 2N\left[v(t) - \phi_n(t) - 2y_2(t) \right],$$

$$\dot{\phi}_n(t) = -(2N+1)\phi_n(t) - \phi_n(0) - 4\sum_{m=1}^{N} m y_{2(N-m+1)}(t) + 2(N+1)v(t).$$

Thus the Nth approximation to the $1/\mathfrak{f}$ noise process is an output of a Markovian system of $N+1$ linear stochastic differential equations of Itô type. Finally, because the expected value of the $1/\mathfrak{f}$ noise process is zero, we find the initial conditions by taking the expectation of (1.177)

$$\phi_n(0) = y_2(0) = y_4(0) = \cdots = y_{2N}(0) = 0. \tag{1.178}$$

An interesting feature of our model is that for $N = 0$, (1.177) becomes an Ornstein–Uhlenbeck process. This type of process is commonly used for colored Gaussian noise models. In Figure 1.5 the frequency response [128] of the truncated transfer

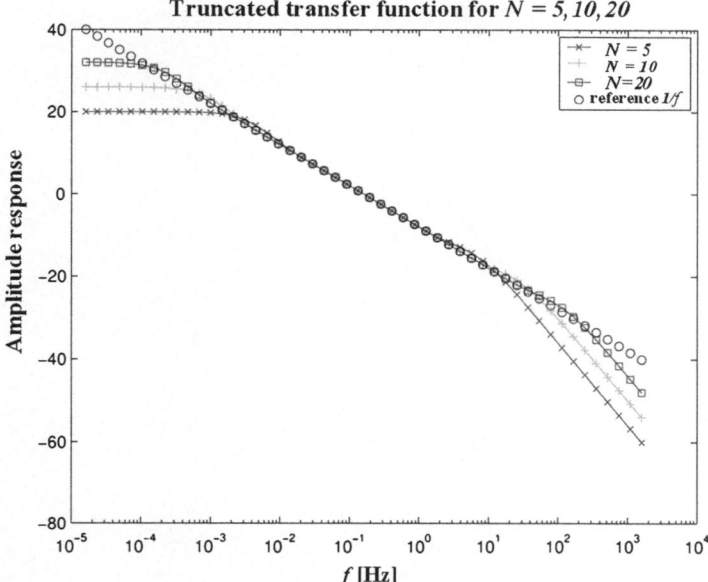

Fig. 1.5 Frequency response of continued fraction approximations of $1/\mathfrak{f}$ with $2\pi\mathfrak{f}_0 = 1$, truncated at $N = 5$ (×), $N = 10$ (+), and $N = 20$ (□). The reference $1/\mathfrak{f}$ is marked by circles.

function $H(s)$ (1.173) is given for the approximation of $1/\mathfrak{f}$, with $2\pi\mathfrak{f} = 1$, truncated at $N = 5$ (×), $N = 10$ (+), and $N = 20$ (□). The reference $1/\mathfrak{f}$ is marked by circles. The figure shows how the range of validity of the approximation expands with increasing N, and that the approximation is without ripples. The influence of $1/\mathfrak{f}$ noise on loss of lock in phase estimation is discussed in Section 6.3.3.

Chapter 2
Euler's Simulation Scheme and Wiener's Measure

The path integral (or equivalently, Wiener's measure) interpretation of stochastic differential equations is useful for both the conceptual understanding of stochastic differential equations and for deriving differential equations that govern the evolution of the pdfs of their solutions. A simple illustration of the computational usefulness of the Wiener probability measure is the easy derivation of the explicit expression (1.63) for the pdf of the MBM. Unfortunately, no explicit expressions exist in general for the pdf of the solution to (2.1). The best alternative to such an explicit expression is a (deterministic) differential equation for the pdf, whose solution can be studied both analytically and numerically directly from the differential equation. A case in point is the diffusion equation and the initial condition (1.64) that the pdf of the MBM satisfies.

The discrete approach to SDEs provides insight into the behavior of the random trajectories of the SDE that is not contained in the FPE. Thus, for example, the probability flux density in the FPE is net flux and cannot be separated into its unidirectional components. The need for such a separation arises in connecting discrete simulations to the continuum. Also the boundary behavior of the random trajectories is not easily expressed in terms of boundary conditions for the FPE. These problems are handled in a natural way by the discrete simulation and by its limit.

2.1 The Euler Scheme and its Convergence

Itô's definition of the stochastic integral on the lattice $t_k = t_0 + k\Delta t$, with $\Delta t = T/N$ and $\Delta w(t) = w(t + \Delta t) - w(t)$, defines the solution of the SDE

$$dx = a(x,t)\,dt + b(x,t)\,dw, \quad x(0) = x_0, \tag{2.1}$$

or equivalently, of the Itô integral equation

$$x(t) = x_0 + \int_0^t a(x(s),s)\,ds + \int_0^t b(x(s),s)\,dw(s), \tag{2.2}$$

Z. Schuss, *Nonlinear Filtering and Optimal Phase Tracking*, Applied
Mathematical Sciences 180, DOI 10.1007/978-1-4614-0487-3_2,
© Springer Science+Business Media, LLC 2012

as the limit $\Delta t \to 0$ of the solution of the Euler scheme

$$x_N(t + \Delta t) = x_N(t) + a(x_N(t), t)\Delta t + b(x_N(t), t)\,\Delta w(t),$$
$$x_N(0) = x_0. \tag{2.3}$$

The increments $\Delta w(t)$ are independent random variables $\Delta w(t) = n(t)\sqrt{\Delta t}$, where the random variables $n(t)$, for each t on the numerical mesh, are independent standard Gaussian variables $\mathcal{N}(0, 1)$. According to the recursive scheme (2.3), at any time t (on the numerical mesh) the process $x_N(t)$ depends on the sampled trajectory $w(s)$ for $s \leq t$, so it is adapted. The existence of the limit $x(t) = \lim_{N\to\infty} x_N(t)$ is the content of the following theorem.

Theorem 2.1.1 (Skorokhod). *If $a(x, t)$ and $b(x, t)$ are uniformly Lipschitz continuous functions in $x \in \mathbb{R}$, $t \in [t_0, T]$, and there is a constant C such that $|a(x, t)| + |b(x, t)| \leq C(1 + |x|)$, than the limit $x(t) \stackrel{\text{Pr}}{=} \lim_{N\to\infty} x_N(t)$ exists and is the solution of* (2.2).

(see proof in [142], [137, Theorem 5.1.1]).

Exercise 2.1 (Killing measure*). How can dynamics with killing be simulated? What if the killing rate is negative at some points? □

2.2   The pdf of Euler's Scheme in $\mathbb{R}$ and the FPE

We assume that the coefficients $a(x, t)$ and $b(x, t)$ are smooth functions in $\mathbb{R} \times \mathbb{R}^+$, with $b(x, t) > \delta > 0$ for some constant δ. The coefficients can be allowed to be random in a way such that for each $x \in \mathbb{R}$, the stochastic processes $a(x, t, \mathfrak{w})$ and $b(x, t, \mathfrak{w})$ are adapted in the sense of Definition 1.2.6. We assume for now that $a(x, t)$ and $b(x, t)$ are deterministic.

Theorem 2.2.1. *The pdf $p_N(x, t \mid x_0)$ of the solution $x_N(t, \mathfrak{w})$ of (2.3) converges to the solution $p(x, t \mid x_0)$ of the initial value problem (1.164), (1.165) as $N \to \infty$, uniformly on every compact subset of the half-plane $[x \in \mathbb{R}, t > t_0]$.*

Proof. The pdf of $x_N(t)$ can be expressed explicitly for t on the lattice, because (2.3), written as

$$\Delta w(t) = \frac{x_N(t + \Delta t) - x_N(t) - a(x_N(t), t)}{b(x_N(t), t)}, \tag{2.4}$$

means that for all t on the lattice the expressions on the right-hand side of (2.4) are i.i.d. Gaussian variables. It follows, as in (1.83), that the pdf of the entire Euler trajectory is the product

$$p\left(x_1, t_1; x_2, t_2; \ldots; x_n, t_n\right)$$

$$= \prod_{k=1}^{n} \left[2\pi b^2(x_{k-1}, t_{k-1})\Delta t\right]^{-1/2} \exp\left\{-\frac{[x_k - x_{k-1} - a(x_{k-1}, t_{k-1})\Delta t]^2}{2b^2(x_{k-1}, t_{k-1})\Delta t}\right\}. \tag{2.5}$$

Setting $x_n = x$ and integrating over $\mathbb{R}$ with respect to all intermediate points $x_1, x_2, \ldots, x_{n-1}$, we find from (2.5) that the transition pdf of the trajectory satisfies on the lattice the recurrence relation

$$p_N(x, t + \Delta t \mid x_0) = \int_{\mathbb{R}} \frac{p_N(y, t \mid x_0)\, dy}{\sqrt{2\pi\Delta t}\, b(y, t)} \exp\left\{-\frac{[x - y - a(y, t)\Delta t]^2}{2b^2(y, t)\Delta t}\right\}. \tag{2.6}$$

The solution of the integral equation (2.6) is called Wiener's *discrete path integral*. Its limit as $N \to \infty$ is called Wiener's *path integral*.

To prove convergence as $N \to \infty$, we first extend $x_N(t)$ off the lattice as follows. If $0 < t \leq \Delta t$, we define

$$\tilde{x}_N(t) = x_0 + a(x_0, 0)\, t + b(x_0, 0)\, w(t), \quad x_N(0) = x_0. \tag{2.7}$$

If $k\Delta t < t \leq (k+1)\Delta t$, where $k = 1, 2, \ldots$, we define $\tilde{t} = k\Delta t$, $\Delta\tilde{t} = t - \tilde{t}$, $\Delta\tilde{w}(t) = w(t) - w(\tilde{t})$, and

$$\tilde{x}_N(t) = \tilde{x}_N(\tilde{t}) + a(\tilde{x}_N(\tilde{t}), \tilde{t})\Delta\tilde{t} + b(\tilde{x}_N(\tilde{t}), \tilde{t})\, \Delta\tilde{w}(t). \tag{2.8}$$

Obviously, for each realization of $w(t)$, we have $\tilde{x}_N(t) = x_N(t)$ at lattice points t. The pdf $\tilde{p}_N(x, t \mid x_0)$ of $\tilde{x}_N(t)$ is identical to the pdf $p_N(x, t \mid x_0)$ of $x_N(t)$ on lattice points t, and it satisfies on the lattice the recurrence relation (2.6). Off the lattice we have the recurrence

$$\tilde{p}_N(x, t \mid x_0) = \int_{\mathbb{R}} \frac{\tilde{p}_N(y, \tilde{t} \mid x_0)\, dy}{\sqrt{2\pi\Delta\tilde{t}}\, b(y, \tilde{t})} \exp\left\{-\frac{[x - y - a(y, \tilde{t})\Delta\tilde{t}]^2}{2b^2(y, \tilde{t})\Delta\tilde{t}}\right\}, \tag{2.9}$$

where $\tilde{p}_N(x, \tilde{t} \mid x_0) = p_N(x, \tilde{t} \mid x_0)$. Note that $\tilde{p}_N(x, t \mid x_0)$ is differentiable with respect to t and twice differentiable with respect to x. Therefore the analysis of (2.9) applies to (2.6) as well. We observe that integrating $\tilde{p}_N(x, t \mid x_0)$ with respect to x_0 against a bounded sufficiently smooth initial function $p_0(x_0)$ results in a sequence of bounded and twice continuously differentiable functions

$$\tilde{p}_N(x, t) = \int_{\mathbb{R}} \tilde{p}_N(x, t \mid x_0) p_0(x_0)\, dx_0 \tag{2.10}$$

that satisfy the recurrence (2.9), the initial condition

$$\lim_{t \to 0} \tilde{p}_N(x,t) = p_0(x), \tag{2.11}$$

uniformly on finite intervals, and whose partial derivatives up to second-order are uniformly bounded (see Exercise 2.2 below).

Differentiation with respect to t at off-lattice points is equivalent to differentiation with respect to $\Delta\tilde{t}$. Differentiating and expanding all functions in powers of $\sqrt{\Delta\tilde{t}}$, we obtain (see Exercises 2.3–2.6 below)

$$\frac{\tilde{p}_N(x,t)}{\partial t} = \frac{1}{2}\frac{\partial^2\left[b^2(x,t)\tilde{p}_N(x,t)\right]}{\partial x^2} - \frac{\partial\left[a(x,t)\tilde{p}_N(x,t)\right]}{\partial x} + O(\Delta t), \tag{2.12}$$

uniformly for $x \in \mathbb{R}$ and $t > 0$. At lattice points, we use the change of variables

$$y = x - a(y,t)\Delta t + \eta b(y,t)\sqrt{\Delta t} \tag{2.13}$$

in (2.6), and expanding in powers of Δt, we obtain (2.12) again. If $p(x,t)$ is the (unique) solution of the initial value problem (2.11) for the FPE (1.164), then $p(x,t) - p_N(x,t)$ satisfies the inhomogeneous FPE with homogeneous initial value and right-hand side that is uniformly $O(\Delta t)$. It follows from the maximum principle for parabolic initial value problems [54], [129] that the difference converges uniformly to zero.

Remark 2.2.1. There are many types of convergence of the Euler scheme [89], [88]. Theorems 2.1.1 and 2.2.1 concern convergence in probability and of probability density and therefore cannot be used as measures for the error of the Euler numerical scheme in a given simulation. Such estimates depend on the sample size and are the subject of numerical analysis of stochastic differential equations.

Exercise 2.2 (Regularity of $p_N(x,t)$). Use the recurrence (2.6) to prove that the functions $p_N(x,t)$ and their partial derivatives with respect to x, up to second-order, are uniformly bounded and the convergence (2.11) is uniform on finite intervals. $\square$

Exercise 2.3 (The differential). Prove that the differential of the transformation (2.13) is given by

$$
\begin{aligned}
dy &= \frac{\sqrt{\Delta t}\,b(y,t)}{1 - \sqrt{\Delta t}\,\eta b_y(y,t) + a_y(y,t)\Delta t}\,d\eta \\
&= d\eta\left\{1 + \eta\sqrt{\Delta t}\,b_x(x,t) - a_x(x,t)\Delta t + \eta^2\Delta t\left[(b_x(x,t))^2 + b_{xx}(x,t)b(x,t)\right]\right\} \\
&\quad \times \sqrt{\Delta t}\,b(x,t)\left[1 + O\left(\sqrt{\Delta t}\right)\right],
\end{aligned} \tag{2.14}
$$

where subscripts denote partial derivatives. $\square$

Exercise 2.4 (The exponent).

(i) Prove that the exponent in (2.6) is expanded as

$$\frac{\left[\eta b(y,t)\sqrt{\Delta t} + a\left(x + \eta b(y,t)\sqrt{\Delta t}, t\right)\Delta t\right]^2}{2b^2(y,t)\Delta t}$$

$$= \eta^2 \left(\frac{a(x,t)}{b(x,t)}\right)_x b(x,t) + \frac{a^2(x,t)}{2b^2(x,t)} + O\left(\Delta t^{3/2}\right).$$

(ii) Show that the exponential function in (2.6) can be expanded as

$$\exp\left\{-\frac{[x - y - a(y,t)\Delta t]^2}{2b^2(y,t)\Delta t}\right\} = \exp\left\{-\frac{\eta^2}{2}\right\}$$

$$\times \left\{1 - \eta\sqrt{\Delta t}\frac{a(x,t)}{b(x,t)} - \Delta t\left[\eta^2\left(\frac{a(x,t)}{b(x,t)}\right)_x b(x,t) + (1 - \eta^2)\frac{a^2(x,t)}{2b^2(x,t)}\right]\right\}$$

$$+ O\left(\Delta t^{3/2}\right). \tag{2.15}$$

(HINT: Use Maple or Mathematica). $\qquad\qquad\square$

Exercise 2.5 (The density $p_N(y,t)$).

(i) Expand

$$p_N(y,t) = p_N(x + \eta b(y,t)\sqrt{\Delta t}, t)$$

$$= p_N(x,t) + p_{N,x}(x,t)\eta\left[b(x,t) + b_x(x,t)\eta b(x,t)\sqrt{\Delta t}\right]\sqrt{\Delta t}$$

$$+ \frac{1}{2}p_{N,xx}(x,t)\eta^2 b^2(x,t)\Delta t + O\left(\Delta t^{3/2}\right).$$

(ii) Show that the pre-exponential factor in (2.6), up to $O\left(\Delta t^{3/2}\right)$, has the form

$$\left\{p_N(x,t) + p_{N,x}(x,t)b(x,t)\eta\sqrt{\Delta t} + \frac{1}{2}\left[p_{N,xx}(x,t)b^2(x,t)\right.\right.$$

$$\left.\left. + \left(b^2(x,t)\right)_x p_{N,x}(x,t)\right]\eta^2\Delta t\right\}$$

$$\times \left\{\left[1 - \frac{a^2(x,t)}{2b^2(x,t)}\Delta t\right] - \eta\sqrt{\Delta t}\frac{a(x,t)}{b(x,t)} - \eta^2\Delta t\left[\left(\frac{a(x,t)}{b(x,t)}\right)_x b(x,t)\right.\right.$$

$$\left.\left. - \frac{a^2(x,t)}{2b^2(x,t)}\right]\right\}$$

$$= \left[1 - \frac{a^2(x,t)}{2b^2(x,t)} \Delta t \right] p_N(x,t) + \eta \sqrt{\Delta t} \left\{ \left[b(x,t) - \frac{a^2(x,t)}{2b(x,t)} \Delta t \right] p_{N,x}(x,t) \right.$$

$$\left. - \frac{a(x,t)}{b(x,t)} p_N(x,t) \right\} + \eta^2 \Delta t \left\{ \frac{1}{2} \left[b^2(x,t) p_{N,xx}(x,t) \right. \right.$$

$$+ \left(b^2(x,t) \right)_x p_{N,x}(x,t) \right] \left[1 - \frac{a^2(x,t)}{2b^2(x,t)} \Delta t \right]$$

$$\left. - a(x,t) p_{N,x}(x,t) - \left[\left(\frac{a(x,t)}{b(x,t)} \right)_x b(x,t) - \frac{a^2(x,t)}{2b^2(x,t)} \right] p(x) \right\}. \qquad (2.16)$$

(iii) Use eqs.(2.14)–(2.16) in (2.6) to obtain

$$p_N(x, t + \Delta t)$$

$$= \frac{1}{\sqrt{2\pi}} \int_{\mathbb{R}} \left\{ \left\{ 1 - a_x(x,t)\Delta t + \eta^2 \Delta t \left[(b_x)^2(x,t) + b_{xx}(x,t)b(x,t) \right] \right\} \right.$$

$$\times \left[1 - \frac{a^2(x,t)}{2b^2(x,t)} \Delta t \right] p_N(x,t) + \eta^2 \Delta t b_x(x,t) \left[b(x,t) p_{N,x}(x,t) \right.$$

$$\left. - \frac{a(x,t)}{b(x,t)} p_N(x,t) \right] + \frac{\eta^2 \Delta t}{2} \left\{ b^2(x,t) p_{N,xx}(x,t) + \left(b^2(x,t) \right)_x p_{N,x}(x,t) \right.$$

$$\left. - 2 \left[\left(\frac{a(x,t)}{b(x,t)} \right)_x b(x,t) - \frac{a^2(x,t)}{2b^2(x,t)} \right] p_N(x,t) \right\} \right\} \exp \left\{ -\frac{\eta^2}{2} \right\} d\eta. \quad (2.17)$$

□

Exercise 2.6 (The FPE). Evaluate the Gaussian integrals in (2.17) and show that

$$\frac{p_N(x, t + \Delta t) - p_N(x, t)}{\Delta t}$$

$$= p_N(x,t) \left[(b_x)^2(x,t) + b_{xx}(x,t)b(x,t) \right] + \frac{1}{2} p_N(x,t) b^2(x,t)$$

$$+ \left(b^2(x,t) \right)_x p_{N,x}(x,t) - \left[a(x,t) p_N(x,t) \right]_x + O\left(\Delta t^{1/2} \right)$$

$$= \frac{1}{2} \left[b^2(x,t) p_N(x,t) \right]_{xx} - \left[a(x,t) p_N(x,t) \right]_x + O\left(\Delta t^{1/2} \right),$$

and hence (2.12). Can the last estimate be improved?   □

Exercise 2.7 (The initial condition). Prove that $p(x,t) - p_N(x,t) \to 0$ uniformly for finite intervals in x and $0 < t < T$.   □

Exercise 2.8 (The Feynman–Kac formula). Prove that if the recurrence (2.6) is modified to

$$p_N(x, t + \Delta t \mid x_0)$$

$$= \int_{\mathbb{R}} \frac{p_N(y, t \mid x_0)\, dy}{\sqrt{2\pi \Delta t}\, b(y, t)} \exp\left\{-\frac{[x - y - a(y, t)\Delta t]^2}{2b^2(y, t)\Delta t} + g(y, t)\Delta t\right\}, \quad (2.18)$$

where $g(y, t)$ is a sufficiently regular function, then $\lim_{N \to \infty} p_N(x, t \mid x_0, s) = p(x, t \mid x_0, s)$, where $p(x, t \mid x_0, s)$ is the solution of the initial value problem

$$p_t = \frac{1}{2}(b^2 p)_{xx} - (ap)_x + gp$$

with the initial condition $\lim_{t \downarrow s} p = \delta(x - x_0)$. $\qquad \square$

Exercise 2.9 (Simulation of the Feynman–Kac formula*). How should the Euler scheme (2.3) be modified so that the corresponding pdf satisfies the recurrence (2.18), in case that $g(x, t)$ is nonpositive? (see Section 1.5.1). What is the interpretation of $p_N(x, t \mid x_0, s)$ and $p(x, t \mid x_0, s)$ if $g(x, t)$ can be positive? How should the Euler scheme (2.3) be modified for this case? $\qquad \square$

Exercise 2.10 (The backward Kolmogorov equation). Derive a partial differential equation with respect to the backward variables for the pdf of the solution of a stochastic equation.

(i) First, prove that

$$p_N(y, t \mid x, s) = \int_{\mathbb{R}} \frac{p_N(y, t \mid z, s + \Delta s)}{\sqrt{2\pi \Delta s}\, b(x, s)} \exp\left\{-\frac{[z - x - a(x, s)\Delta s]^2}{2b^2(x, s)\Delta s}\right\} dz.$$

(ii) Then prove that the transition pdf $p(y, t \mid x, s) = \lim_{N \to \infty} p_N(y, t \mid x, s)$ satisfies the terminal value problem for the backward Kolmogorov equation

$$p_s + a(x, s)p_x + \frac{1}{2}b^2(x, s)p_{xx} = 0,$$

$$\lim_{t \uparrow s} p = \delta(y - x),$$

in $\mathbb{R}$ with respect to the backward variables (x, s). (HINT: Change the variable of integration to $z = b(x, s)\xi\sqrt{\Delta s} + x + a(x, s)\Delta s$ and expand everything in sight in powers of $\sqrt{\Delta s}$, as above. Finally, prove convergence using the maximum principle.)

(iii) What is the Feynman–Kac formula for the backward equation? $\qquad \square$

2.2.1 Euler's Scheme in $\mathbb{R}^d$

We consider the d-dimensional stochastic dynamics

$$dx = a(x,t)\,dt + \sqrt{2}B(x,t)\,dw, \quad x(0) = x_0, \qquad (2.19)$$

where $a(x,t) : \mathbb{R}^d \times [0,\infty) \mapsto \mathbb{R}^d$ is a vector of smooth functions for all $x \in \mathbb{R}$, $t \geq 0$, $B(x,t) : \mathbb{R}^d \times [0,\infty) \mapsto \mathbb{M}_{n \times m}$ is a smooth $n \times m$ matrix of smooth functions, and $w(t) : [0,\infty) \mapsto \mathbb{R}^m$ is a vector of m independent MBMs. We assume that the diffusion tensor $\sigma(x,t) = B(x,t)B^T(x,t)$ is uniformly positive definite in $\mathbb{R}^d$. The Euler scheme for (2.19) is

$$x(t+\Delta t) = x(t) + a(x(t),t)\Delta t + \sqrt{2}B(x(t),t)\,\Delta w(t), \quad x_N(0) = x_0. \quad (2.20)$$

Exercise 2.11 (Convergence of trajectories*). Generalize the proof of Skorokhod's theorem (Theorem 2.1.1) to the d-dimensional case.   □

2.2.2 The Convergence of the pdf in Euler's scheme

We consider now the convergence of the pdf of the trajectories of (2.20). Setting

$$\mathcal{B}(x,y,t) = [y - x - a(x,t)\Delta t]^T \sigma^{-1}(x,t)\,[y - x - a(x,t)\Delta t], \qquad (2.21)$$

we see that the pdf of the trajectories of (2.20) satisfies the d-dimensional version of the recurrence relation (2.6),

$$p_N(y,t+\Delta t) = \int_{\mathbb{R}^d} \frac{p_N(x,t)\,dx}{(4\pi\Delta t)^{d/2}\sqrt{\det\sigma(x,t)}} \exp\left\{-\frac{\mathcal{B}(x,y,t)}{4\Delta t}\right\}. \qquad (2.22)$$

Theorem 2.2.2. *Under the above assumptions, if the initial point x_0 is chosen from a smooth bounded density $p_0(x_0)$, then the pdf $p_N(y,t)$ of the solution $x_N(t)$ of (2.20) converges as $N \to \infty$ to the solution $p(y,t)$ of the initial value problem*

$$\frac{\partial p(y,t)}{\partial t} = \sum_{i=1}^{d}\sum_{j=1}^{d} \frac{\partial^2 \sigma^{ij}(y,t)p(y,t)}{\partial y^i \partial y^j} - \sum_{i=1}^{d} \frac{\partial a^i(y,t)p(y,t)}{\partial y^i}, \qquad (2.23)$$

$$\lim_{t \downarrow 0} p(y,t) = p_0(y), \qquad (2.24)$$

uniformly on compact subset of the half-space $[y \in \mathbb{R}^d, t > 0]$.

Proof. As above, we change variables in (2.22) to $z = \sigma^{-1/2}(x,t)(x - y + a(x,t)\Delta t)/\sqrt{\Delta t}$ and expand the integrand in powers of $\sqrt{\Delta t}$. First, we need to expand the Jacobian of the transformation. Differentiating the identity $\sigma^{1/2}\sigma^{-1/2} = I$, we write

$$\sigma^{1/2}\nabla\left(\sigma^{-1/2}\right) + \nabla\left(\sigma^{1/2}\right)\sigma^{-1/2} = 0, \tag{2.25}$$

or

$$\sigma^{1/2}\nabla\left(\sigma^{-1/2}\right)\sigma^{1/2} = -\nabla(\sigma^{1/2}),$$

from which it follows that the Jacobian matrix is

$$\frac{\partial z}{\partial x} = \frac{\sigma^{-1/2}}{\sqrt{\Delta t}}\left[I - \sqrt{\Delta t}\,\nabla\left(\sigma^{1/2}\right)\cdot z + O(\Delta t)\right]$$

and that the Jacobian of the transformation is

$$\mathcal{J} = \left|\det\left(\frac{\partial z}{\partial x}\right)\right| = \frac{1 - \sqrt{\Delta t}\,\operatorname{tr}\left[\nabla\left(\sigma^{1/2}\right)\cdot z\right] + O(\Delta t)}{(\Delta t)^{d/2}\sqrt{\det\sigma}}. \tag{2.26}$$

Expanding the transformed integrand about y in powers of $\sqrt{\Delta t}$, we note that terms linear in z vanish, because they give rise to Gaussian integrals with an odd integrand. We end up with the approximate Fokker–Planck equation for $p_N(y,t)$,

$$\frac{\partial p_N(y,t)}{\partial t} = \sum_{i=1}^{d}\sum_{j=1}^{d}\frac{\partial^2\sigma^{ij}(y,t)p_N(y,t)}{\partial y^i\,\partial y^j} - \sum_{i=1}^{d}\frac{\partial a^i(y,t)p_N(y,t)}{\partial y^i} + O(\sqrt{\Delta t}).$$

The uniform convergence of $p_N(y,t)$ to the solution $p(y,t)$ of the initial value problem (2.23), (2.24) is proved as in the one-dimensional case above.

2.2.3 Unidirectional and Net Probability Flux Density

The flux density in continuum diffusion theory (Fick's law) is the net flux through a given point (or surface, in higher dimensions). Unidirectional fluxes are not defined in the diffusion or Fokker–Planck equations, because velocity is not a state variable, so that the equations cannot separate unidirectional fluxes. However, it is often necessary to evaluate the unidirectional probability flux across a given interface in simulations of diffusive trajectories of particles. This is the case, for example, if a simulation of diffusing particles is connected to a region, where only a coarse-grained continuum description of the particles is used. In this case, the exchange of trajectories between the two regions, across the interface, requires the calculation of the unidirectional diffusion flux from the continuum region into the simulated region. This situation is encountered in simulations of ionic motion through protein channels of biological membranes, where the number of ions in the salt solution away from the channel is too large to simulate. This issue is discussed further in [137, Section 2.2.3]. In this section we keep the notation of the previous one.

Definition 2.2.1 (Unidirectional probability flux density). *The unidirectional probability current (flux) density at a point x_1 is the probability of trajectories that propagate from the ray $x < x_1$ into the ray $x > x_1$ in unit time. It is given by*

$$J_{LR}(x_1, t) = \lim_{\Delta t \to 0} J_{LR}(x_1, t, \Delta t), \tag{2.27}$$

where

$$J_{LR}(x_1, t, \Delta t)$$

$$= \frac{1}{\Delta t} \int_{x_1}^{\infty} dx \int_{-\infty}^{x_1} \frac{dy}{\sqrt{4\pi \Delta t \sigma(y,t)}} \exp\left\{ -\frac{[x - y - a(y,t)\Delta t]^2}{4\sigma(y,t)\Delta t} \right\} p_N(y,t). \tag{2.28}$$

Remark 2.2.2. Note that the dependence of p_N on the initial point has been suppressed in (2.28).

Theorem 2.2.3 (Unidirectional and net fluxes in one dimension). *The discrete unidirectional fluxes at a point x_1 are given by*

$$J_{LR,RL}(x_1, t, \Delta t) = \sqrt{\frac{\sigma(x_1,t)}{\pi \Delta t}} p_N(x_1,t) \pm \frac{1}{2} J(x_1,t) + O(\sqrt{\Delta t}), \tag{2.29}$$

where the net flux is

$$J(x_1, t) = \lim_{\Delta t \to 0} [J_{LR}(x_1,t) - J_{RL}(x_1,t)]$$

$$= \left\{ -\frac{\partial [\sigma(x,t)p(x,t)]}{\partial x} + a(x,t)p(x,t) \right\}_{x=x_1}. \tag{2.30}$$

Remark 2.2.3. It is clear from (2.29) that the unidirectional fluxes in Definition 2.2.1 are infinite, but the net flux is finite.

Proof. The integral (2.28) can be calculated by the Laplace method [12] at the saddle point $x = y = x_1$. First, we change variables in (2.28) to $x = x_1 + \xi\sqrt{\Delta t}$ and $y = x_1 - \eta\sqrt{\Delta t}$ to obtain

$$J_{LR}(x_1, t, \Delta t) = \int_0^\infty d\xi \int_0^\infty \frac{p_N\left(x_1 - \eta\sqrt{\Delta t}, t\right) d\eta}{\sqrt{4\pi \Delta t \sigma(x_1 - \eta\sqrt{\Delta t}, t)}}$$

$$\times \exp\left\{ -\frac{\left[\xi + \eta - a(x_1 - \eta\sqrt{\Delta t}, t)\sqrt{\Delta t}\right]^2}{4\sigma(x_1 - \eta\sqrt{\Delta t}, t)} \right\},$$

and changing the variable in the inner integral to $\eta = \zeta - \xi$, we get

$$J_{LR}(x_1, t, \Delta t) = \int_0^\infty d\xi \int_\xi^\infty \frac{p_N\left(x_1 - (\zeta - \xi)\sqrt{\Delta t}, t\right) d\zeta}{\sqrt{4\pi \Delta t \sigma(x_1 - (\zeta - \xi)\sqrt{\Delta t}, t)}}$$

$$\times \exp\left\{-\frac{\left[\zeta - a(x_1 - (\zeta - \xi)\sqrt{\Delta t}, t)\sqrt{\Delta t}\right]^2}{4\sigma(x_1 - (\zeta - \xi)\sqrt{\Delta t}, t)}\right\}. \qquad (2.31)$$

Next, we expand the exponent in powers of $\sqrt{\Delta t}$ to obtain

$$\frac{\left[\zeta - a(x_1 - (\zeta - \xi)\sqrt{\Delta t}, t)\sqrt{\Delta t}\right]^2}{4\sigma(x_1 - (\zeta - \xi)\sqrt{\Delta t}, t)}$$

$$= \frac{\zeta^2}{4\sigma(x_1, t)} + \left[\frac{\zeta^2(\zeta - \xi)\sigma_x(x_1, t)}{4\sigma^2(x_1, t)} - \frac{\zeta a(x_1, t)}{2\sigma(x_1, t)}\right]\sqrt{\Delta t} + O(\Delta t), \qquad (2.32)$$

the pre-exponential factor

$$\frac{1}{\sqrt{\sigma\left(x_1 - (\zeta - \xi)\sqrt{\Delta t}, t\right)}} = \frac{1}{\sqrt{\sigma(x_1, t)}}\left[1 + \frac{\sigma_x(x_1, t)}{2\sigma(x_1, t)}(\zeta - \xi)\sqrt{\Delta t} + O(\Delta t)\right],$$

$$(2.33)$$

and the pdf

$$p_N\left(x_1 - (\zeta - \xi)\sqrt{\Delta t}, t\right) = p_N(x_1, t) - \frac{\partial p_N(x_1, t)}{\partial x}(\zeta - \xi)\sqrt{\Delta t} + O(\Delta t).$$

Using the above expansions in (2.31), we obtain

$$J_{LR}(x_1, t, \Delta t) = \int_0^\infty d\xi \int_\xi^\infty \frac{p_N(x_1, t) d\zeta}{\sqrt{4\pi \Delta t \sigma(x_1, t)}} \exp\left\{-\frac{\zeta^2}{4\sigma(x_1, t)}\right\}$$

$$\times \left\{1 - \sqrt{\Delta t}\left[\frac{\zeta^2(\zeta - \xi)\sigma_x(x_1, t)}{4\sigma^2(x_1, t)} - \frac{\zeta a(x_1, t)}{2\sigma(x_1, t)}\right.\right.$$

$$\left.\left. - \frac{\sigma_x(x_1, t)(\zeta - \xi)}{2\sigma(x_1, t)} + (\zeta - \xi)\frac{p_{N,x}(x_1, t)}{p_N(x_1, t)}\right] + O(\Delta t)\right\}.$$

$$(2.34)$$

Similarly, $J_{RL}(x_1, t) = \lim_{\Delta t \to 0} J_{RL}(x_1, t, \Delta t)$, where

$$J_{RL}(x_1, t, \Delta t)$$

$$= \frac{1}{\Delta t} \int_{-\infty}^{x_1} dx \int_{x_1}^{\infty} \frac{dy}{\sqrt{4\pi \Delta t \sigma(y, t)}} \exp\left\{-\frac{[x - y - a(y, t)\Delta t]^2}{4\sigma(y, t)\Delta t}\right\} p_N(y, t). \tag{2.35}$$

The change of variables $x = x_1 - \xi\sqrt{\Delta t}$, $y = x_1 + \eta\sqrt{\Delta t}$ in (2.35) gives

$$J_{RL}(x_1, t, \Delta t) = \int_0^{\infty} d\xi \int_{\xi}^{\infty} \frac{p_N(x_1, t)\, d\zeta}{\sqrt{4\pi \Delta t \sigma(x_1, t)}} \exp\left\{-\frac{\zeta^2}{4\sigma(x_1, t)}\right\}$$

$$\times \left\{1 + \sqrt{\Delta t}\left[\frac{\zeta^2(\zeta - \xi)\sigma_x(x_1, t)}{4\sigma^2(x_1, t)} - \frac{\zeta a(x_1, t)}{2\sigma(x_1, t)}\right.\right.$$

$$\left.\left. - \frac{\sigma_x(x_1, t)(\zeta - \xi)}{2\sigma(x_1, t)} - (\zeta - \xi)\frac{p_{N,x}(x_1, t)}{p_N(x_1, t)}\right] + O(\Delta t)\right\}. \tag{2.36}$$

Because $p_N(x_1, t) > 0$, both $J_{LR}(x_1, t)$ and $J_{RL}(x_1, t)$ are infinite. Using the identities of Exercise 2.12 below, we find that the net flux density is, however, finite and is given by

$$J_{\text{net}}(x_1, t) = \lim_{\Delta t \to 0} \{J_{LR}(x_1, t, \Delta t) - J_{RL}(x_1, t, \Delta t)\}$$

$$= -2 \int_0^{\infty} d\xi \int_{\xi}^{\infty} \frac{d\zeta}{\sqrt{4\pi \Delta t \sigma(x_1, t)}} \exp\left\{-\frac{\zeta^2}{4\sigma(x_1, t)}\right\}$$

$$\times \left[\left(\frac{\zeta^2(\zeta - \xi)\sigma_x(x_1, t)}{4\sigma^2(x_1, t)} - \frac{\zeta a(x_1, t)}{2\sigma(x_1, t)} - \frac{\sigma_x(x_1, t)(\zeta - \xi)}{2\sigma(x_1, t)}\right)\right.$$

$$\left. \times p_N(x_1, t) + (\zeta - \xi)\, p_{N,x}(x_1, t)\right]$$

$$= \left\{-\frac{\partial[\sigma(x, t)p(x, t)]}{\partial x} + a(x, t)p(x, t)\right\}_{x=x_1}, \tag{2.37}$$

as asserted.   □

Exercise 2.12 (Identities). Prove the following identities (by changing the order of integration),

$$\int_0^{\infty} d\xi \int_{\xi}^{\infty} \frac{\zeta^2(\zeta - \xi)\, d\zeta}{\sqrt{4\pi\sigma}} \exp\left\{-\frac{\zeta^2}{4\sigma}\right\} = \int_0^{\infty} \frac{\zeta^4\, d\zeta}{4\sqrt{\pi\sigma}} \exp\left\{-\frac{\zeta^2}{4\sigma}\right\} = 3\sigma^2,$$

$$\int_0^\infty d\xi \int_\xi^\infty \frac{\zeta \, d\zeta}{\sqrt{4\pi\sigma}} \exp\left\{-\frac{\zeta^2}{4\sigma}\right\} = \int_0^\infty \frac{\zeta^2 \, d\zeta}{\sqrt{4\pi\sigma}} \exp\left\{-\frac{\zeta^2}{4\sigma}\right\} = \sigma,$$

$$\int_0^\infty d\xi \int_\xi^\infty \frac{(\zeta - \xi) \, d\zeta}{\sqrt{4\pi\sigma}} \exp\left\{-\frac{\zeta^2}{4\sigma}\right\} = \frac{1}{4} \int_0^\infty \frac{\zeta^2 \, d\zeta}{\sqrt{\pi\sigma}} \exp\left\{-\frac{\zeta^2}{4\sigma}\right\} = \frac{\sigma}{2}. \qquad \square$$

Equation (2.30) is the classical expression for the probability (or heat) current in diffusion theory [58]. The FPE (1.164) can be written in terms of the flux density function $J(x, t)$ in the conservation law form

$$\frac{\partial p(x, t)}{\partial t} = -\frac{\partial J(x, t)}{\partial x}. \qquad (2.38)$$

The unidirectional flux in $\mathbb{R}^d$ is the probability density of trajectories that propagate per unit time from a domain D across its boundary, ∂D, into the complementary part of space, D^c. It is given by $J_{\text{out}}(\partial D, t) = \lim_{\Delta t \to 0} J_{\text{out}}(\partial D, t, \Delta t)$, where

$$J_{\text{out}}(\partial D, t, \Delta t) = \frac{1}{\Delta t} \int_{D^c} d\mathbf{x} \int_D \frac{p_N(\mathbf{y}, t) \, d\mathbf{y}}{(4\pi\Delta t)^{d/2} \sqrt{\det \boldsymbol{\sigma}(\mathbf{y}, t)}}$$

$$\times \exp\left\{-\frac{(\mathbf{x} - \mathbf{y} - \mathbf{a}(\mathbf{y}, t)\Delta t)^T \boldsymbol{\sigma}^{-1}(\mathbf{y}, t)(\mathbf{x} - \mathbf{y} - \mathbf{a}(\mathbf{y}, t)\Delta t)}{4\Delta t}\right\}. \qquad (2.39)$$

Similarly, the unidirectional flux into the domain is defined as the limit $J_{\text{in}}(\partial D, t) = \lim_{\Delta t \to 0} J_{\text{in}}(\partial D, t, \Delta t)$, where

$$J_{\text{in}}(\partial D, t, \Delta t) = \frac{1}{\Delta t} \int_D d\mathbf{x} \int_{D^c} \frac{p_N(\mathbf{y}, t) \, d\mathbf{y}}{(4\pi\Delta t)^{d/2} \sqrt{\det \boldsymbol{\sigma}(\mathbf{y}, t)}}$$

$$\times \exp\left\{-\frac{(\mathbf{x} - \mathbf{y} - \mathbf{a}(\mathbf{y}, t)\Delta t)^T \boldsymbol{\sigma}^{-1}(\mathbf{y}, t)(\mathbf{x} - \mathbf{y} - \mathbf{a}(\mathbf{y}, t)\Delta t)}{4\Delta t}\right\}. \qquad (2.40)$$

The net flux from the domain is defined as the limit

$$J_{\text{net}}(\partial D, t) = \lim_{\Delta t \to 0} J_{\text{net}}(\partial D, t, \Delta t),$$

where $J_{\text{net}}(\partial D, t, \Delta t) = J_{\text{out}}(\partial D, t, \Delta t) - J_{\text{in}}(\partial D, t, \Delta t)$.

Theorem 2.2.4 (Unidirectional and net fluxes in $\mathbb{R}^d$). *The discrete unidirectional flux densities at a boundary point $\mathbf{x}_B$ are given by*

$$\mathbf{J}_{\text{out,in}}(\mathbf{x}_B, t) \cdot \mathbf{n}(\mathbf{x}_B) = \sqrt{\frac{\sigma_n(\mathbf{x}_B, t)}{\pi\Delta t}} \, p(\mathbf{x}_B, t)$$

$$\pm \frac{1}{2} \mathbf{J}_{\text{net}}(\mathbf{x}_B, t) \cdot \mathbf{n}(\mathbf{x}_B) + O(\sqrt{\Delta t}), \qquad (2.41)$$

where $n(x)$ is the unit outer normal at a boundary point x, $\sigma_n(x_B, t) = n(x_B)^T \sigma(x_B, t) n(x_B)$, and the components of the net flux density vector, $i = 1, 2, \ldots, d$, are

$$J_{net}^i(x_B, t) = -\left\{ \sum_{j=1}^{d} \frac{\partial \sigma^{ij}(x, t) p(x, t)}{\partial x^j} + a^i(x, t) p(x, t) \right\}_{x=x_B}. \qquad (2.42)$$

The total net flux is

$$J_{net}(\partial D, t) = \oint_D J_{net}(x, t) \cdot n(x) \, dS_x. \qquad (2.43)$$

Proof. To evaluate the unidirectional and net fluxes, we define near a boundary point x_B the vector $v(x_B) = \sigma^{1/2}(x_B, t) n(x_B)$, where $n(x_B)$ is the unit outer normal at x_B, and map a two-sided neighborhood $\mathcal{N}$ of the boundary by the transformation

$$x = x_B + \sigma^{1/2}(x_B, t) \left[x^\perp - \xi v(x_B) \right] \sqrt{\Delta t}, \qquad (2.44)$$

where $x^\perp$ is a $(d-1)$-dimensional vector orthogonal to $\xi v(x_B)$. Here $\xi < 0$ for $x \in D$ and $\xi > 0$ for $x \in D^c$ (this applies to both x and y in the integrals (2.39) and (2.40)). The boundary is then mapped into the hyperplane $\xi = 0$. We may confine the domain of integration in the double integral (2.39) to $\mathcal{N}$, because the contribution of integration outside $\mathcal{N}$ decays exponentially fast as $\Delta t \to 0$. We partition the boundary into patches $\mathcal{P}_B$ about a finite set of boundary points x_B and freeze the coefficients at x_B inside the slice $\{(x^\perp, \xi) \in \mathcal{N} \mid x^\perp \in \mathcal{P}_B\}$. We expand first

$$\frac{(x - y - a(y,t)\Delta t)^T \sigma^{-1}(x_B, t)(x - y - a(y,t)\Delta t)}{\Delta t}$$
$$= (y^\perp - x^\perp)^T \sigma^{-1}(x, t)(y - x) - 2a^T(x, t)\sigma^{-1}(x, t)(y - x)\Delta t + O(\Delta t^2),$$

and then about x_B in the variables $(x^\perp, \xi)$. The transformation (2.44) maps each side of the slice onto a half-space. The variables $x^\perp$ integrate out in the double integrals (2.39), (2.40), expressed in the variables $x^\perp, \xi$ (in both integrals) and the calculation of the unidirectional flux density reduces to that in the one-dimensional case. We obtain the unidirectional flux densities in the form (2.41)–(2.43), as asserted.   □

Exercise 2.13 (Details of the proof). Fill in the missing details of the proof.   □

Exercise 2.14 (The FPE is a conservation law). Prove that in analogy with (2.38), the FPE (2.23) in $\mathbb{R}^d$ can also be written as the conservation law

$$\frac{\partial p}{\partial t} = -\nabla \cdot J,$$

where

$$J^i = \sum_{j=1}^d \frac{\partial^2 \sigma^{ij}(y,t) p(y,t)}{\partial y^i \partial y^j} + \frac{\partial a^i(y,t) p(y,t)}{\partial y^i}, \text{ for } i = 1, 2, \ldots, d.$$

□

2.3 The Wiener Measure Induced by SDEs

The solution of $x(t)$ of the SDE

$$dx = a(x,t)\,dt + b(x,t)\,dw, \quad x(0) = x_0, \tag{2.45}$$

is a Markov process, so its multidimensional density is determined uniquely by the transition probability density function $p(y,t \mid x,s)$, which is the solution of the FPE

$$\frac{\partial p(y,t \mid x,s)}{\partial t} = \frac{1}{2} \frac{\partial^2 \left[b^2(y,t) p(y,t \mid x,s) \right]}{\partial y^2} - \frac{\partial \left[a(y,t) p(y,t \mid x,s) \right]}{\partial y} \tag{2.46}$$

with the initial condition

$$\lim_{t \downarrow s} p(y,t \mid x,s) = \delta(y - x). \tag{2.47}$$

We can use it to construct a Wiener probability measure on the space of continuous functions (trajectories) in analogy to that constructed in Section 1.2.4. The cylinder sets are defined as

$$C\left(t_1, \ldots, t_K; I_1, I_2, \ldots, I_K\right) \tag{2.48}$$

$$= \left\{ \mathfrak{w} \in \Omega \mid x(t_1, \mathfrak{w}) \in I_1, x(t_2, \mathfrak{w}) \in I_2, \ldots, x(t_K, \mathfrak{w}) \in I_K \right\}.$$

These are the same cylinder sets as in Section 1.2.3, but they are assigned different probabilities. Specifically, we define

$$\Pr\left\{ C\left(t_1, \ldots, t_K; I_1, \ldots, I_K\right) \right\} = \int_{I_1} \cdots \int_{I_K} \prod_{j=1}^K p(y_j, t_j \mid y_{j-1}, t_{j-1})\, dy_j. \tag{2.49}$$

The transition probability density function $p(y,t \mid x,s)$ satisfies the Chapman–Kolmogorov equation (1.80), so the consistency condition

$$C\left(t_1, \ldots, t_K; I_1, I_2, \ldots, I_j = \mathbb{R}, \ldots, I_K\right)$$

$$= C\left(t_1, \ldots, t_{j-1}, t_{j+1}, \ldots, t_K; I_1, I_2, \ldots, I_{j-1}, I_{j+1}, \ldots, I_K\right)$$

is satisfied.

Using Theorem 2.2.1, we can write each factor $p\left(y_j, t_j \mid y_{j-1}, t_{j-1}\right)$ in (2.49) as a limit. More specifically, we partition each interval $[t_{j-1}, t_j]$ $(j = 1, 2, \ldots, K)$ by the points $t_{j-1} = t_j^{(0)} < t_j^{(1)} < \cdots < t_j^{(N_j)} = t_j$ such that $\Delta t_{k,N_j} = t_j^{(k)} - t_j^{(k-1)} = (t_j - t_{j-1})/N_j \to 0$ as $N_j \to \infty$ and write

$$p\left(y_j, t_j \mid y_{j-1}, t_{j-1}\right) = \lim_{N_j \to \infty} \underbrace{\int_{\mathbb{R}} \int_{\mathbb{R}} \cdots \int_{\mathbb{R}}}_{N_j - 1} \prod_{k=1}^{N_j - 1} d z_k$$

$$\times \prod_{k=1}^{N_j} \frac{1}{\sqrt{2\pi \Delta t_{k,N_j}}\, b\left(z_{k-1}, t_{k-1,N_j}\right)}$$

$$\times \exp\left\{ \frac{-\left[z_k - z_{k-1} - a\left(z_{k-1}, t_{k-1,N_j}\right) \Delta t_{k-1,N_j}\right]^2}{2 b^2\left(z_{k-1}, t_{k-1,N_j}\right) \Delta t_{k,N_j}} \right\},$$

$$\tag{2.50}$$

with $z_{N_j} = y_j$, $z_0 = y_{j-1}$, which can be used in (2.49). We denote by $\mathrm{Pr}_{a,b}\{A\}$ the extension of this probability measure from cylinders to any set A in $\mathcal{F}$. The case in which $a(x, t)$ and $b(x, t)$ are adapted stochastic processes is handled in a similar manner [60], [52].

Definition 2.3.1 (Brownian filtration). *The sets of events $\mathcal{F}_t$ generated by cylinder sets confined to times $0 \leq t_i \leq t$, for any fixed t, is called* Brownian filtration *and is said to be generated by Brownian events up to time t.*

Obviously, $\mathcal{F}_s \subset \mathcal{F}_t \subset \mathcal{F}$ if $0 \leq s < t < \infty$. Note that the elementary events of a Brownian filtration $\mathcal{F}_t$ are continuous functions in the entire time range, not just the initial segments in the time interval $[0, t]$. However, only the initial segments of Brownian paths in $\mathcal{F}_t$ that occur by time t are observed and so can be used to define the filtration. The pairs $(\Omega, \mathcal{F}_t)$ are different probability spaces for different values of t.

2.4 Brownian Simulations at Boundaries

Diffusion processes often model particles confined to a given domain in space, for example ions in biological cells. The behavior of the diffusion paths at the boundary of the domain is often determined by physical laws; for example, ions cannot penetrate biological cell membranes due to the much lower dielectric constant of the lipid cell membrane (about $\varepsilon = 2$) being much lower than that of the intracellular salt solution (about $\varepsilon = 80$). Sometimes, diffusing trajectories that cross the boundary of a domain cannot return for a long time and can be considered instantaneously terminated then and there. This can happen, for example, in modeling the diffusive motion of an atom inside a molecule that collides thermally with other molecules. Due to the collisions, the atom, held by the chemical bond, can be displaced to a distance at which the chemical bond is broken, thus dissociating from the molecule permanently. In other situations the diffusing paths can be terminated at the boundary with a given probability, for example, a diffusing protein can stick to a receptor on the cell membrane, or continue its diffusive motion inside the cell. There are many more modes of boundary behavior of diffusion processes inside bounded domains (see, e.g., [45], [110], [81]), so a theory of diffusion inside bounded domains with a variety of boundary behaviors is needed.

The easiest way to define a diffusion process inside a given domain with a prescribed boundary behavior is to run discrete computer simulations. The relevant mathematical problems are the question of convergence, determining the partial differential equations that the transition probabilities and their functionals satisfy, identifying the boundary conditions for the partial differential equations, and determining the probability measures defined in function space by the confined diffusions. The imposed boundary conditions on the simulated trajectories are reflected in the pdf, in boundary conditions for the FPE, but sometimes more complicated connections show up. And conversely, often boundary conditions imposed on the FPE to express physical processes that occur at the boundary, for example, a reactive boundary condition that expresses a possible binding of a molecule, can be expressed in terms of the boundary behavior of simulated trajectories of an SDE. The Wiener path integral is a convenient tool for the study of the duality between the boundary behavior of trajectories and boundary (and other) conditions for the FPE, as discussed below.

2.5 Absorbing Boundaries

The simplest simulation of the Itô dynamics

$$dx = a(x, t)\, dt + b(x, t)\, dw \ \text{ for } t > s, \ x(s) = x_0, \tag{2.51}$$

is the Euler scheme

$$x_N(t + \Delta t) = x_N(t) + a(x_N(t), t)\Delta t + b(x_N(t), t)\, \Delta w(t), \ \ x_N(s) = x_0. \tag{2.52}$$

If the trajectories of $x_N(t, \mathfrak{w})$ that start at $x_0 > 0$ (and are determined by (2.52)) are truncated at the first time they cross the origin, we say that the origin is an *absorbing boundary*.

Exercise 2.15 (Convergence of trajectories*). Generalize the proof of Skorokhod's theorem (Theorem 2.1.1) to Euler's scheme with an absorbing boundary (see [142]). □

The path integral corresponding to this situation is defined on the subset of trajectories that *never cross a* from left to right. Thus the integration in the definition (2.6) of the pdf does not extend over $\mathbb{R}$, but rather is confined to the ray $[0, \infty)$. That is, the pdf is given by

$$p_N(x,t \mid x_0, s) = \underbrace{\int_0^\infty dy_1 \int_0^\infty dy_2 \cdots \int_0^\infty dy_{N-1}}_{N-1} \prod_{j=1}^N \frac{1}{\sqrt{2\pi \Delta t} \, b(y_{j-1}, t_{j-1})}$$

$$\times \exp \left\{ -\frac{\left[y_j - y_{j-1} - a(y_{j-1}, t_{j-1}) \Delta t) \right]^2}{2 b^2(y_{j-1}, t_{j-1}) \Delta t} \right\}, \tag{2.53}$$

where $t_0 = s$, $y_0 = x_0$ and $t_N = t$, $y_N = x$. As in (2.10), we define $p_N(x, t \mid s) = \int_0^\infty p_N(x, t \mid x_0, s) p_0(x_0) \, dx_0$, where $p_0(x_0)$ is a sufficiently smooth test density with compact support on the positive axis.

Theorem 2.5.1. *For every $T > 0$, the Wiener integral $p_N(x, t \mid s)$ converges to the solution $p(x, t \mid s)$ of the initial–boundary value problem*

$$\frac{\partial p(y, t \mid s)}{\partial t} = \frac{1}{2} \frac{\partial^2 \left[b^2(y, t) p(y, t \mid s) \right]}{\partial y^2} - \frac{\partial \left[a(y, t) p(y, t \mid s) \right]}{\partial y}, \tag{2.54}$$

$$\lim_{t \downarrow s} p(y, t \mid s) = p_0(y), \tag{2.55}$$

$$p(0, t \mid s) = 0 \ \text{for } t > s, \tag{2.56}$$

uniformly for all $x > 0$, $s < t < T$.

Proof. If $x > 0$, then the change of variables $y = x - a(y, t)\Delta t + \eta b(y, t)\sqrt{\Delta t}$ changes the domain of integration from $0 < y < \infty$ to the ray $-(x - a(y, t)\Delta t)/b(y, t)\sqrt{\Delta t} < \eta < \infty$, so integration can be extended to $\mathbb{R}$ with exponentially decaying error as $\Delta t \to 0$. The proof of Theorem 2.2.1 then shows that the limit function $p(x, t)$ satisfies (2.46), (2.47). If, however, we set $x = 0$ in the expansion of the path integral (2.53) that leads to (2.17), the change of variables maps the domain of integration onto only the half-line $0 \le \eta < \infty$, rather than onto the entire line. The value of the Gaussian integral over this domain is $\frac{1}{2}$, so assuming that the limit of $p_N(x, t \mid s) \to p(x, t \mid s)$ as $N \to \infty$ exists, we obtain the identity $p(0, t \mid s) = \frac{1}{2} p(0, t \mid s)$, which apparently implies that $p(y, t \mid s)$ satisfies the boundary condition (2.56).

The pdf $p_N(y, t \mid s)$, however, does not necessarily converge to the solution $p(y, t \mid s)$ of (2.46), (2.47) with the boundary condition (2.56), uniformly up to the boundary. More specifically, it is not clear that

$$\lim_{y \to 0} \lim_{N \to \infty} p_N(y,t \mid s) = \lim_{N \to \infty} \lim_{y \to 0} p_N(y,t \mid s), \qquad (2.57)$$

because, as is typical for diffusion approximations of Markovian jump processes that jump over the boundary [90], [92], [91], [42], the convergence is not necessarily uniform, and typically, a boundary layer is formed. A boundary layer expansion is needed to capture the boundary phenomena. To examine the convergence of $p_N(y,t \mid s)$ near $y = 0$, we rewrite (2.53) as the integral equation

$$p_N(y, t + \Delta t \mid x_0) = \int_0^\infty \frac{p_N(x,t \mid s)}{\sqrt{4\pi\sigma(x,t)\Delta t}} \exp\left\{ -\frac{[y - x - a(x,t)\Delta t]^2}{4\sigma(x,t)\Delta t} \right\} dx,$$

where $\sigma(x,t) = \frac{1}{2}b^2(x,t)$, and introduce the local variable $y = \eta\sqrt{\Delta t}$ and the boundary layer solution $p_{\text{bl}}(\eta, t) = p_N(\eta\sqrt{\Delta t}, t \mid x_0)$. Changing variables $x = \xi\sqrt{\Delta t}$ in the integral gives the integral equation

$$p_{\text{bl}}(\eta, t + \Delta t \mid s) = \int_0^\infty \frac{p_{\text{bl}}(\xi, t \mid s)}{\sqrt{4\pi\sigma(\xi\sqrt{\Delta t}, t)}} \exp\left\{ -\frac{\left[\eta - \xi - a(\xi\sqrt{\Delta t}, t)\sqrt{\Delta t}\right]^2}{4\sigma(\xi\sqrt{\Delta t}, t)} \right\} d\xi.$$

The boundary layer solution has an asymptotic expansion in powers of $\sqrt{\Delta t}$

$$p_{\text{bl}}(\eta, t \mid s) \sim p_{\text{bl}}^{(0)}(\eta, t \mid s) + \sqrt{\Delta t}\, p_{\text{bl}}^{(1)}(\eta, t \mid s) + \Delta t\, p_{\text{bl}}^{(2)}(\eta, t \mid s) + \dots . \quad (2.58)$$

Expanding all functions in the integral equation in powers of $\sqrt{\Delta t}$ and equating similar orders, we obtain integral equations that the asymptotic terms of (2.58) must satisfy. The leading-order $O(1)$ term gives the Wiener–Hopf-type equation [120]

$$p_{\text{bl}}^{(0)}(\eta, t \mid s) = \int_0^\infty \frac{p_{\text{bl}}^{(0)}(\xi, t \mid s)}{\sqrt{4\pi\sigma(0,t)}} \exp\left\{ -\frac{(\eta - \xi)^2}{4\sigma(0,t)} \right\} d\xi \qquad (2.59)$$

on the half-line for $\eta > 0$. Integrating (2.59) with respect to η over $\mathbb{R}^+$, changing the order of integration, and changing variables to $\eta = \xi + z$ on the right-hand side, we obtain

$$\int_0^\infty p_{\text{bl}}^{(0)}(\eta, t \mid s)\, d\eta = \int_0^\infty \frac{p_{\text{bl}}^{(0)}(\xi, t \mid s)}{\sqrt{4\pi\sigma(0,t)}} \int_{-\xi}^\infty \exp\left\{ -\frac{z^2}{4\sigma(0,t)} \right\} dz\, d\xi$$

$$= \int_0^\infty p_{\text{bl}}^{(0)}(\xi, t \mid s)\, d\xi$$

$$- \int_0^\infty \frac{p_{\text{bl}}^{(0)}(\xi, t \mid s)}{\sqrt{4\pi\sigma(0,t)}} \int_\xi^\infty \exp\left\{ -\frac{z^2}{4\sigma(0,t)} \right\} dz\, d\xi;$$

hence

$$\int_0^\infty \frac{p_{\mathrm{bl}}^{(0)}(\xi, t \mid s)}{\sqrt{4\pi\sigma(0,t)}} \int_\xi^\infty \exp\left\{-\frac{z^2}{4\sigma(0,t)}\right\} dz\, d\xi = 0. \qquad (2.60)$$

It follows that $p_{\mathrm{bl}}^{(0)}(\xi, t \mid s) = 0$, because all functions in (2.60) are continuous and nonnegative.

Away from the boundary layer, the solution admits an outer expansion

$$p_{\mathrm{out}}(y, t \mid s) \sim p_{\mathrm{out}}^{(0)}(y, t \mid s) + \sqrt{\Delta t}\, p_{\mathrm{out}}^{(1)}(y, t \mid s) + \dots, \qquad (2.61)$$

where $p_{\mathrm{out}}^{(0)}(y, t \mid x_0)$ is an as yet undetermined function that satisfies (2.12). The leading-order matching condition of the boundary layer and the outer solutions is $\lim_{\eta\to\infty} p_{\mathrm{bl}}^{(0)}(\eta, t \mid s) = p_{\mathrm{out}}^{(0)}(0, t \mid s)$, so that $p_{\mathrm{out}}^{(0)}(0, t \mid s) = 0$. Because

$$\lim_{y\to 0}\lim_{N\to\infty} p_N(y, t \mid s) = p_{\mathrm{out}}^{(0)}(0, t \mid s) = 0,$$

$$\lim_{N\to\infty}\lim_{y\to 0} p_N(y, t \mid s) = p_{\mathrm{bl}}^{(0)}(0, t \mid s) = 0,$$

the limits are interchangeable and (2.57) holds, and so does the boundary condition (2.56).

The remainder of the proof follows that of Theorem 2.2.1. We extend $p_N(x, t \mid s)$ to t off the lattice by an interpolation $\tilde{p}_N(x, t \mid s)$, as in (2.7) and (2.8). The boundary layer expansion of $\tilde{p}_N(x, t \mid s)$ is similar to that of $p_N(x, t \mid s)$ and implies that for every $\varepsilon > 0$ and $T > 0$ there is $\delta > 0$ such that if $0 \leq x < \delta$ and $s < t < T$, then $\tilde{p}_N(x, t \mid s) < \varepsilon$, $p(x, t \mid s) < \varepsilon$, $|\tilde{p}_N(\delta, t \mid s) - p(\delta, t \mid s)| < 2\varepsilon$, and $\tilde{p}_N(x, s \mid s) - p(x, s \mid s) = p_0(x) - p_0(x) = 0$. The maximum principle implies that $|\tilde{p}_N(x, t \mid s) - p(x, t \mid s)| < 2\varepsilon$ for all $x > \delta$, $s \leq t \leq T$. Because δ is arbitrarily small, the convergence is uniform. □

Exercise 2.16 (Diffusion in an interval with absorbing boundaries). Generalize Theorem 2.5.1 to diffusion in a finite interval with absorption at both boundaries. Generalize Exercises 2.2–2.9 to this case. □

Exercise 2.17 (Convergence of trajectories in d dimensions*). Generalize the proof of Skorokhod's theorem (Theorem 2.1.1) to Euler's scheme in a domain $D \subset \mathbb{R}^d$ with an absorbing boundary. □

Theorem 2.5.2. *For every $T > s \geq 0$ the pdf $p_{\Delta t}(y, t \mid x, s)$ of the Euler scheme*

$$x(t + \Delta t) = x(t) + a(x(t), t)\Delta t + \sqrt{2}B(x(t), t)\Delta w(t, \Delta t), \quad x_N(s) = x_0,$$

where all trajectories are instantaneously terminated when they exit D, converges as $\Delta t \to 0$ to the solution $p(y, t \mid x, s)$ of the initial value problem for the FPE, (2.23), (2.24), with the absorbing (Dirichlet) boundary condition

$$p(y, t \mid x, s) = 0 \text{ for } y \in \partial D, \ x \in D. \qquad (2.62)$$

Exercise 2.18 (Proof of Theorem 2.5.2). Prove Theorem 2.5.2 by following these steps:

(i) Derive the Chapman–Kolmogorov equation

$$p_{\Delta t}(y, t + \Delta t\, x_0, s) = \int_D \frac{p_{\Delta t}(x, t \mid x_0, s)\, dx}{(4\pi \Delta t)^{d/2} \sqrt{\det \sigma(x, t)}} \exp\left\{-\frac{\mathcal{B}(x, y, t)}{4\Delta t}\right\},$$

where $\mathcal{B}(x, y, t) = [y - x - a(x, t)\Delta t]^T \sigma^{-1}(x, t)[y - x - a(x, t)\Delta t]$, as in (2.21).

(ii) Generalize Exercises 2.2–2.9 to the integral in (i).
(iii) Show that there is no boundary layer.
(iv) Use the maximum principle to prove convergence.

□

2.5.1 Unidirectional Flux and the Survival Probability

The trajectories absorbed at the boundary give rise to a unidirectional probability flux from the domain into the boundary. The absorbing boundary condition (2.56) implies that the pdf vanishes for all $x \geq 0$, so that its right derivatives at the origin vanish. It follows from (2.35) that $J_{RL}(0, t \mid s) = 0$. On the other hand, (2.28) and (2.30) give

$$J(0, t \mid s) = J_{LR}(0, t \mid s) = -\left.\frac{\partial \sigma(x, t) p(x, t \mid s)}{\partial x}\right|_{x=0}.$$

Because $\sigma(x, t) > 0$ and $p(x, t \mid s) > 0$ for $x < 0$, but $p(0, t \mid s) = 0$, it follows that $J(0, t \mid s) > 0$. This means that there is positive flux into the absorbing boundary, so that the probability of trajectories that survive in the region to the left of the absorbing boundary, $\int_{-\infty}^{0} p(x, t \mid s)\, dx$, must be a decreasing function of time. This can be seen directly from (2.38) by integrating it with respect to x over the ray $(-\infty, 0)$ and using the fact that $\lim_{x \to -\infty} J(x, t) = 0$,

$$\frac{d}{dt} \int_{-\infty}^{0} p(x, t \mid s)\, dx = -J(0, t \mid s) < 0. \tag{2.63}$$

Equation (2.63) means that the total population of trajectories in the domain $x < 0$ decreases with time, so that the transition pdf $p(x, t \mid s)$ is defective in the sense that it does not integrate to 1 over the domain.

To clarify the meaning of the defective pdf, we note that in fact for any subset A of the domain,

$$\int_A p(x,t \mid s)\, dx = \Pr\{x(t) \in A, \tau > t \mid s\},$$

because $x(t)$ can be in A only if it has not been absorbed in the boundary by time t, that is, $p(x,t \mid s)$ is actually the joint density of $x(t)$ and (complementary) PDF of the first passage time τ,

$$p(x,t \mid s) = \Pr\{x(t) = x, \tau > t \mid s\}.$$

Definition 2.5.1 (The survival probability). *The survival probability at time t of trajectories in the domain $x < 0$ that started at time $s < t$ at a point $x_s < 0$, denoted by $S(t \mid x_s, s)$, is the conditional probability that the first passage time to the absorbing boundary $x = 0$, denoted by τ, does not exceed t. That is,*

$$S(t \mid x_s, s) = \Pr\{\tau > t \mid x_s, s\} = \int_{-\infty}^{0} p\,(x,t \mid x_s, s)\, dx. \qquad (2.64)$$

Analogous definitions apply to the multidimensional dynamics

$$dx = a(x,t)\, dt + \sqrt{2} B(x,t)\, dw \qquad (2.65)$$

in a domain D in $\mathbb{R}^d$ with an absorbing boundary. The flux density vector $J(y,t \mid x,s)$ in (2.42) reduces to

$$J^i(y,t \mid x,s) = -\sum_{j=1}^{n} \frac{\partial \left[\sigma^{ij}(y,t) p\,(y,t \mid x,s)\right]}{\partial y^j}, \qquad (2.66)$$

where $\sigma(x,t) = B^T(x,t) B(x,t)$. The probability per unit time of trajectories that are absorbed into a given surface $S \subset \partial D$ is given by $F = \int_S J(y,t \mid x,s) \cdot n(y)\, dS_y$, which can be interpreted as follows.

Theorem 2.5.3 (Normal flux density at an absorbing boundary). *The normal flux density $J(y,t \mid x,s) \cdot n(y)$ at the absorbing boundary is the conditional probability per unit surface area and per unit time that passes through the surface at the boundary point y at time t. Thus it is the conditional probability density (per unit area) of stochastic trajectories absorbed at the boundary point y at a given instance of time $t > s$, given that they started at x at time s.*

The survival probability and the probability distribution function of the first passage time τ to the boundary ∂D are related by the following theorem.

Theorem 2.5.4 (Survival probability and the first passage time).

$$\Pr\{\tau > t \mid x(s) = x\} = S(t \mid x, s) = \int_D p(y, t \mid x, s)\, dy, \qquad (2.67)$$

where $p(y, t \mid x, s)$ is the solution of the initial–boundary value problem for the Fokker–Planck equation (2.23), (2.24), (2.62). The mean first passage time to the boundary, after time s, is

$$\mathbb{E}[\tau \mid x, s] = \int_s^\infty S(t \mid x, s)\, dt = \int_s^\infty \int_D p(y, t \mid x, s)\, dy\, dt. \qquad (2.68)$$

The proofs of Theorems 2.5.3 and 2.5.4 are straightforward. The MFPT $\mathbb{E}[\tau \mid x, s]$ is the mean first passage time after the initial time s.

Example 2.1 (Flux in 1D). The one-dimensional Fokker–Planck equation has the form $p_t = -J_y(y, t \mid x, s)$, where the one-dimensional flux is given by $J(y, t \mid x, s) = a(y, t)p(y, t \mid x, s) - [\sigma(y, t)p(y, t \mid x, s)]_y$. At an absorbing boundary, $J(y, t \mid x, s) = -[\sigma(y, t)p(y, t \mid x, s)]_y$ for $x \in D$ and $y \in \partial D$, because $p(y, t)\big|_{y \in \partial D} = 0$. $\qquad\qquad \square$

Chapter 3
Nonlinear Filtering and Smoothing of Diffusions

Filtering theory is concerned with the extraction of information from noisy measurements of a signal. For example, in radio communications the signal may be speech, music, or data, which are converted by a microphone or a computer into a variable voltage $x(t)$ or a vector of variable voltages $x(t)$. The signal is often assumed to be a stationary random process and is often characterized by its power spectral density function. Linear filtering theory is by now a classical subject that has been thoroughly discussed in the literature. Nonlinear filtering, however, is still a subject of intensive research.

3.1 Diffusion Model of Random Signals

In many applications the signal statistics are modeled as those of a diffusion process defined by a system of Itô stochastic differential equations of the form

$$dx(t) = m(x(t), t)\,dt + \sigma(x(t), t)\,dw(t), \qquad (3.1)$$

where $m(x, t)$ is the drift vector, $\sigma(x, t)$ is the diffusion matrix, and $w(t)$ is a vector of independent standard Brownian motions. The units of the components of $x(t)$ depend on the type of the signal. It can be volts, radians, hertz, meters (on the oscilloscope screen), and so on. Keeping in mind that the units of the MBM $w(t)$ are $\sqrt{\sec}$, equation (3.1) defines the units of the coefficients in terms of the units of the signal and of the MBM. We assume therefore throughout this chapter that all variables are dimensionless. The statistics of the trajectories of (3.1) represent those of the physical signals that are transmitted in a given channel, for example, the statistics of all voltages that the antennas of all FM radio stations around the globe emit, classical music, jazz, rock, news, political gobbledygook, commercials, and so on. Not all components of the signal are necessarily transmitted.

Z. Schuss, *Nonlinear Filtering and Optimal Phase Tracking*, Applied
Mathematical Sciences 180, DOI 10.1007/978-1-4614-0487-3_3,
© Springer Science+Business Media, LLC 2012

Before transmission the signal usually undergoes modulation by the transmitter and is converted into the amplitude of a carrier wave (AM transmission), or into the phase or frequency of a transmitted wave (PM or FM transmissions, respectively), or any other form of modulation. The modulated signal is the voltage (or voltages) on the transmitter's antenna. The modulation is a memoryless transformation of the signal.

For example, in *amplitude modulated* (AM) transmission on carrier frequency ω_0 (usually measured in kHz) the modulated signal on the transmitter's antenna is the voltage

$$h(x(t), t) = \sqrt{2} x(t) \sin \omega_0 t. \tag{3.2}$$

Usually the original random signal is filtered before it is modulated by (3.2). This means that it is first fed into a linear or nonlinear system of differential equations and the output is modulated by (3.2). This means that the filtered signal is a component of the output of a system of differential equations of the form (3.1).

In *phase modulated* (PM) transmission with carrier frequency ω_0 (usually measured in MHz) the modulated signal on the antenna is the voltage

$$h(x(t), t) = \sqrt{2} \sin[\omega_0 t + \beta x(t)]. \tag{3.3}$$

For a signal with $\mathrm{Var}\, x(t) = 1$, we call β the *modulation index*. In *frequency modulated* (FM) transmission on carrier frequency ω_0 (usually in the range 88 MHz–105 MHz), the signal $x(t)$ is converted into frequency by the transformation

$$h(x(t), t) = \sqrt{2} \sin \left(\omega_0 t + d_f \int_0^t x(s)\, ds \right), \tag{3.4}$$

where the parameter d_f is called *frequency deviation*. The modulation in (3.4) is not a memoryless transformation of $x(t)$, because the integral contains all the past trajectory of the signal up to time t. The modulation (3.4) can, however, be viewed as a memoryless transformation of the output of a system of the form (3.1) if we define the two-dimensional signal $x(t) = (x_1(t), x_2(t))^T$ as the output of the Itô system

$$dx_1(t) = m(x_1(t), t)\, dt + \sigma(x_1(t), t)\, dw, \quad dx_2(t) = d_f x_1(t)\, dt, \tag{3.5}$$

and then (3.4) can be written as the memoryless transformation of $x(t)$

$$h(x(t), t) = \sqrt{2} \sin \left(\omega_0 t + x_2(t) \right). \tag{3.6}$$

The modulated signal can also have several components, that is, $h(x(t), t)$ can be a vector. Components of the modulated signal (not necessarily all of them) are sent to the transmitter and are picked up by the receiver in a usually noisy transmission channel. There are many sources of noise in a transmission channel. These may include Johnson noise in the electronic components, atmospheric noise, jamming, interchannel interference, and so on.

3.2 Diffusion Model of the Received Signal

The noisy output of the receiver's antenna, denoted by $y(t)$, is usually modeled as the sum of the transmitted signal and the acquired noise. That is, the received signal can be written as the output of the Itô system

$$dy(t) = h(x(t), t)\, dt + \rho\, dv, \tag{3.7}$$

where $v(t)$ is a vector of independent standard Brownian motions, independent of $w(t)$, and ρ is the noise matrix. Usually ρ is assumed independent of $x(t)$, because otherwise the signal can be detected from the variance of the measurements noise $\rho\dot{v}$. Using white noise as a carrier is not an efficient method of modulation. However, ρ can be a function of t or even of $y(t)$ and t. We denote by y_0^t the trajectory of the measurements up to time t. All the information about the signal available at time t is contained in y_0^t. The filtration generated by the process $y(t)$ is denoted by $\mathcal{G}_t$. We confine our attention to one-dimensional models of the signal and the measurements. More general cases can be found in [105].

3.3 Small Noise and Reduction to Standard Form

Often, the measurements noise is assumed to be small, after appropriate scaling of the model. The assumption of small noise is often valid, because if the noise is not small, one may as well decide on the value of the signal by flipping a coin. When the measurements noise is small and the signal is linear, the system (3.1), (3.7), can be reduced to a standard form.

Thus, if $\|\rho\| \ll \|\sigma(x(t), t)\|$ and $\|\rho\sigma(x(t), t)\| \ll 1$ (e.g., in the maximum norm), and (3.1) is linear, the problem of estimating $x(t)$ with an observation process $y(t) \in \mathbb{R}^2$ that satisfies (3.7), can be reduced to the standard form

$$\dot{x} = A\,x + \varepsilon\,B\,\dot{w}, \quad x(0) = x_0, \tag{3.8}$$

$$\dot{y} = h(x) + \varepsilon\,\dot{v}, \quad y(0) = 0, \tag{3.9}$$

where $\varepsilon \ll 1$.

We illustrate the reduction by the benchmark first-order problem of filtering the phase-modulated Brownian motion

$$\dot{x} = \sigma\dot{w}, \quad \dot{y} = h(x) + \rho\dot{v}, \tag{3.10}$$

where
$$h(x) = \begin{bmatrix} \sin x(t) \\ \cos x(t) \end{bmatrix},$$

with small measurements noise ($\rho \ll \sigma$). To reduce (3.10) to the form (3.8), (3.9), we scale $t = at'$ and use the Brownian scaling $dw(t) = \sqrt{a}\, dw'(t')$ (and a similar scaling of $v(t)$). Writing $x(t) = x'(t')$, $y(t) = by'(t')$, the scaled system (3.10) becomes
$$dx' = \sqrt{a}\sigma dw', \quad dy' = \frac{a}{b}h(x')\, dt' + \frac{\rho\sqrt{a}}{b} v'.$$

Setting
$$a = b = \frac{\rho}{\sigma}, \quad \varepsilon^2 = \rho\sigma, \quad t' = \left(\frac{\sigma}{\rho}\right) t,$$

we obtain the scaled system
$$\dot{x}' = \varepsilon\dot{w}', \quad \dot{y}' = h(x') + \varepsilon\dot{v}'. \tag{3.11}$$

Similarly, in the second-order problem of filtering the frequency modulated Brownian motion in a low-noise channel,
$$\begin{bmatrix} \dot{x}_1 \\ \dot{x}_2 \end{bmatrix} = \begin{bmatrix} 0 & 1 \\ 0 & 0 \end{bmatrix}\begin{bmatrix} x_1 \\ x_2 \end{bmatrix} + \sigma\begin{bmatrix} 0 \\ 1 \end{bmatrix}\dot{w},$$
$$\dot{y} = h(x_1) + \rho\dot{v},$$

we set
$$a = \gamma = \sqrt{\frac{\rho}{\sigma}}, \quad b = \sqrt{\frac{\sigma}{\rho}}, \quad t' = bt,$$

and obtain the scaled system
$$\begin{bmatrix} \dot{x}'_1 \\ \dot{x}'_2 \end{bmatrix} = \begin{bmatrix} 0 & ab \\ 0 & 0 \end{bmatrix}\begin{bmatrix} x'_1 \\ x'_2 \end{bmatrix} + \begin{bmatrix} 0 \\ \dfrac{\sigma\sqrt{a}}{b} \end{bmatrix}\dot{w}'$$

$$\tag{3.12}$$

$$\dot{y}' = \frac{a}{\gamma}h(x'_1) + \frac{\rho\sqrt{a}}{\gamma}\dot{v}'.$$

In the notation of (3.8), (3.9) the matrices of the scaled system (3.12) and the small parameter are, respectively,
$$A = \begin{bmatrix} 0 & 1 \\ 0 & 0 \end{bmatrix}, \quad B = \begin{bmatrix} 0 \\ 1 \end{bmatrix}, \quad \varepsilon = \sqrt[4]{\sigma\rho^3}.$$

Exercise 3.1 (Standard form). To reduce to standard form:

(i) Show that the above scaling method reduces the general problem of low-noise filtering of a linear signal $d\boldsymbol{x}(t) = A\boldsymbol{x}(t)\,dt + \sigma\boldsymbol{B}\,d\boldsymbol{w}(t)$ (with $\|\boldsymbol{B}\| = 1$ in some norm) to the standard form

$$d\boldsymbol{x}'(t') = \frac{\varepsilon^2}{\sigma^2}A\boldsymbol{x}'(t')\,dt' + \varepsilon\boldsymbol{B}\,d\boldsymbol{w}'(t'), \quad d\boldsymbol{y}'(t') = \boldsymbol{h}(\boldsymbol{x}'(t')) + \varepsilon\,\boldsymbol{v}'(t').$$

(ii) Find classes of nonlinear signals measured in a low-noise channel that can be reduced to the standard form (3.1), (3.7), with σ and ρ replaced by $\varepsilon\tilde{\sigma}$ and $\varepsilon\tilde{\rho}$, where $\varepsilon \ll 1$ and $\tilde{\sigma}, \tilde{\rho} = O(1)$. □

3.4 Optimal Filtering and a Posteriori Density

A *causal estimator* of $\boldsymbol{x}(t)$ is a stochastic process $\tilde{\boldsymbol{x}}(t)$, measurable with respect to the filtration $\mathcal{G}_t$ generated by the process $\boldsymbol{y}(t)$, whose value is an estimator of $\boldsymbol{x}(t)$. This means that a causal estimator depends at time t only on the measurements up to time t and not on any future information. There are many different optimality criteria for choosing a causal estimator. The *optimal filtering problem* is to find a causal estimator that satisfies a given optimality criterion.

Note that the filtering problem calls for a *real-time* decision rule that continuously estimates the signal instantly after its noisy measurement. The instantaneous decision is needed for automatic control of devices that cannot tolerate delay, and often for speech and music (though some delay can be tolerated here). If the decision can be delayed, the estimation problem is called the *smoothing* problem. If the decision has to be made about the future value of the signal given its past measurements, the estimation problem is called the *prediction* problem. The latter is useful in trying to predict stock prices (good luck!), or to anticipate the location of a moving craft, and so on.

We discuss below three conventional optimality criteria for the filtering problem and the optimal estimators they define. We say that an optimal filter is *realizable* if its dynamics is computable, given the measurements. For example, if the measurements process can be used as an input into a finite system of computable ordinary differential equations (ODEs) whose output is the optimal estimator, the optimal filter is realizable. This means that the ODEs can be solved either numerically or by an analog circuit. Similarly, if a partial differential equation (PDE) can be found such that the optimal filter is a computable functional of its solution, the optimal filter is realizable, provided the solution can be computed in real time. There aren't, however, too many analog circuits that solve partial differential equations, and the class of parabolic PDEs whose solutions can be expressed in terms of a finite number of known elementary or transcendental special functions is too meagre to satisfy the needs of filtering theory. The existing algorithms for solving parabolic PDEs are too slow to be of much use in solving filtering problems. Therefore explicit realizations of optimal estimators for the filtering problem of diffusions are feasible only in special cases, but the general case is considered nonrealizable.

The simplest realizable case is that of linear theory that assumes linear models for both signal and measurements. The resulting estimator is the so-called *Kalman–Bucy* filter [77]. The nonlinear optimal filtering problem has realizable exact solution only in exceptional cases that are of little practical use [34]. It has, however, approximate solutions if the measurements noise is small, as discussed above.

We consider next the model (3.1), (3.7).

Definition 3.4.1 (The minimum conditional error variance estimator). *The conditional mean square estimation error (CMSEE) of an estimator $\tilde{x}(t)$ is*

$$\widehat{|e|^2}(t) = \mathbb{E}\left[\,|x(t) - \tilde{x}(t)|^2 \,\middle|\, y_0^t\right], \tag{3.13}$$

where y_0^t is the measured trajectory and the mean square estimation error (MSEE) is the unconditional expectation

$$\overline{|e|^2}(t) = \mathbb{E}\widehat{|e|^2}(t) = \mathbb{E}\left[\,|x(t) - \tilde{x}(t)|^2\right]. \tag{3.14}$$

Conditional averaging is denoted by $\widehat{\cdot}$, whereas unconditional averaging is denoted by $\overline{\cdot}$. Specifically, the CMSEE $\widehat{|e|^2}(t)$ is the average of the estimation error of $\tilde{x}(t)$ over all possible trajectories of the process $x(\cdot)$ up to time t, given the measured trajectory y_0^t. The MSEE $\overline{|e|^2}(t)$ is the estimation error of $\tilde{x}(t)$ averaged over all possible trajectories $x(t)$ of the signal and all possible trajectories y_0^t of the measurements.

Definition 3.4.2 (The optimal filtering problem). *The optimal filtering problem is to find an estimator that minimizes the CMSEE over all causal ($\mathcal{G}_t$-measurable) estimators of the signal.*

Theorem 3.4.1 (The minimum conditional mean square error estimator). *Among all estimators that are $\mathcal{G}_t$-measurable, the one with the minimal CMSEE is the conditional expectation $\hat{x}(t)$ of the signal, given the measurements:*

$$\hat{x}(t) = \mathbb{E}\left[x(t) \,|\, y_0^t\right]. \tag{3.15}$$

Proof. Indeed, assume that $\tilde{x}(t)$ is $\mathcal{G}_t$-measurable. Then

$$\mathbb{E}\left[\,|x(t) - \tilde{x}(t)|^2 \,\middle|\, y_0^t\right] = \mathbb{E}\left[\,|x(t) - \hat{x}(t)|^2 \,\middle|\, y_0^t\right]$$
$$+ 2\mathbb{E}\left[\,[x(t) - \hat{x}(t)] \cdot [\hat{x}(t) - \tilde{x}(t)] \,|\, y_0^t\right]$$
$$+ \mathbb{E}\left[\,|\tilde{x}(t) - \hat{x}(t)|^2 \,\middle|\, y_0^t\right].$$

Because $\hat{x}(t)$ and $\tilde{x}(t)$ are $\mathcal{G}_t$-measurable, they are no longer random in $\mathcal{G}_t$. Thus

$$\mathbb{E}\left[[x(t)-\hat{x}(t)]\cdot[\hat{x}(t)-\tilde{x}(t)]\mid y_0^t\right] = \mathbb{E}\left[[x(t)-\hat{x}(t)]\mid y_0^t\right]\cdot[\hat{x}(t)-\tilde{x}(t)] = 0,$$

because

$$\mathbb{E}\left\{[x(t)-\hat{x}(t)]\mid y_0^t\right\} = \hat{x}(t)-\hat{x}(t) = 0.$$

It follows that

$$\mathbb{E}\left[|x(t)-\tilde{x}(t)|^2\ \middle|\ y_0^t\right] = \mathbb{E}\left[|x(t)-\hat{x}(t)|^2\ \middle|\ y_0^t\right] + \mathbb{E}\left[|\tilde{x}(t)-\hat{x}(t)|^2\ \middle|\ y_0^t\right]$$

$$\geq \mathbb{E}\left\{|x(t)-\hat{x}(t)|^2\ \middle|\ y_0^t\right\}.$$

$\square$

Definition 3.4.3 (The a posteriori density of the signal). *The conditional probability distribution function of the signal, given the trajectory $y_0^t = \{y(s), 0 \leq s \leq t\}$ of the measurements, is defined for every measurable set A by*

$$P(A, t \mid y_0^t) = \Pr\{x(t) \in A \mid y(s), 0 \leq s \leq t\}.$$

It has a probability density function $p(x, t \mid y_0^t)$ such that

$$P(A, t \mid y_0^t) = \int_A p(x, t \mid y_0^t)\, dx. \tag{3.16}$$

Note that $p(x, t \mid y_0^t)$ is a stochastic process defined on the probability space $\mathcal{G}_t$ of trajectories y_0^t.

The estimator $\hat{x}(t)$ can be expressed in terms of the a posteriori probability density function as

$$\hat{x}(t) = \int_{-\infty}^{\infty} x\, p(x, t \mid y_0^t)\, dx \tag{3.17}$$

and its CMSEE as

$$\widehat{|e|^2}(t) = \int_{-\infty}^{\infty} |x - \hat{x}(t)|^2 p(x, t \mid y_0^t)\, dx. \tag{3.18}$$

Definition 3.4.4 (The maximum a posteriori probability estimator). *The maximum a posteriori probability (MAP) estimator is a causal estimator that maximizes the a posteriori probability density function at each time t, that is, $x_{MAP}(t)$ is defined by the relation*

$$p(x_{MAP}, t \mid y_0^t) = \max_{x} p(x, t \mid y_0^t). \tag{3.19}$$

A dynamic programming deterministic approach to the filtering problem was proposed in [101], [117] and further elaborated in [70], [50], [51]. Instead of a stochastic model (3.1), (3.7), the signal is modeled as a deterministic trajectory with unknown error and so is the measured signal. The errors have some measure of magnitude, σ and ρ, respectively, but practically nothing is assumed about their randomness. Instead of the a posteriori probability density $p(x, t \mid y_0^t)$, a functional $-J(x(\cdot))$ of the signal $x(t)$ is introduced, analogous to the signal information. Filtering is achieved by minimizing the L^2 norm of the errors (or their "energy") in the class of all causal (independent of future measurements) trajectories $x(t)$ of the signal in every time interval $[0, t]$, thereby maximizing the chosen measure of information. The maximizing trajectory $x_{\text{MNE}}(t)$ is called the minimum noise energy (MNE) filter. The following is its formal definition.

Definition 3.4.5 (The minimum noise energy estimator). *The minimum noise energy estimator, denoted by $x_{\text{MNE}}(t)$, is the end value of the trajectory $x(s)$, $0 \leq s \leq t$, that minimizes the energy functional*

$$J(x(\cdot)) = \frac{1}{2} \int_0^t \left\{ \frac{[\dot{y}(s) - h(x(s), s)]^2}{\rho^2} + \frac{[\dot{x}(s) - m(x(s), s)]^2}{\sigma^2} \right\} ds \quad (3.20)$$

in the class of causal processes $x(\cdot)$.

We have to assume that the expression in the braces in (3.20) is finite, that is, that the expression is integrable. It cannot represent the energy of white noise $\int_0^t \left[|\dot{w}(t)|^2 + |\dot{v}(t)|^2 \right] dt$, because white noises are not square integrable. One can imagine, however, a model in which the white noises in (3.1) and (3.7) are replaced with square integrable wide-band noises, and after the MNE filter is found (by minimizing $J(x(\cdot))$), the white noise limit of infinite bandwidth is taken.

The function

$$S(x, t, T) = J(x_{\text{MNE}}(T)), \quad (3.21)$$

where $x_{\text{MNE}}(t) = x$, satisfies for $t < T$ the Hamilton–Jacobi–Bellman (HJB) equation [14]

$$S_t + \frac{\sigma^2}{2\rho} |\nabla_x S|^2 + m(x, t) \cdot \nabla_x S = \frac{1}{\rho} \left[\frac{1}{2} |h(x, t)|^2 - h(x, t) \cdot \dot{y}_S(t) \right].$$

The relationship between the MNE and the MMSEE filters is elaborated in Section 4.1.3 below. It is shown that for small measurements noise ρ their error variances are asymptotically the same. Their stability as phase trackers is, however, not nearly the same: the mean time for the MNE tracker to lose lock on the signal is longer by many orders of magnitude than that of all MMSEE-type trackers (see Chapters 5 and 7).

3.5   The Zakai Equation

We consider the one-dimensional version of (3.1) and (3.7) in the standard form of Section 3.3,

$$dx(t) = m(x, t)\, dt + \varepsilon\sigma\, dw(t), \tag{3.22}$$

$$dy(t) = h(x, t)\, dt + \varepsilon\rho\, dv(t), \tag{3.23}$$

where $m(x, t)$ and $h(x, t)$ are possibly nonlinear, sufficiently smooth functions. The processes $w(t)$ and $v(t)$ are independent standard Brownian motions, σ and ρ are constants, and ε can be 1 or can represent a small parameter. An unnormalized version of the a posteriori density $p(x, t \mid y_0^t)$ is a function $\varphi(x, t)$ such that

$$p(x, t \mid y_0^t) = \frac{\varphi(x, t)}{\displaystyle\int_{\mathbb{R}} \varphi(x, t)\, dx}. \tag{3.24}$$

Our purpose here is to find an unnormalized version $\varphi(x, t)$ that satisfies a linear partial differential equation.

Theorem 3.5.1 (Zakai [164]). *The solution of the initial value problem for Zakai's equation in Stratonovich form,*

$$d_S\varphi(x, t) = \left\{ \mathcal{L}\varphi(x, t) - \frac{h^2(x, t)\varphi(x, t)}{2\varepsilon^2\rho^2} \right\} dt + \frac{h(x, t)\varphi(x, t)}{\varepsilon^2\rho^2}\, d_S y(t), \tag{3.25}$$

$$\lim_{t \to 0} \varphi(x, 0) = \lim_{t \to 0} p(x, t \mid y_0^t),$$

where

$$\mathcal{L}\varphi(x, t) = -[m(x, t)\varphi(x, t)]_x + \frac{1}{2}[\varepsilon^2\sigma^2\varphi(x, t)]_{xx}, \tag{3.26}$$

is an unnormalized version of the a posteriori density in the sense of (3.24). *The Itô form of Zakai's equation* (3.25) *is given by*

$$d\varphi(x, t) = \mathcal{L}\varphi(x, t)\, dt + \frac{h(x, t)\varphi(x, t)}{\varepsilon^2\rho^2}\, dy(t). \tag{3.27}$$

Proof. The joint transition pdf $p(x, y, t \mid \xi, \eta, s)$ of a trajectory $(x(t), y(t))$ of (3.22) and (3.23) is the solution of the initial value problem for the FPE for $t > s$

$$\frac{\partial p\,(x,y,t\mid\xi,\eta,s)}{\partial t}$$

$$= -\frac{\partial m(x,t)\,p\,(x,y,t\mid\xi,\eta,s)}{\partial x} - \frac{\partial h(x,t)\,p\,(x,y,t\mid\xi,\eta,s)}{\partial y}$$

$$+ \frac{(\varepsilon\sigma)^2}{2}\frac{\partial^2 p\,(x,y,t\mid\xi,\eta,s)}{\partial x^2} + \frac{(\varepsilon\rho)^2}{2}\frac{\partial^2 p\,(x,y,t\mid\xi,\eta,s)}{\partial y^2},\qquad (3.28)$$

$$\lim_{t\downarrow s} p\,(x,y,t\mid\xi,\eta,s) = \delta(x-\xi)\delta(y-\eta).\qquad (3.29)$$

It is the limit of the joint density of the Euler scheme for the simulation of the filtering problem on a finite interval $0 \le s \le t \le T$ (see Chapter 2 and [137, Chapter 3]). Specifically, discretizing (3.22), (3.23) on a sequence of grids

$$\left\{ t_i = s + i\Delta t, \quad i = 0, 1, \dots, N, \quad \Delta t = \frac{t}{N} \right\},$$

we define discrete trajectories by the Euler scheme

$$x_N(t_{i+1}) = x_N(t_i) + \Delta t\, m(x_N(t_i), t_i) + \varepsilon\sigma\,\Delta w(t_i), \quad x_N(t_0) = \xi, \qquad (3.30)$$

$$y_N(t_{i+1}) = y_N(t_i) + \Delta t\, h\,(x_N(t_i), t_i) + \varepsilon\rho\,\Delta v(t_i), \quad y_N(t_0) = \eta, \qquad (3.31)$$

for $i = 0, 1, \dots, N - 1$, where $\Delta w(t_i)$ and $\Delta v(t_i)$ are independent zero-mean Gaussian random variables with variance Δt.

As in (2.5), the pdf of an entire Euler trajectory $(x_N(s), y_N(s))$ $(0 \le s \le t)$ is the Gaussian

$$p_N(x_1, \dots, x_N; y_1, \dots, y_N; t_1, \dots, t_N) = \prod_{k=1}^{N}\left[\frac{\exp\left\{-\dfrac{\mathcal{B}_k(x_k, x_{k-1})}{2\varepsilon^2\Delta t}\right\}}{2\pi\varepsilon^2\rho\sigma\Delta t}\right],$$

$$(3.32)$$

where the exponent is the quadratic form

$$\mathcal{B}_k(x_k, x_{k-1}) = [x_k - x_{k-1} - \Delta t a_{k-1}]^T\, B\,[x_k - x_{k-1} - \Delta t a_{k-1}],$$

where

$$x_k = \begin{bmatrix} x_k \\ y_k \end{bmatrix}, \quad a_k = \begin{bmatrix} m(x_k, t_k) \\ h(x_k, t_k) \end{bmatrix}, \quad B = \begin{bmatrix} \sigma^{-2} & 0 \\ 0 & \rho^{-2} \end{bmatrix}.$$

The transition probability density, as in (2.6), is therefore

$$p(x, y, t \mid \xi, \eta, s)$$

$$= \lim_{N \to \infty} \underbrace{\int_{\mathbb{R}} dx_1 \int_{\mathbb{R}} dx_2 \cdots \int_{\mathbb{R}} dx_{N-1}}_{N-1} \underbrace{\int_{\mathbb{R}} dy_1 \int_{\mathbb{R}} dy_2 \cdots \int_{\mathbb{R}} dy_{N-1}}_{N-1}$$

$$\times \prod_{k=1}^{N} \left[\frac{\exp\left\{ -\dfrac{\mathcal{B}_k(\boldsymbol{x}_k, \boldsymbol{x}_{k-1})}{2\varepsilon^2 \Delta t} \right\}}{2\pi \varepsilon^2 \rho \sigma \Delta t} \right], \tag{3.33}$$

where $x_N = x$, $y_N = y$, $x_0 = \xi$, $y_0 = \eta$. It is the solution of the FPE (3.28) with the initial condition (3.29). The pdf (3.32) can be written as

$$p_N(x_1, \ldots, x_N; y_1, \ldots, y_N; t_1, \ldots, t_N) \tag{3.34}$$

$$= \prod_{k=1}^{N} \left[\frac{1}{\sqrt{2\pi \Delta t}\, \varepsilon \sigma} \exp\left\{ -\frac{[x_k - x_{k-1} - m(x_{k-1}, t_{k-1})\, \Delta t]^2}{2\varepsilon^2 \sigma^2 \Delta t} \right\} \right.$$

$$\left. \times \exp\left\{ \frac{1}{\varepsilon^2 \rho^2} h(x_{k-1}, t_{k-1})(y_k - y_{k-1}) - \frac{1}{2\varepsilon^2 \rho^2} h^2(x_{k-1}, t_{k-1}) \Delta t \right\} \right]$$

$$\times \left[\prod_{k=1}^{N} \frac{\exp\left\{ -\dfrac{(y_k - y_{k-1})^2}{2\varepsilon^2 \rho^2 \Delta t} \right\}}{\sqrt{2\pi \Delta t}\, \varepsilon \rho} \right].$$

The proof of Theorem 2.2.1 and the Feynman–Kac formula (1.142) (see Exercise 2.9) show that the first product, integrated with respect to all intermediate points $x_1, x_2, \ldots, x_{N-1}$, converges to the function

$$\varphi(x, t \mid s) \tag{3.35}$$

$$= \lim_{N \to \infty} \underbrace{\int_{\mathbb{R}} dx_1 \int_{\mathbb{R}} dx_2 \cdots \int_{\mathbb{R}} dx_{N-1}}_{N-1}$$

$$\times \prod_{k=1}^{N} \left[\frac{1}{\sqrt{2\pi \Delta t}\, \varepsilon \sigma} \exp\left\{ -\frac{[x_k - x_{k-1} - m(x_{k-1}, t_{k-1})\, \Delta t]^2}{2\varepsilon^2 \sigma^2 \Delta t} \right\} \right.$$

$$\left. \times \exp\left\{ \frac{1}{\varepsilon^2 \rho^2} h(x_{k-1}, t_{k-1})(y_k - y_{k-1}) - \frac{1}{2\varepsilon^2 \rho^2} h^2(x_{k-1}, t_{k-1}) \Delta t \right\} \right],$$

which is the solution of Zakai's equation in Stratonovich form (3.25). Because $x_0 = \xi$, the product (3.35) satisfies the initial condition

$$\lim_{t \downarrow s} \varphi(x, t \mid s) = \delta(x - \xi). \tag{3.36}$$

The Itô form of Zakai's equation is obtained from (3.25) by subtracting the Wong–Zakai correction (see Theorem 1.3.1 and [137, Theorem 4.2.1]). Setting $\varphi(x, t) = \varphi(x, t \mid 0)$, we note that the stochastic process defined by (3.25) is $\varphi(x, t)$, so the correction has to be done with respect to $\varphi(x, t)$. Note that the white-noise term driving (3.25) is $[h(x, t)\varphi(x, t)/\varepsilon\rho]\, dv$, so that the noise coefficient is $[h(x, t)\varphi(x, t)/\varepsilon\rho]$. This term has to be differentiated with respect to $\varphi(x, t)$ in order to find the Wong–Zakai correction. The correction term in the drift is given by

$$-\frac{1}{2}\frac{h(x, t)}{\varepsilon\rho}\frac{h(x, t)\varphi(x, t)}{\varepsilon\rho} = -\frac{h^2(x, t)\varphi(x, t)}{2\varepsilon^2\rho^2},$$

so the Itô form of Zakai's equation is

$$d\varphi(x, t) = \mathcal{L}\varphi(x, t)\, dt + \frac{h(x, t)\varphi(x, t)}{\varepsilon^2\rho^2}\, dy(t). \tag{3.37}$$

We assume henceforward that the Zakai equation (3.27) has a unique solution in the strong sense. That is, we assume that $\varphi(x, t)$ is twice differentiable with respect to x and satisfies the equation in the Itô sense.

The joint density

$$p_N(x_N, t_N;\, y_1, y_2, \ldots, y_N)$$
$$= \Pr\{x_N(t_N) = x_N, y_N(t_1) = y_1, y_N(t_2) = y_2, \ldots, y_N(t_N) = y_N\}$$

can now be written at $t = t_N$, $x_N = x$ as

$$p_N(x, t;\, y_1, y_2, \ldots, y_N) \tag{3.38}$$
$$= [\varphi(x, t) + o(1)]\prod_{k=1}^{N}\frac{1}{\sqrt{2\pi\Delta t\varepsilon\rho}}\exp\left\{-\frac{(y_k - y_{k-1})^2}{2\varepsilon^2\rho^2\Delta t}\right\},$$

where $o(1) \to 0$ as $N \to \infty$. Equivalently,

$$\varphi(x, t \mid s) = \frac{p_N(x, t;\, y_1, y_2, \ldots, y_N)}{\displaystyle\prod_{k=1}^{N}\frac{1}{\sqrt{2\pi\Delta t\varepsilon\rho}}\exp\left\{-\frac{(y_k - y_{k-1})^2}{2\varepsilon^2\rho^2\Delta t}\right\}} + o(1), \tag{3.39}$$

which can be interpreted as follows: $\varphi(x, t \mid s)$ is the conditional density of $x_N(t)$, given the entire trajectory $\{y_N(t_i)\}_{i=0}^N$. However, the probability density of the trajectories $\{y_N(t_i)\}_{i=0}^N$,

$$p_N^B(y_s^t) = \prod_{k=1}^N \frac{1}{\sqrt{2\pi\Delta t}\varepsilon\rho} \exp\left\{-\frac{(y_k - y_{k-1})^2}{2\varepsilon^2\rho^2\Delta t}\right\},$$

is Brownian, rather than the a priori density

$$p_N(y_s^t) = \int_{-\infty}^\infty p_N(x, t; y_1, y_2, \ldots, y_N\} \, dx, \qquad (3.40)$$

imposed by (3.22), (3.23).

Next, we show that $\varphi(x, t) = \varphi(x, t \mid 0)$ is an unnormalized a posteriori density. The a posteriori density of the discretized process is

$$p_N(x, t \mid y_0^t\} = \frac{p_N(x, t; y_1, y_2, \ldots, y_N\}}{p_N(y_0^t)} = \frac{p_N^B(y_0^t)}{p_N(y_0^t)} \{\varphi(x, t) + o(1)\}. \quad (3.41)$$

As $N \to \infty$, both sides of (3.41) converge to a finite limit, which we write as

$$p(x, t \mid y_0^t) = \alpha(t)\varphi(x, t), \qquad (3.42)$$

where

$$\alpha(t) = \lim_{N \to \infty} \frac{p_N^B(y_0^t)}{p_N(y_0^t)} \qquad (3.43)$$

is a function independent of x. Because $\int_\mathbb{R} p(x, t \mid y_0^t) \, dx = 1$, we have

$$\int_\mathbb{R} \varphi(x, t) \, dx = \frac{1}{\alpha(t)}; \qquad (3.44)$$

hence (3.24). Note that (3.36) implies that $\alpha(0) = 1$.

$\square$

Lemma 3.5.1. *The Stratonovich and Itô forms of $\alpha(t)$ are, respectively,*

$$\alpha(t) = \exp\left\{\frac{1}{2\varepsilon^2\rho^2}\int_0^t \widehat{h}^2(t) \, dt - \frac{1}{\varepsilon^2\rho^2}\int_0^t \widehat{h}(t) \, d_S y(t)\right\} \qquad (3.45)$$

$$= \exp\left\{\frac{1}{2\varepsilon^2\rho^2}\int_0^t \widehat{h}^2(t) \, dt - \frac{1}{\varepsilon^2\rho^2}\int_0^t \widehat{h}(t) \, dy(t)\right\}, \qquad (3.46)$$

where

$$\widehat{h}(t) = \mathbb{E}[h(x(t), t) \mid \mathcal{G}_t] = \int_{-\infty}^\infty h(x, t) \, p(x, t \mid y_0^t) \, dx \qquad (3.47)$$

is the conditional moment of $h(x(t), t)$.

Proof. Equation (3.44) implies that $\alpha(t)$ satisfies the stochastic differential equation

$$\frac{d_S\alpha(t)}{\alpha(t)} = -\frac{\int_{-\infty}^{\infty} d_S\varphi(x,t)\,dx}{\int_{-\infty}^{\infty} \varphi(x,t)\,dx} = -\alpha(t)\int_{-\infty}^{\infty} d_S\varphi(x,t)\,dx; \qquad (3.48)$$

hence, using the Zakai–Stratonoich equation (3.25), we obtain

$$-\frac{d_S\alpha(t)}{\alpha(t)} = \alpha(t)\int_{-\infty}^{\infty} \left\{ \left[\mathcal{L}\varphi(x,t) - \frac{h^2(x,t)\varphi(x,t)}{2\varepsilon^2\rho^2} \right] dt \right.$$
$$\left. + \frac{h(x,t)\varphi(x,t)}{\varepsilon^2\rho^2} d_S y(t) \right\} dx$$
$$= -\frac{\widehat{h^2}(t)}{2\varepsilon^2\rho^2} dt + \frac{\hat{h}(t)}{\varepsilon^2\rho^2} d_S y(t),$$

because $\int_{-\infty}^{\infty} \mathcal{L}\varphi(x,t)\,dx = 0$. Thus $\alpha(t)$ is the solution of the differential equation

$$d_S\alpha(t) = \frac{\widehat{h^2}(t)}{2\varepsilon^2\rho^2}\alpha(t)\,dt - \frac{\hat{h}(t)}{\varepsilon^2\rho^2}\alpha(t)\,d_S y(t), \quad \alpha(0) = 1, \qquad (3.49)$$

whose solution is (3.45).

To prove (3.46), we note that $\hat{h}(t)$ is a stochastic process that depends on y_0^t, so instead of using the Wong–Zakai correction to convert the Stratonovich integral in (3.45) to Itô form, we derive first a stochastic equation for $\gamma(t) = \int_{-\infty}^{\infty} \varphi(x,t)\,dx$. According to (3.27),

$$d\gamma(t) = \int_{-\infty}^{\infty} d\varphi(x,t)\,dx = \int_{-\infty}^{\infty} \left\{ \mathcal{L}\varphi(x,t)\,dt + \frac{h(x,t)\varphi(x,t)}{\varepsilon^2\rho^2} dy(t) \right\} dx$$
$$= \frac{\gamma(t)\hat{h}(t)}{\varepsilon^2\rho^2} dy(t). \qquad (3.50)$$

The solution of the Itô equation (3.50) is found by setting $\eta = \log\gamma$ and applying Itô's formula. We get

$$\gamma(t) = \exp\left\{ -\frac{1}{2\varepsilon^2\rho^2}\int_0^t \hat{h}^2(t)\,dt + \frac{1}{\varepsilon^2\rho^2}\int_0^t \hat{h}(t)\,dy(t) \right\};$$

hence (3.46).   □

Exercise 3.2 (The Itô form of (3.49)). Use Itô's formula to prove that

$$d\alpha(t) = \frac{\alpha(t)\hat{h}^2(t)}{\varepsilon^2\rho^2} dt - \frac{\alpha(t)\hat{h}(t)}{\varepsilon^2\rho^2} dy. \qquad (3.51)$$

□

The optimal filtering can be accomplished for each realization of y_0^t by feeding the rate of change of the measured noisy signal, $\dot{y}(t)$, into the Zakai–Itô equation (3.27) and getting the output $\varphi(x, t)$. Thus, in order to filter optimally the Zakai equation (3.27) has to be solved continuously in time: whenever a new measurement is taken, the a posteriori pdf $p(x, t \mid y_0^t)$ has to be recalculated instantly. Unfortunately, explicit solutions of Zakai's equation are not readily available, so that approximate solutions are called for [165]. The minimum variance estimator $\hat{x}(t)$ and its CMSEE $\widehat{e^2}(t)$ (see (3.17) and (3.18)) are given, respectively, by

$$\hat{x}(t) = \frac{\int_{-\infty}^{\infty} x \varphi(x, t)\, dx}{\int_{-\infty}^{\infty} \varphi(x, t)\, dx}, \tag{3.52}$$

$$\widehat{e^2}(t) = \frac{\int_{-\infty}^{\infty} [x - \hat{x}(t)]^2\, \varphi(x, t)\, dx}{\int_{-\infty}^{\infty} \varphi(x, t)\, dx}. \tag{3.53}$$

The MAP estimator x_{MAP} maximizes $p(x, t \mid y_0^t)$ at each time t. Thus the optimal filtering problem can be solved if a computable scheme for the evaluation of the integral (3.52) can be found.

Exercise 3.3 (Zakai's equation in higher dimensions). Derive Zakai's equation for an unnormalized a posteriori density of the signal, given the measurements, for a multidimensional model of the signal and measurements. □

3.5.1 Zakai's Equations for Smoothing Problems

In fixed-interval smoothing the measurements y_0^T are given in a fixed interval $[0, T]$, and the smoothing problem is to estimate $x(t)$ in this interval so that the MSEE is minimal. Also in this case, the MMSEE estimator is

$$\hat{x}(t) = \mathbb{E}[x(t) \mid y_0^T] = \int_{\mathbb{R}} x p(x, t \mid y_0^T)\, dx \quad \text{for } 0 < t < T. \tag{3.54}$$

The a posteriori pdf $p(x, t \mid y_0^T)$ in this case can be found from the solution of two Zakai–Stratonovich equations, one running from 0 to T and the other from T to 0. Specifically, we may state the following result.

Theorem 3.5.2 (Zakai's equations for fixed-interval smoothing). *The a posteriori pdf $p(x, t \mid y_0^T)$ is given by*

$$p\left(x,t\mid y_0^T\right) = \frac{\varphi_+(x,t)\varphi_-(x,t)}{\displaystyle\int_{\mathbb{R}} \varphi_+(x,t)\varphi_-(x,t)\, dx}, \tag{3.55}$$

where $\varphi_+(x,t)$ is the solution of the initial value problem

$$\lim_{t\downarrow 0} \varphi_+(x,t) = \delta(x-\xi) \tag{3.56}$$

for Zakai's equation (3.25) (or (3.27)) and $\varphi_-(x,t)$ is the solution of the terminal value problem for the backward equation

$$-d_S\varphi_-(x,t) = \left\{ \mathcal{L}_x^*\varphi_-(x,t) - \frac{h^2(x,t)\varphi_-(x,t)}{2\varepsilon^2\rho^2} \right\} dt + \frac{h(x,t)\varphi_-(x,t)}{\varepsilon^2\rho^2} d_S\, y(t),$$

$$\lim_{t\to T} \varphi_-(x,t) = \delta(x-\eta), \tag{3.57}$$

for $0 < t < T$, where $\mathcal{L}_x^*$ is the backward Kolmogorov operator defined in (1.106).

Proof. Fixing $x_j = x$, $t_j = t$, we break the product in (3.35) into $\prod_{k=1}^{j} \times \prod_{k=j+1}^{N}$ and suppress integration with respect to x_j. As in the proof of Zakai's equation in Theorem 3.5.1, the repeated integral of the product $\prod_{k=1}^{j}$ is the solution of the initial value problem for the Zakai–Stratonovich equation. The proof of the Feynman–Kac formula and [137, Exercise 5.10] show that the integrated product $\prod_{k=j+1}^{N}$ satisfies the backward Zakai–Stratonovich equation (3.57). □

The fixed-delay filtering-smoothing problem is to find the MMSEE estimate of both $x(t)$ and $x(t+\tau)$ simultaneously for a fixed-delay $\tau > 0$, given the measurements $y(s)$ in the interval $0 \le s \le t+\tau$.

Theorem 3.5.3 (Zakai's equations for fixed-delay filtering-smoothing). *The a posteriori pdf $p\left(x,t,u,t+\tau \mid y_0^{t+\tau}\right)$ of the fixed-delay filtering-smoothing problem is given by*

$$p\left(x,t,u,t+\tau \mid y_0^{t+\tau}\right) = \frac{\varphi_+(x,t)\varphi^-(x,t,u,t+\tau)}{\displaystyle\int_{\mathbb{R}}\int_{\mathbb{R}} \varphi_+(x,t)\varphi^-(x,t,u,t+\tau)\, dx\, du}, \tag{3.58}$$

where $\varphi_+(x,t)$ is the solution of the initial value problem (3.56) for the Zakai–Stratonovich equation (3.25) (or (3.27)) and $\varphi^-(x,t,u,s)$ is the solution of the forward–backward Stratonovich stochastic equation

$$d_{S,t}\varphi^- + \left[\mathcal{L}_{x,t}^*\varphi^- - \frac{h^2(x,t)\varphi^-}{2\varepsilon^2\rho^2} \right] dt + \frac{h(x,t)\varphi^-}{\varepsilon^2\rho^2} d_S\, y(t)$$

$$- d_{S,s}\varphi^-\big|_{s=t+\tau} + \left[\mathcal{L}_{u,t+\tau}\varphi^- - \frac{h^2(u,t+\tau)\varphi^-}{2\varepsilon^2\rho^2} \right] dt$$

$$+ \frac{h(u,t+\tau)\varphi^-}{\varepsilon^2\rho^2} d_{S,t}\, y(t+\tau) = 0 \quad \text{for } t > 0. \tag{3.59}$$

We may assume that $\varphi^-(x, 0, u, \tau)$ is known (for example, by initial filtering in the time interval $[0, \tau]$ with the initial condition $\varphi^-(x, 0, u, 0) = \delta(u - x)$).

Proof. To find the joint filtering–smoothing a posteriori density, we fix $x_j = x$, $t_j = t$, $x_N = u$, $t_N = t + \tau$, break the product

$$\varphi(x, t, u, t+\tau) = \lim_{N \to \infty} \underbrace{\int_{\mathbb{R}} dx_1 \int_{\mathbb{R}} dx_2 \cdots \int_{\mathbb{R}} dx_{N-1}}_{N-1}$$

$$\times \prod_{k=1}^{N} \left[\frac{1}{\sqrt{2\pi \Delta t}\, \varepsilon \sigma} \exp \left\{ -\frac{[x_k - x_{k-1} - m(x_{k-1}, t_{k-1})\, \Delta t]^2}{2\varepsilon^2 \sigma^2 \Delta t} \right\} \right.$$

$$\left. \times \exp \left\{ \frac{1}{\varepsilon^2 \rho^2} h(x_{k-1}, t_{k-1})(y_k - y_{k-1}) - \frac{1}{2\varepsilon^2 \rho^2} h^2(x_{k-1}, t_{k-1}) \Delta t \right\} \right]$$

into $\prod_{k=1}^{j} \times \prod_{k=j+1}^{N}$, and suppress integration with respect to x_j (note that j depends on N). The integral of the first product,

$$\varphi_+(x, t) = \lim_{N \to \infty} \underbrace{\int_{\mathbb{R}} dx_1 \int_{\mathbb{R}} dx_2 \cdots \int_{\mathbb{R}} dx_{j-1}}_{j-1}$$

$$\times \prod_{k=1}^{j} \left[\frac{1}{\sqrt{2\pi \Delta t}\, \varepsilon \sigma} \exp \left\{ -\frac{[x_k - x_{k-1} - m(x_{k-1}, t_{k-1})\, \Delta t]^2}{2\varepsilon^2 \sigma^2 \Delta t} \right\} \right.$$

$$\left. \times \exp \left\{ \frac{1}{\varepsilon^2 \rho^2} h(x_{k-1}, t_{k-1})(y_k - y_{k-1}) - \frac{1}{2\varepsilon^2 \rho^2} h^2(x_{k-1}, t_{k-1}) \Delta t \right\} \right],$$

is that defined in Zakai's theorem (Theorem 3.5.2). The second product,

$$\varphi^-(x, t, u, t+\tau) = \lim_{N \to \infty} \underbrace{\int_{\mathbb{R}} dx_{j+1} \int_{\mathbb{R}} dx_{j+2} \cdots \int_{\mathbb{R}} dx_{N-1}}_{N-j-1}$$

$$\times \prod_{k=j+1}^{N} \left[\frac{1}{\sqrt{2\pi \Delta t}\, \varepsilon \sigma} \exp \left\{ -\frac{[x_k - x_{k-1} - m(x_{k-1}, t_{k-1})\, \Delta t]^2}{2\varepsilon^2 \sigma^2 \Delta t} \right\} \right.$$

$$\left. \times \exp \left\{ \frac{1}{\varepsilon^2 \rho^2} h(x_{k-1}, t_{k-1})(y_k - y_{k-1}) - \frac{1}{2\varepsilon^2 \rho^2} h^2(x_{k-1}, t_{k-1}) \Delta t \right\} \right],$$

is the limit as $\Delta t \to 0$ of the solution of the integral equation

$$\varphi_{\Delta t}^- (x,t,u,t+\tau) = \int_{\mathbb{R}} d\xi \int_{\mathbb{R}} d\eta \, \varphi_{\Delta t}^- (\xi, t+\Delta t, \eta, t+\tau-\Delta t)$$

$$\times \exp \left\{ -\frac{[\xi - x - m(x,t)\Delta t]^2}{2\varepsilon^2 \sigma^2 \Delta t} \right\} \exp \left\{ -\frac{[u - \eta - m(\eta, t+\tau-\Delta t)\Delta t]^2}{2\varepsilon^2 \sigma^2 \Delta t} \right\}$$

$$\times \exp \left\{ \frac{1}{\varepsilon^2 \rho^2} h(\xi, t+\Delta t)\Delta y(t+\Delta t) - \frac{1}{2\varepsilon^2 \rho^2} h^2(\xi, t+\Delta t)\Delta t \right\}$$

$$\times \exp \left\{ \frac{1}{\varepsilon^2 \rho^2} h(\eta, t+\tau-\Delta t)\Delta y(t+\tau) - \frac{1}{2\varepsilon^2 \rho^2} h^2(\eta, t+\tau-\Delta t)\Delta t \right\}.$$

Expanding the integrals as above, we find that

$$\varphi^- = \varphi^- (x,t,u,s) = \lim_{\Delta t \to 0} \varphi_{\Delta t}^- (x,t,u,s)$$

is the solution of the forward–backward Stratonovich stochastic equation

$$d_{S,t}\varphi^- + \left[\mathcal{L}_{x,t}^* \varphi^- - \frac{h^2(x,t)\varphi^-}{2\varepsilon^2\rho^2} \right] dt + \frac{h(x,t)\varphi^-}{\varepsilon^2\rho^2} d_S \, y(t)$$

$$- d_{S,s}\varphi^- |_{s=t+\tau} + \left[\mathcal{L}_{u,t+\tau}\varphi^- - \frac{h^2(u,t+\tau)\varphi^-}{2\varepsilon^2\rho^2} \right] dt$$

$$+ \frac{h(u,t+\tau)\varphi^-}{\varepsilon^2\rho^2} d_{S,t} \, y(t+\tau) = 0 \quad \text{for } t > 0; \tag{3.60}$$

hence (3.58) follows. □

A separated solution

$$\varphi^-(x,t,u,s) = \varphi_1(x,t)\varphi_2(u,s)$$

means that first $\varphi_2(u,t+\tau)$ is constructed by filtering forward from time $t=0$ to time $t+\tau$ and then $\varphi_1(x,t)$ is constructed by filtering backward from time $t+\tau$ to time t with terminal condition $\varphi_1(u,t+\tau) = \varphi_2(u,t+\tau)$. A separated approximation to the solution exists if ε is sufficiently small (see Section 4.1 below).

The structure (3.58) of the a posteriori fixed-delay filtering–smoothing density means that the fixed-delay estimator is constructed by filtering forward from time 0 to time t and to time $t+\tau$ and then backward from time $t+\tau$ to time t. The estimator is obtained by averaging the forward and backward filters with appropriate weight (see Section 4.1). Different approaches to the nonlinear fixed-delay filtering–smoothing problem are given in [166] and [104].

3.5.2   Kushner's Equation for the a Posteriori Density

The a posteriori density $p(x, t \mid y_0^t)$ satisfies a nonlinear stochastic partial differential equation.

Theorem 3.5.4 (Kushner [95]). *The a posteriori density is the solution of the nonlinear initial value problem for Kushner's equation in Stratonovich form*

$$d_S p(x, t \mid y_0^t) = \left\{ \mathcal{L} p(x, t \mid y_0^t) - \frac{\left[h^2(x, t) - \widehat{h^2}(t) \right] p(x, t \mid y_0^t)}{2\varepsilon^2 \rho^2} \right\} dt$$

$$+ \frac{\left[h(x, t) - \hat{h}(t) \right] p(x, t \mid y_0^t)}{\varepsilon^2 \rho^2} d_S y, \qquad (3.61)$$

where $\hat{h}(t)$ is given by (3.47). The Itô form of (3.61) is given by

$$dp(x, t \mid y_0^t) = \left[\mathcal{L} p(x, t \mid y_0^t) - \frac{\hat{h}(t)[h(x, t) - \hat{h}(t)]}{\varepsilon^2 \rho^2} p(x, t \mid y_0^t) \right] dt$$

$$+ \frac{[h(x, t) - \hat{h}(t)]}{\varepsilon^2 \rho^2} p(x, t \mid y_0^t) \, dy(t). \qquad (3.62)$$

Proof. Differentiating (3.42) with respect to t in the Stratonovich sense and using (3.49), (3.25), we obtain

$$d_S p(x, t \mid y_0^t) = \varphi(x, t) \, d_S \alpha(t) + \alpha(t) d_S \varphi(x, t)$$

$$= \left[\frac{\widehat{h^2}(t)}{2\varepsilon^2 \rho^2} \alpha(t) \, dt - \frac{\hat{h}(t)}{\varepsilon^2 \rho^2} \alpha(t) \, d_S y(t) \right] \varphi(x, t)$$

$$+ \alpha \left\{ \mathcal{L} \varphi(x, t) - \frac{h^2(x, t) \varphi(x, t)}{2\varepsilon^2 \rho^2} \right\} dt + \frac{h(x, t) \varphi(x, t)}{\varepsilon^2 \rho^2} d_S y(t)$$

$$= \left[\frac{\widehat{h^2}(t)}{2\varepsilon^2 \rho^2} dt - \frac{\hat{h}(t)}{\varepsilon^2 \rho^2} d_S y(t) \right] p(x, t \mid y_0^t)$$

$$+ \left\{ \mathcal{L} p(x, t \mid y_0^t) - \frac{p(x, t \mid y_0^t) h^2(x, t)}{2\varepsilon^2 \rho^2} \right\} dt$$

$$+ \frac{p(x, t \mid y_0^t) h(x, t)}{\varepsilon^2 \rho^2} d_S y(t),$$

which after regrouping gives Kushner's equation (3.61) in Stratonovich form. To convert (3.61) to Itô form, we apply Itô's formula to the product of two processes $p(x, t \mid y_0^t) = \alpha(t)\varphi(x, t)$. Using the Itô equations (3.27) and (3.51), we obtain (3.62).

Exercise 3.4 (Time-variable noises). What changes in the Zakai and Kushner equations if $\sigma = \sigma(t)$ and $\rho = \rho(t)$? $\qquad\square$

Exercise 3.5 (Conditional moments). Use the Kushner equation to derive equations for conditional moments $\hat{f}(x(t))$, in particular for $\hat{x}(t)$, $\widehat{x^2}(t)$, etc. Note that the moment equations give a finite system only in the linear case. $\qquad\square$

Exercise 3.6 (Initial conditions). Use (3.36) to define solutions of Zakai's and Kushner's equations with other initial conditions. $\qquad\square$

Exercise 3.7 (Kushner's equation for smoothers*). Are there Kushner-type equations for fixed-interval and fixed-delay smoothers? $\qquad\square$

Exercise 3.8 (Change of measure [164]*). Use Girsanov's theorem [137, Theorem 6.6.1] to show:

(i) Changing the measure induced by (3.22), (3.23) with the Radon–Nikodym derivative

$$X(t) = \exp\left\{ \frac{1}{\varepsilon^2 \rho^2} \int_0^t h(x(t), t)\, dy(t) - \frac{1}{2\varepsilon^2 \rho^2} \int_0^t h^2(x(t), t)\, dt \right\} \quad (3.63)$$

converts the measurements process into Brownian motion, independent of the signal $x(t)$.

(ii) Due to independence, the conditional (a posteriori) probability density function and moments of $x(t)$ are the same as the unconditional (a priori) ones, with respect to the new measure. How can they be converted to those with respect to the original measure? $\qquad\square$

3.6 The Kalman–Bucy Linear Filter

The linear filtering model in one dimension is given by the system (3.22), (3.23) [77] with

$$m(x, t) = m(t)x, \quad h(x, t) = h(t)x. \quad (3.64)$$

We can write

$$m'(x, t) = m(t), \quad h'(x, t) = h(t). \quad (3.65)$$

The Zakai–Stratonovich equation for the system (3.22), (3.23) with the linear coefficients (3.64) is given by

$$d_S\varphi = \left[-m(t)\varphi - m(t)x\varphi_x + \frac{1}{2}\varepsilon^2\sigma^2(t)\varphi_{xx} - \frac{h^2(t)x^2\varphi}{2\varepsilon^2\rho^2}\right] dt + \frac{h(t)x\varphi}{\varepsilon^2\rho^2} d_S y.$$

(3.66)

Seeking a solution of the form

$$\varphi = \gamma(t)\exp\left\{-\frac{[x - \hat{x}(t)]^2}{2P(t)}\right\},$$

(3.67)

with unknown functions $\gamma(t)$, $\hat{x}(t)$, and $P(t)$, we obtain the differential equations

$$d\hat{x}(t) = \hat{x}(t)m'(\hat{x}((t),t) dt + \frac{P(t)h'(\hat{x}(t),t)}{\varepsilon^2\rho^2}\left[dy(t) - \hat{x}(t)h'(\hat{x}(t),t) dt\right]$$

(3.68)

$$dP(t) = 2P(t)m'(\hat{x}(t),t) dt + \frac{\varepsilon^4\sigma^2\rho^2 - P^2(t)h'^2(\hat{x}(t),t)}{\varepsilon^2\rho^2} dt.$$

Equations (3.52) and (3.67) show that $\hat{x}(t)$, as defined in the first equation in (3.68), is indeed the conditional expectation of $x(t)$, given the measurements, and therefore it is the minimum variance estimator for the linear model. Equations (3.53) and (3.67) give

$$P(t) = \frac{\displaystyle\int_{-\infty}^{\infty} [x - \hat{x}(t)]^2\, \varphi(x,t)\, dx}{\displaystyle\int_{-\infty}^{\infty} \varphi(x,t)\, dx} = \int_{-\infty}^{\infty} [x - \hat{x}(t))]^2 p(x,t)\, dx = \hat{e}^2(t).$$

It follows that equations (3.68) are the minimum variance filter for the signal and the CMSEE.

Definition 3.6.1 (The Kalman–Bucy filter).   *The system (3.68) is called the Kalman–Bucy filter, and the factor*

$$\frac{P(t)h'(\hat{x}(t),t)}{\varepsilon^2\rho^2}$$

(3.69)

is called the Kalman gain.

In this model, the Wong–Zakai correction vanishes, so the Itô and the Stratonovich forms of the filter (3.68) are the same. The driving factor $[dy(t) - \hat{x}(t)(t)h(t) dt]$ in (3.68) is called the *innovation process* and has properties similar to those of white noise [71].

Exercise 3.9 (Derivation of the filtering equations from the Zakai–Stratonovich equation). Derive the filter equations (3.68) from the Zakai–Stratonovich equation (3.66). Consider time-variable noises as well.   □

Exercise 3.10 ($\gamma(t)$). Calculate the pre-exponential factor $\gamma(t)$ in (3.67).   □

Exercise 3.11 (Filtering at small noise). Show that the filtering error vanishes in the limit of vanishing measurements noise ($\rho \to 0$).   □

Exercise 3.12 (Perfect filtering*). Find the Kalman–Bucy minimum-noise-variance filter for a linear multidimensional filtering problem. Find necessary and sufficient conditions under which the CMSEEs of all the signal components vanish with the measurements noise (this case is called *perfect filtering* [96], [139]). Find the components whose CMSEEs do not vanish with the noise when these conditions are not satisfied.   □

Exercise 3.13 (AM, PM, and FM). Write the Zakai–Stratonovich equations for AM (3.2), PM (3.3), and FM (3.4) filtering problems.   □

Exercise 3.14 (Linear smoothing theory [75], [59]). Use Theorems 3.5.2 and 3.5.3 to develop a linear smoothing theory. (HINT: Assume that all densities are Gaussian).   □

Chapter 4
Low-Noise Analysis of Zakai's Equation

Zakai's equation is a stochastic linear parabolic initial value problem that except for the linear case, has a closed-form solution only in exceptional and not very useful cases [165]. To understand the difficulties in applying the Zakai equation to the filtering problem, we consider again the simplified one-dimensional filtering problem (3.22), (3.23),

$$dx(t) = m(x(t))\, dt + \sigma\, dw(t), \tag{4.1}$$

$$dy(t) = h(x(t))\, dt + \rho\, dv(t). \tag{4.2}$$

Assume that $m(x)$ and $h(x)$ are analytic functions of x, $h'(x) > 0$, and

$$\mathbb{E}\left[\frac{1}{h'(x(t))}\right] < \infty. \tag{4.3}$$

According to (3.52), the optimal estimator of the signal, $\hat{x}(t)$, given the measurements y_0^t, is

$$\hat{x}(t) = \frac{\displaystyle\int_{-\infty}^{\infty} x\varphi(x,t)\, dx}{\displaystyle\int_{-\infty}^{\infty} \varphi(x,t)\, dx}. \tag{4.4}$$

The estimator (4.4) is realizable (see Section 3.4) if it is computable in the sense that it is an output of a finite-dimensional dynamical system whose input is the measured signal y_0^t. Thus it is realizable if the solution $\varphi(x,t)$ of Zakai's equation (3.27), or of its Stratonovich form (3.25), is an output of a finite-dimensional dynamical system, that is, if the solution $\varphi(x,t)$ is realizable.

In the linear case all of the above are known and the Kalman–Bucy filter (3.68) is the solution to the optimal filtering problem. If the system (4.1), (4.2) is nonlinear, there is no realizable solution in general [34]. If, however, the measurements noise

Z. Schuss, *Nonlinear Filtering and Optimal Phase Tracking*, Applied Mathematical Sciences 180, DOI 10.1007/978-1-4614-0487-3_4,
© Springer Science+Business Media, LLC 2012

is small, realizable approximations to $\hat{x}(t)$ can be constructed with any degree of accuracy, as described below.

4.1 The Wentzel–Kramers-Brillouin Method

The low-measurement-noise assumption means that ρ in (4.2) is a small parameter. The simplest approach to nonlinear filtering in this case is to linearize all equations about the unknown MMSEE filter [71], that is, to expand everything in sight in powers of ρ. The resulting filter is called the extended Kalman filter (EKF). The underlying assumption in this approach is that $\hat{x}(t)$ has a power series expansion [125] and all coefficients in the expansion satisfy stochastic differential equations that can be truncated at any finite power of ρ. It turns out, however, that this is not the case.

The structure of the unnormalized a posteriori pdf $\varphi(x, t)$ can be discerned from that in the linear case (3.67). The solution of Zakai's equation has an essential singularity at $\rho = 0$, so it cannot have a representation in positive powers of ρ and its Laurent series expansion has an infinite number of negative powers of ρ (see Wikipedia). We should, therefore, expect an essential singularity in the solution of Zakai's equation for the general case as well. Equations whose solutions have essential singularities at certain values of a parameter are called singular perturbation problems, and they are ubiquitous in mathematical physics. The hallmark of singular perturbation problems is a change in the order of the equation at the singular value of the parameter or a loss of boundary conditions. This is the case of Zakai's equation, which is reduced from a second-order partial differential equation to a first-order ordinary equation in the limit $\rho \to 0$. In this section, the WKB method for constructing approximate solutions to singularly perturbed PDEs, originally developed for the Schrödinger equation, is explained and applied to Zakai's equation (see also [137]).

To emphasize that all functions depend on the low-noise parameter ρ, we choose $\varepsilon = 1$ and introduce ρ as one of the variables. The WKB method for constructing approximate solutions with an essential singularity as a function of a small parameter is to resolve the essential singularity by the substitution

$$\varphi(x, t, \rho) = \exp\left\{ -\frac{\Psi(x, t, \rho)}{\rho} \right\}, \tag{4.5}$$

where the *eikonal function* $\Psi(x, t, \rho)$ is a regular function. Given the structure (4.5), we can use the Laplace method to evaluate the integrals in (4.4). To this end, we need to determine the trajectory $\tilde{x}(t, \rho)$ that minimizes $\Psi(x, t, \rho)$ with respect to x for each t and also to determine the partial derivatives $\partial^n \Psi(\tilde{x}(t, \rho), t, \rho)/\partial x^n$ on this trajectory. Then, expanding $\Psi(x, t, \rho)$ in a Taylor series about $\tilde{x}(t, \rho)$, we write

$$\Psi(x,t,\rho) = \Psi(\tilde{x}(t,\rho),t,\rho) + \frac{(x-\tilde{x}(t))^2}{2P(t,\rho)} \sum_{k=3}^{\infty} q_k(t,\rho)\frac{(x-\tilde{x}(t,\rho))^k}{k!}, \quad (4.6)$$

where

$$P(t,\rho) = \frac{1}{\Psi_{xx}(\tilde{x}(t,\rho),t,\rho)} \quad (4.7)$$

and

$$q_k(t,\rho) = \frac{\partial^n \Psi(\tilde{x}(t,\rho),t,\rho)}{\partial x^n}, \qquad k = 3,4,\dots . \quad (4.8)$$

Now the Laplace expansion of the integrals in (4.4) gives

$$\hat{x}(t) = \tilde{x}(t,\rho) - \rho\frac{P^2(t,\rho)q_3(t,\rho)}{2} + o\left(\rho^2\right). \quad (4.9)$$

The conditional mean square "distance" between $\hat{x}(t)$ and $x(t)$ is the error (3.53), whose expansion is

$$\widehat{e^2}(t) = \frac{\displaystyle\int_{-\infty}^{\infty} [x - \hat{x}\,(t,\rho)]^2\,\varphi\,(x,t,\rho)\,dx}{\displaystyle\int_{-\infty}^{\infty} \varphi\,(x,t,\rho)\,dx} \quad (4.10)$$

$$= \rho P(t,\rho) + \rho^2\left[P^4(t,\rho)q_3^2(t,\rho) - \frac{1}{2}P^3(t,\rho)q_4(t,\rho)\right] + O\left(\rho^3\right). \quad (4.11)$$

Thus, realizable approximations of $\varphi(x,t,\rho)$ and of $\hat{x}(t)$ can be obtained by truncating all the series in powers of ρ in sight. Specifically, we construct the functions $\tilde{x}(t,\rho)$, $P(t,\rho)$, $q_k(t,\rho)$ $(k = 3,4,\dots)$ in the asymptotic series form

$$\tilde{x}(t,\rho) \sim x_0(t) + \rho x_1(t) + \rho^{3/2}x_2(t) + \cdots ,$$

$$P(t,\rho) \sim P_0(t) + \rho^{1/2}P_1(t) + \rho P_2(t) + \cdots , \quad (4.12)$$

$$q_k(t,\rho) \sim q_{k,0}(t) + \rho^{1/2}q_{k,1}(t) + \rho q_{k,2}(t) + \cdots .$$

At this point the functions $\tilde{x}(t,\rho)$, $P(t,\rho)$, and $q_k(t,\rho)$, $(k = 3,4,\dots)$ are governed by an infinite series of coupled stochastic differential equations driven by the measurement process $y(t)$, so this approximation process does not yet provide a realizable filter. If, however, the infinite system can be truncated, the resulting finite-dimensional system of stochastic differential equations driven by $y(t)$ is a realizable approximation to the optimal filter. The degree of approximation can be estimated by comparing the error of the approximate filter to the error (4.10) of the optimal filter.

4.1.1 An Asymptotic Solution of Zakai's Equation

The transformation (4.5) converts Zakai's equation in Stratonovich form (3.25) into the nonlinear equation

$$\Psi_t(x,t,\rho) = -m'(x)\rho + m(x)\Psi_x(x,t,\rho) - \frac{1}{2}\sigma^2\Psi_{xx}(x,t,\rho) \tag{4.13}$$

$$+ \frac{1}{\rho}\left[\frac{\sigma^2}{2}\Psi_x^2(x,t,\rho) - h(x)\left(\frac{1}{2}h(x) - \dot{y}(t)\right)\right].$$

Note that (4.13) is a stochastic partial differential equation and $\Psi(x,t,\rho)$ is a random function due to the driving random term $\dot{y}(t)$ (the derivative is in the Stratonovich sense).

The assumed condition $h'(x) > 0$ ensures that $\Psi(x,t,\rho)$ has a unique minimum $\tilde{x}(t,\rho)$ for every $t > 0$ (see (4.30) below). We proceed to construct $\Psi(x,t,\rho)$ by expanding it in a Taylor series about $\tilde{x}(t,\rho)$. Obviously,

$$\Psi_x(\tilde{x}(t,\rho),t,\rho) = 0; \tag{4.14}$$

hence, by (4.13),

$$\frac{d}{dt}\Psi(\tilde{x}(t,\rho),t,\rho) = m'(\tilde{x}(t,\rho))\rho + \frac{1}{2}\sigma^2\Psi_{xx}(\tilde{x}(t,\rho),t,\rho)$$

$$+ \frac{1}{\rho}\left[\frac{1}{2}h^2(\tilde{x}(t,\rho)) - h(\tilde{x}(t,\rho))\dot{y}\right]. \tag{4.15}$$

Note that the value of $\Psi(\tilde{x}(t,\rho),t,\rho)$ does not affect the conditional density $p(x,t,\rho)$ (see (3.24)), because it is independent of x and is canceled in the numerator and denominator.

Next, we calculate the partial derivatives of $\Psi(\tilde{x}(t,\rho),t,\rho)$ with respect to x on the trajectory $\tilde{x}(t,\rho)$. To this end, we differentiate (4.13) with respect to x, and using the identity

$$\Psi_{x,t}(\tilde{x}(t,\rho),t,\rho) = \frac{d}{dt}\Psi_x(\tilde{x}(t,\rho),t,\rho) - \Psi_{xx}(\tilde{x}(t,\rho),t,\rho)\dot{\tilde{x}}(t)$$

$$= -\Psi_{xx}(\tilde{x}(t,\rho),t,\rho)\dot{\tilde{x}}(t), \tag{4.16}$$

where $\dot{\tilde{x}}(t)$ is the Stratonovich derivative of $\tilde{x}(t,\rho)$, we obtain

$$\Psi_{xx}(\tilde{x}(t,\rho),t,\rho)\dot{\tilde{x}}(t) = -m''(\tilde{x}(t,\rho))\rho + m(\tilde{x}(t,\rho))\Psi_{xx}(\tilde{x}(t,\rho),t,\rho)$$

$$- \frac{1}{2}\sigma^2\Psi_{xxx}(\tilde{x}(t,\rho),t,\rho)$$

$$- \frac{1}{\rho}h'(\tilde{x}(t,\rho))[h(\tilde{x}(t,\rho)) - \dot{y}(t)]. \tag{4.17}$$

Using the notation (4.7), (4.8), we rewrite (4.17) as

$$\dot{\tilde{x}}(t) = -\frac{1}{2}\sigma^2 P(t,\rho)q_3(t,\rho) + m(\tilde{x}(t,\rho)) - \rho m''(\tilde{x}(t,\rho))P(t,\rho)$$

$$+ \frac{P(t,\rho)}{\rho}h'(\tilde{x}(t,\rho))[\dot{y}(t) - h(\tilde{x}(t,\rho))]. \tag{4.18}$$

Differentiating (4.13) with respect to x twice and arguing as above, we obtain for $P(t,\rho)$ the equation

$$\dot{P}(t,\rho) = \frac{1}{\rho}\left[\sigma^2 - P^2(t,\rho)h'^2(\tilde{x}(t,\rho))\right] + 2m'(\tilde{x}(t,\rho))P(t,\rho) \tag{4.19}$$

$$-\frac{\sigma^2}{2}P^2(t,\rho)[q_4(t,\rho) - P(t,\rho)q_3^2(t,\rho)]$$

$$-\rho P^2(t,\rho)[m'''(\tilde{x}(t,\rho)) - P(t,\rho)q_3(t,\rho)m''(\tilde{x}(t,\rho))]$$

$$+\frac{P^2(t,\rho)}{\rho}[h''(\tilde{x}(t,\rho)) - P(t,\rho)q_3(t,\rho)h'(\tilde{x}(t,\rho))][\dot{y}(t) - h(\tilde{x}(t,\rho))].$$

Proceeding as above, we obtain an infinite system of differential equations for $q_k(t,\rho)$,

$$\dot{q}_3(t,\rho) = -\frac{3}{\rho}\left[\sigma^2\frac{q_3(t,\rho)}{P(t,\rho)} - h'(\tilde{x}(t,\rho))h''(\tilde{x}(t,\rho))\right] \tag{4.20}$$

$$+\frac{1}{2}\sigma^2 q_5(t,\rho) - 3m'(\tilde{x}(t,\rho))q_3(t,\rho) - \frac{3m''(\tilde{x}(t,\rho))}{P(t,\rho)}$$

$$-\sigma^2 P(t,\rho)q_3(t,\rho)q_4(t,\rho)$$

$$-\rho[m^{(iv)}(\tilde{x}(t,\rho)) - m''(\tilde{x}(t,\rho))P(t,\rho)q_4(t,\rho)]$$

$$+\frac{1}{\rho}[P(t,\rho)q_4(t,\rho)h'(\tilde{x}(t,\rho)) - h''(\tilde{x}(t,\rho))][\dot{y}(t) - h(\tilde{x}(t,\rho))].$$

In general, for $k \geq 3$,

$$\dot{q}_k(t,\rho) = -\frac{1}{\rho}\left\{k\left[\sigma^2\frac{q_k(t,\rho)}{P(t,\rho)} - h'(\tilde{x}(t,\rho))h^{(k-1)}(\tilde{x}(t,\rho))\right]\right. \tag{4.21}$$

$$\left.+\frac{1}{2}\sum_{i=2}^{k-2}\binom{k}{i}\left[\sigma^2 q_{i+1}(t,\rho)q_{k+1-i}(t,\rho) - h^{(i)}(\tilde{x}(t,\rho))h^{(k-i)}(\tilde{x}(t,\rho))\right]\right\}$$

$$-\frac{1}{2}\left[q_{k+2}(t,\rho) - q_{k+1}(t,\rho)\sigma^2 P(t,\rho)q_3(t,\rho)\right] - \frac{km^{(k-1)}(\tilde{x}(t,\rho))}{P(t,\rho)}$$

$$- \sum_{i=1}^{k-2} \binom{k}{i} m^{(i)}(\tilde{x}(t,\rho)) q_{k+1-i}(t,\rho)$$

$$+ \rho \left[m^{(k+1)}(\tilde{x}(t,\rho)) - q_{k+1}(t,\rho) m''(\tilde{x}(t,\rho)) P(t,\rho) \right]$$

$$+ \frac{1}{\rho} \left[P(t,\rho) q_{k+1}(t,\rho) h'(\tilde{x}(t,\rho)) - h^{(k)}(\tilde{x}(t,\rho)) \right] [\dot{y}(t) - h(\tilde{x}(t,\rho))].$$

Note that the equations in the infinite system (4.21) are coupled.

Next, we proceed to develop a self-consistent sequence of asymptotic finite-dimensional approximations to the solutions of the system (4.18)–(4.21) for small ρ. More specifically, we expand $\tilde{x}(t,\rho))$, $P(t,\rho)$, and $q_k(t,\rho)$ in powers of ρ and establish a *truncation rule*. First, we postulate the following expansions:

$$\tilde{x}(t,\rho)) \sim x_0(t) + \sum_{i=1}^{\infty} \rho^{i/2} x_i(t), \tag{4.22}$$

$$P(t,\rho) \sim P_0(t) + \sum_{i=1}^{\infty} \rho^{i/2} P_i(t), \tag{4.23}$$

$$q_k(t,\rho) \sim q_{k,0}(t) + \sum_{i=1}^{\infty} \rho^{i/2} q_{k,i}(t), \tag{4.24}$$

where $x_i(t)$, $P_i(t)$, and $q_{k,i}(t)$ are stochastic processes such that Var $x_i(t)$, Var $P_i(t)$, and Var $q_{k,i}(t)$ are bounded uniformly with respect to t and ρ, for small ρ [84], [85]. Now we substitute (4.22)–(4.24) in (4.18) and to leading-order in ρ, we obtain for $x_0(t)$ the equation

$$\dot{x}_0(t) = \frac{P_0(t) h'(x_0(t))}{\rho} [\dot{y}(t) - h(x_0(t))]. \tag{4.25}$$

Next, we estimate the conditional error of $x_0(t)$:

$$\mathbb{E}\left[(x(t,\rho) - x_0(t))^2 \,\Big|\, y_0^t \right] = \frac{\displaystyle\int_{-\infty}^{\infty} (\xi - x_0(t))^2 \, \varphi(\xi,t,\rho) \, d\xi}{\displaystyle\int_{-\infty}^{\infty} \varphi(\xi,t,\rho) \, d\xi}. \tag{4.26}$$

Equations (4.5), (4.6), and the Laplace expansions (4.22)–(4.24) of the integrals in (4.26) give

$$\mathbb{E}\left[(x(t,\rho) - x_0(t))^2 \,\Big|\, y_0^t \right] = O(\rho). \tag{4.27}$$

The normalized error

$$\varepsilon(t) = \frac{x(t,\rho) - x_0(t)}{\sqrt{\rho}}$$

satisfies

$$\widehat{\varepsilon^2}(t) = O(1)$$

as $\rho \to 0$. We also have

$$\frac{\dot{y}(t) - h(x_0(t))}{\rho} = \frac{[h(x(t,\rho)) + \rho\dot{v} - h(x_0(t))]}{\rho} \tag{4.28}$$

$$= \dot{v} + \frac{h'(x_0(t))\varepsilon(t)}{\sqrt{\rho}} + \sum_{k=2}^{\infty} \frac{h^{(k)}(x_0(t))\varepsilon^k(t)\rho^{k/2-1}}{k!}.$$

Next, we note that in balancing terms in stochastic differential equations, terms of order ρ^α in the drift coefficient should be balanced with terms of order $\rho^{\alpha/2}$ in the noise coefficient. This is due to the fact that the noise coefficient is squared in the Fokker–Planck equation. Thus, it can easily be shown that $x_1(t) \to 0$ as $t \to \infty$. Therefore, we disregard the transient term $x_1(t)$ in the expansion (4.22) and re-expand $\tilde{x}(t,\rho)$ in the form

$$\tilde{x}(t,\rho) \sim x_0(t) + \sum_{i=1}^{\infty} \rho^{(i+1)/2}x_i(t), \tag{4.29}$$

where in this expansion $x_1(t)$ does not decay.

The leading terms in eqs.(4.23), (4.24) are found using these expansions in (4.19) and (4.20) as

$$P_0(t) = \frac{\sigma}{h'(x_0(t)} > 0 \tag{4.30}$$

and

$$q_{k,0}(t) = \frac{h^{(k-1)}(x_0(t))}{\sigma} \qquad (k \geq 3). \tag{4.31}$$

Exercise 4.1 (First-order approximate optimal filter). Show that $x_1(t)$ and $P_1(t)$ satisfy the equations

$$\dot{x}_1(t) = \frac{1}{\rho}\left[-\frac{1}{2}\sigma^2 P_0(t)q_{3,0}t) + m(x_0(t)) - P_0(t)h'^2(x_0(t))x_1(t) \right] \tag{4.32}$$

$$+ \frac{1}{\rho^{3/2}} P_1(t)h'(x_0(t))[\dot{y}(t) - h(x_0(t))]$$

and

$$\dot{P}_1(t) = \frac{-2P_0(t)h'^2(x_0(t))P_1(t)}{\rho} - \frac{\sigma^2 h''(x_0(t))[\dot{y}(t) - h(x_0(t))]}{\rho^{3/2}h'^2(x_0(t))}. \qquad (4.33)$$

$\square$

Exercise 4.2 ($\tilde{x}(t, \rho)$ is also the MAP filter and the minimum-noise-energy filter). Show that $\tilde{x}(t, \rho)$ is asymptotically also the MAP filter x_{MAP} (3.19) and the minimum-noise-energy filter $x_{\mathrm{MNE}}(t)$ (3.20) (see Section 4.1.3 below). $\square$

Exercise 4.3 (The second-order approximate optimal filter). Write down equations for the next-order terms, x_2, $P_2(t)$, $q_{3,1}(t)$, and $q_{4,0}(t)$ [84], [85]. $\square$

Example 4.1 (An asymptotically optimal fixed-interval smoother). The solutions $\varphi_+(x, t, \rho)$ and $\varphi_-(x, t, \rho)$ of the forward and backward Zakai equations in Theorem 3.5.2 can be constructed by the WKB method as above. The eikonal functions (see (4.5)) have forward and backward minimizers $\tilde{x}_+(t, \rho)$ and $\tilde{x}_-(t, \rho)$, respectively, that can be approximated by asymptotically optimal filters, as above. According to (4.5) and (3.55), the optimal fixed-interval smoother is given therefore by

$$\hat{x}(t, \rho) = \int_{-\infty}^{\infty} x p\left(x, t \mid y_0^T\right) dx = \frac{\displaystyle\int_{-\infty}^{\infty} x \varphi_+(x, t, \rho)\varphi_-(x, t, \rho)\, dx}{\displaystyle\int_{-\infty}^{\infty} \varphi_+(x, t, \rho)\varphi_-(x, t, \rho)\, dx}$$

$$= \frac{\displaystyle\int_{-\infty}^{\infty} x \exp\left\{-\frac{\Psi_+(x, t, \rho) + \Psi_-(x, t, \rho)}{\rho}\right\} dx}{\displaystyle\int_{-\infty}^{\infty} \exp\left\{-\frac{\Psi_+(x, t, \rho) + \Psi_-(x, t, \rho)}{\rho}\right\} dx}$$

$$\sim \frac{\displaystyle\int_{-\infty}^{\infty} x \exp\left\{-\frac{[x - \tilde{x}_+(t, \rho)]^2}{2P_+(t, \rho)\rho} - \frac{[x - \tilde{x}_-(t, \rho)]^2}{2P_-(t, \rho)\rho}\right\} dx}{\displaystyle\int_{-\infty}^{\infty} \exp\left\{-\frac{[x - \tilde{x}_+(t, \rho)]^2}{2P_+(t, \rho)\rho} - \frac{[x - \tilde{x}_-(t, \rho)]^2}{2P_-(t, \rho)\rho}\right\} dx}$$

$$= \frac{\tilde{x}_+(t, \rho)P_-(t, \rho) + \tilde{x}_-(t, \rho)P_+(t, \rho)}{P_+(t, \rho) + P_-(t, \rho)},$$

where

$$P_+(t, \rho) = \frac{1}{\Psi_{+,xx}(\tilde{x}_+(t, \rho), t, \rho)}, \qquad P_-(t, \rho) = \frac{1}{\Psi_{-,xx}(\tilde{x}_+(t, \rho), t, \rho)}.$$

The a posteriori error variance is given by

$$\mathbb{E}\left[(x(t) - \hat{x}(t, \rho))^2 \mid y_0^T\right] \sim \rho \frac{P_+(t, \rho)P_-(t, \rho)}{2[P_+(t, \rho) + P_-(t, \rho)]}. \qquad (4.34)$$

Equation (4.34) indicates that fixed-interval smoothing reduces the filtering error by a factor of about 2 [59].   □

Exercise 4.4 (An asymptotically optimal fixed-delay smoother). Use Theorem 3.5.3 to construct an asymptotically optimal fixed-delay smoother.   □

4.1.2   Realizable Approximations of $\hat{x}(t)$

We truncate the system (4.12) according to the following *truncation rule*: neglect $x_i(t)$, $P_i(t)$, and $q_{k,i}(t)$ for $i > n$ and $k + j \geq n + 3$ in (4.12). We obtain for each n a finite system of equations for the approximate filter. The number of stochastic differential equations to be solved is $(n + 1)(n + 2)/2$. This number can be reduced by introducing the cumulative variables obtained by truncating the series (4.12) at $i = N$ and denoting the finite sums by $\tilde{x}_N(t)$, $\tilde{P}_N(t)$, and $\tilde{q}_{k,N}(t)$, respectively. They satisfy a reduced system of equations. The leading-order approximation (zeroth order) to the optimal filter is

$$\dot{\tilde{x}}_0(t) = \frac{\tilde{P}_0(t)}{\rho} h'(\tilde{x}_0(t))[\dot{y}(t) - h(\tilde{x}_0(t))], \qquad (4.35)$$

where $\tilde{P}_0(t)$ is given in (4.30), which means that the approximate Kalman gain (3.69) is σ/ρ, so that

$$\dot{\tilde{x}}_0(t) = \frac{\sigma}{\rho}[\dot{y}(t) - h(\tilde{x}_0(t))] \qquad (4.36)$$

(a "constant-gain" one-dimensional filter). The first-order approximation is

$$\dot{\tilde{x}}_1(t) = -\frac{1}{2}\sigma h''(\tilde{x}_1(t))\tilde{P}_1(t) + m(\tilde{x}_1(t) + \frac{\tilde{P}_1(t)}{\rho} h'(\tilde{x}_1(t))[\dot{y}(t) - h(\tilde{x}_1(t))],$$

$$(4.37)$$

$$\dot{\tilde{P}}_1(t) = \frac{\sigma^2 - \tilde{P}_1^2(t)h'^2(\tilde{x}_1(t))}{\rho},$$

which is two-dimensional. Note that the constant-gain filter (4.36) can be understood in either the Itô or Stratonovich sense, because the Wong–Zakai correction vanishes for the constant-noise coefficient ρ^2/σ in the noisy component of $\dot{y}(t)$. The first-order filter (4.37), however, has to be understood in the Stratonovich sense, because the noise coefficient in the first equation is state-dependent for nonlinear $h(x,t)$. The transformation of the equations to Itôs form, which in general is more suitable for digital filter realization, is straightforward, and obviously does not affect the accuracy of the approximation.

Exercise 4.5 (The Itô form of the first-order filter). Convert the first-order filter (4.37) to Itô form. $\square$

Exercise 4.6 (Second-order cumulative approximate optimal filter). Write down equations for the next-order terms $\tilde{x}_2$, $\tilde{P}_2(t)$, $\tilde{q}_{3,1}(t)$, and $\tilde{q}_{4,0}(t)$ [84], [85]. $\square$

Thus the number of equations to be solved at the nth approximation is $n + 1$. The Taylor expansion of $\Psi(x, t, \rho)$ is given by

$$\Psi(x, t, \rho) = \Psi(\tilde{x}(t, \rho), t, \rho) + \frac{[x - \tilde{x}(t, \rho)]^2}{2P(t, \rho)}$$

$$+ \sum_{k=3}^{\infty} q_k(t, \rho) \frac{[x - \tilde{x}(t, \rho)]^k}{k!}. \tag{4.38}$$

The asymptotic expansion of $\Psi(x, t, \rho)$ is obtained by replacing $\tilde{x}(t, \rho)$, $P(t, \rho)$, and $q_k(t, \rho)$ in (4.38) by their asymptotic expansions (4.22)–(4.24), and the asymptotic expansion of $\Psi(\tilde{x}(t, \rho), t, \rho)$ is determined from (4.15). Note again that $\Psi(\tilde{x}(t, \rho), t, \rho)$ does not affect the value of the conditional density $p(x, t, \rho)$.

4.1.3 Applications to the Optimal Filtering Problem

In this section we apply the results of Section 4.1.1 to the three problems of optimal filtering: finding the minimum-noise-variance, MAP, and minimum-noise-energy estimators of the signal $x(t)$, given the measurements y_0^t. These are the most commonly used filters in communication and stochastic control practice.

The Asymptotic MAP Filter

This estimator is defined by (3.19). In view of (3.24) and (4.5), $x_{\mathrm{MAP}}(t)$, which maximizes $p(x, t, \rho)$, maximizes the unnormalized a posteriori density $\varphi(x, t, \rho)$. Thus

$$x_{\mathrm{MAP}}(t) = \tilde{x}(t, \rho). \tag{4.39}$$

It follows that one can use the system (4.36)–(4.37) as finite-dimensional approximations to $x_{\mathrm{MAP}}(t)$.

The Asymptotic Minimum-Error-Variance Filter

The minimum-error-variance estimator $\hat{x}(t)$ is defined in (3.52), and the variance of its error, $\widehat{e^2}(t)$, is defined in (3.53). The first terms in the asymptotic expansion of $\hat{x}(t)$ for small ρ are given in (4.9) and those in the expansion of $\widehat{e^2}(t)$ in (4.11). Using the full asymptotic expansion of $\varphi(x, t, \rho)$ and the Laplace expansion of the integrals in (3.52), (3.53), we obtain

$$\hat{x}(t) = \frac{\displaystyle\int_{-\infty}^{\infty} x\varphi(x, t, \rho)\, dx}{\displaystyle\int_{-\infty}^{\infty} \varphi(x, t, \rho)\, dx} = \tilde{x}(t, \rho) - \rho\frac{P^2(t, \rho)q_3(t, \rho)}{2} \tag{4.40}$$

$$+ \rho^2 \left[-\frac{P^3(t, \rho)q_5(t, \rho)}{8} + \frac{2q_3(t, \rho)q_4(t, \rho)P^4(t, \rho)}{3} - \frac{5P^5(t, \rho)q_3^3(t, \rho)}{8} \right]$$

$$+ O\left(\rho^3\right).$$

The error variance is given by

$$\widehat{e^2}(t) = \frac{\displaystyle\int_{-\infty}^{\infty} [x - \hat{x}(t)]^2 \varphi(x, t, \rho)\, dx}{\displaystyle\int_{-\infty}^{\infty} \varphi(x, t, \rho)\, dx} \tag{4.41}$$

$$= \rho P(t, \rho) + \rho^2 \left[P^4(t, \rho)q_3^2(t, \rho) - \frac{1}{2}P^3(t, \rho)q_4 t, \rho) \right] + O\left(\rho^3\right).$$

Using (4.40) and the truncated expansion

$$x_N^*(t) = x_0(t) + \sum_{i=1}^{N} \rho^{(i+1)/2}\hat{x}_i(t) \rightarrow \hat{x}(t) \quad \text{as } N \rightarrow \infty, \tag{4.42}$$

we can employ the approximations (4.36)–(4.37) in the equivalent Itô form (having used the Wong–Zakai correction) to get the following approximate filters. The zeroth-order approximation is given by

$$dx_0^* = \frac{\sigma}{\rho}[dy - h(x_0^*)\, dt]. \tag{4.43}$$

This is the constant-gain nonlinear filter again. Its error variance is given by

$$\widehat{e_0^{*2}} = \overline{[x(t) - x_0^*(t)]^2} = \frac{\rho\sigma}{h'(x_0^*(t))} + O\left(\rho^{3/2}\right); \tag{4.44}$$

hence, by (4.41) and (4.42),

$$\widehat{e^2} - \widehat{e_0^{*2}} = O\left(\rho^{3/2}\right). \tag{4.45}$$

The first-order approximation is given by

$$dx_1^* = \left[-\frac{\sigma^2 h''(x_1^*)}{2h'(x_1^*)} + m(x_1^*) \right] dt + P_1^*(t)h'(x_1^*)\frac{[dy - h(x_1^*)\,dt]}{\rho}, \quad (4.46)$$

$$dP_1^* = \frac{\sigma^2 - P_1^{*2}h'^2(x_1^*)}{\rho}\,dt,$$

and it can be shown that

$$\widehat{e^2} - \widehat{e_1^{*2}} = O\left(\rho^2\right). \tag{4.47}$$

Exercise 4.7 (Error estimate). Prove (4.47).   □

Exercise 4.8 (Second-order approximate optimal filter). Find the equations of the second-order approximation of the filter x_2^* and its gains P_2^* and $q_{3,1}^*$.   □

Exercise 4.9 (Second-order error estimate). Show that

$$\widehat{e^2} - \widehat{e_2^{*2}} = O\left(\rho^{5/2}\right). \tag{4.48}$$

□

The expansion (4.41) can be used to evaluate the degree of approximation of the error variance of any given estimator to that of the optimal filter for small ρ. In particular, the stationary error variance of the optimal filter is given by

$$\mathbb{E}e^2(t) = \mathbb{E}\widehat{e^2} = \rho\sigma\mathbb{E}\frac{1}{h'(x(t))} + O\left(\rho^{3/2}\right) \tag{4.49}$$

$$= \rho\sigma\int_{-\infty}^{\infty}\frac{p(x)\,dx}{h'(x(t))} + O\left(\rho^{3/2}\right),$$

where $p(x)$ is the stationary probability density function of the signal $x(t)$ and is given by

$$p(x) = \frac{\exp\left\{\dfrac{2}{\sigma^2}\displaystyle\int_0^x m(s)\,ds\right\}}{\displaystyle\int_{-\infty}^{\infty}\exp\left\{\dfrac{2}{\sigma^2}\displaystyle\int_0^s m(t)\,dt\right\}\,ds}, \tag{4.50}$$

assuming the integral exists. Furthermore, $\widehat{e_N^{*2}}$ agrees with $\widehat{e^2}$ up to order $\rho^{1+N/2}$.

The Asymptotic Minimum-Noise-Energy Filter

We recall that according to Definition 3.4.5, the MNE filter $x_{\mathrm{MNE}}(t)$ is the trajectory that for each $0 \leq t \leq T$ minimizes the functional (3.20),

$$J(x(\cdot)) = \frac{1}{2} \int_0^t \left\{ \frac{[\dot{y}(s) - h(x(s), s)]^2}{\rho^2} + \frac{[\dot{x}(s) - m(x(s), s)]^2}{\sigma^2} \right\} \, ds. \qquad (4.51)$$

The Hamilton–Jacobi–Bellman equation for the minimum value [101], [14]

$$S(x, t) = \min_{\{x(\cdot) \in \mathcal{C}^1(\mathbb{R}^+) \mid x(t) = x\}} J(x(\cdot))$$

is given by

$$S_t + \frac{\sigma^2}{2\rho} S_x^2 + m(x) S_x = \frac{1}{\rho} \left[\frac{1}{2} h^2(x) - h(x) \dot{y}_S(t) \right]. \qquad (4.52)$$

Note that the HJB equation (4.52) and the transformed equation (4.13) are identical to the two leading-orders of magnitude $O(\rho^{-1})$ and $O(1)$. Terms of order $O(\rho)$ in (4.13) and (4.52), however, are not the same. Note also that because of the smoothness assumptions of the disturbances, in order to apply the energy performance index to diffusion processes, we must use the Stratonovich form; otherwise, the criterion is meaningless.

To solve for $x_{\text{MNE}}(t)$, we adopt the procedure of Section 4.1.1 to the HJB equation (4.52) and obtain the following equations:

$$\dot{x}_{\text{MNE}}(t) = m\left(x_{\text{MNE}}(t)\right) + \frac{\mathcal{P}(t) h'\left(x_{\text{MNE}}(t)\right)}{\rho} \left[\dot{y}(t) - h\left(x_{\text{MNE}}(t)\right)\right], \qquad (4.53)$$

$$\dot{\mathcal{P}}(t) = \frac{\sigma^2 - \mathcal{P}^2(t) h'^2\left(x_{\text{MNE}}(t)\right)}{\rho} + 2m'\left(x_{\text{MNE}}(t)\right) \mathcal{P}(t) \qquad (4.54)$$

$$+ \frac{\mathcal{P}^2(t) h''\left(x_{\text{MNE}}(t)\right) - \mathcal{P}(t) Q_3(t) h'\left(x_{\text{MNE}}(t)\right)}{\rho} \left[\dot{y}(t) - h\left(x_{\text{MNE}}(t)\right)\right],$$

$$\dot{Q}_3(t) = -\frac{3}{\rho} \left[\sigma^2 \frac{Q_3(t)}{\mathcal{P}(t)} - h'(x_{\text{MNE}}(t)) h''(x_{\text{MNE}}(t)) \right] - 3m'(x_{\text{MNE}}(t)) Q_3(t)$$

$$- \frac{2m''\left(x_{\text{MNE}}(t)\right)}{\mathcal{P}(t)} + \frac{1}{\rho} \left[Q_4(t) \mathcal{P}(t) h'(x_{\text{MNE}}(t)) \right.$$

$$\left. - h'''\left(x_{\text{MNE}}(t)\right) \right] \left[\dot{y}(t) - h\left(x_{\text{MNE}}(t)\right)\right], \qquad (4.55)$$

and so on. Here

$$\mathcal{P}(t) = \frac{1}{S_{xx}\left((x_{\text{MNE}}(t)), t\right)}, \qquad (4.56)$$

$$Q_k(t) = \left. \frac{\partial^k S(x, t)}{\partial x^k} \right|_{x = x_{\text{MNE}}(t)}, \quad k \geq 3. \qquad (4.57)$$

Using the truncation rule of Section 4.1.2, we derive a sequence of finite-dimensional approximations to $x_{MNE}(t)$ as follows. Writing

$$\tilde{x}_{MNE}^N(t) = x_{ME,0}(t) + \sum_{i=1}^{N} \rho^{(i+1)/2} x_{ME,i}(t) \to \tilde{x}_{MNE}(t), \tag{4.58}$$

$$\tilde{\mathcal{P}}_N(t) = \mathcal{P}_0(t) + \sum_{i=1}^{N} \rho^{i/2} \mathcal{P}_i(t) \to \mathcal{P}_0(t), \tag{4.59}$$

$$\tilde{Q}_k^N(t) = Q_{k,0}(t) + \sum_{i=1}^{N} \rho^{i/2} Q_{k,i}(t) \to Q(t), \quad \text{as } N \to \infty \ (k \geq 3), \tag{4.60}$$

we obtain for $N = 0$,

$$\dot{\tilde{x}}_{ME}^0(t) = \frac{\sigma}{\rho} \left[\dot{y}(t) - h\left(\tilde{x}_{MNE}^0(t) \right) \right], \tag{4.61}$$

for $N = 1$,

$$\dot{\tilde{x}}_{ME}^1(t) = m\left(\dot{\tilde{x}}_{ME}^1(t) \right) + \frac{\tilde{\mathcal{P}}_1(t) h'\left(x_{MNE}^1(t) \right)}{\rho} \left[\dot{y}(t) - h\left(\tilde{x}_{MNE}^0(t) \right) \right], \tag{4.62}$$

$$\dot{\tilde{\mathcal{P}}}_1(t) = \frac{\sigma^2 - \tilde{\mathcal{P}}_1(t) h'\left(x_{MNE}^1(t) \right)}{\rho},$$

and so on. Note that the zeroth-order approximations for all three filters are the same, i.e.,

$$\tilde{x}_{MNE}^0(t) = \tilde{x}_0(t) = \hat{x}_0(t), \tag{4.63}$$

while the next-order approximations differ considerably from each other in case $h'(x)$ is not constant. For linear measurements they agree to higher orders.

Comparison of the error energies of the various approximations with that of the optimal one gives

$$J\left(x_{MNE}(t) \right) - J\left(x_{MNE}^0(t) \right) = O(\rho), \tag{4.64}$$

$$J\left(x_{MNE}(t) \right) - J\left(\tilde{x}_{MNE}(t) \right) = O\left(\rho^{3/2} \right),$$

$$J\left(x_{MNE}(t) \right) - J\left(\tilde{x}_{MNE}^1(t) \right) = O\left(\rho^{3/2} \right),$$

$$J\left(x_{MNE}(t) \right) - J\left(\tilde{x}_{MNE}^2(t) \right) = O\left(\rho^2 \right).$$

Exercise 4.10 (Derivation of filter equations). Derive the equations (4.61)–(4.62) by applying the truncation rules to the system (4.58)–(4.60). □

Exercise 4.11 (Derivation of the second-order approximate MNE filter). Use the truncation rules in the system (4.58)–(4.60) to find the equations of the second-order MNE filter $\tilde{x}_{\mathrm{ME}}^2(t)$ and its gains $\tilde{P}_2(t)$, $\tilde{Q}_{3,1}(t)$.   □

Exercise 4.12 (Proof of (4.64)). Use the Hamilton–Jacobi equation (4.52) instead of the eikonal equation and use the Laplace expansion of integrals to prove (4.64).   □

Exercise 4.13 (The linear case). Consider the linear case $dx = -ax\, dt + \sigma\, dw$ with $a > 0$ and $dy = x\, dt + \rho\, dv$. Setting $m(x) = -ax$, $h(x) = x$ in (4.18)–(4.21), get $q_k(t) = 0$ for $k \geq 3$ and obtain the Kalman–Bucy filter

$$\dot{\tilde{x}}(t) = -a\tilde{x}(t) + \frac{\tilde{P}(t)}{\rho}[\dot{y}(t) - \tilde{x}(t)], \quad \dot{\tilde{P}}(t) = \frac{\sigma^2 - \tilde{P}^2(t)}{\rho} - 2a\tilde{P}(t).$$

Conclude that in the linear case, $\hat{x}(t) = \tilde{x}(t)$.   □

Exercise 4.14 (Linear measurements of a nonlinear signal). Consider the case of linear measurements $h(x) = x$ of a nonlinear signal $x(t)$ given by (4.1). Find the filtering equations (4.46) for $x_1^*(t)$ and the equations for $x_2^*(t)$. Show that $q_{3,1}^* \to 0$ as $t \to \infty$ and $P_2^*(t) = \sigma + \rho m'(x_2^*) + O(\rho^{3/2})$.   □

Exercise 4.15 ($x_3^*(t)$ and $x_4^*(t)$ for linear measurements). Find the filtering equations for $x_3^*(t)$ and $x_4^*(t)$ in the case of linear measurements. Show that higher accuracy in this case is achieved at the expense of raising the dimension of the filter.   □

Exercise 4.16 (Error bounds). Upper and lower bounds (UB and LB) on the minimal mean square estimation error for the scalar nonlinear filtering problem (4.1), (4.2) were obtained in [17] in the form

$$\mathrm{LB} = \frac{\rho\sigma}{[\mathbb{E}(h'^2(x))]^{1/2}} \leq [\mathbb{E}(e^2)]^{1/2} \leq \rho\sigma\left[\mathbb{E}\left(\frac{1}{h'^2(x)}\right)\right]^{1/2} = \mathrm{UB}.$$

Use the Jensen and Schwarz inequalities to show that

$$\frac{\rho\sigma}{[\mathbb{E}(h'^2(x))]^{1/2}} \leq \mathbb{E}\left(\frac{1}{h'(x)}\right) \leq \left[\mathbb{E}\left(\frac{1}{h'^2(x)}\right)\right]^{1/2}$$

and show that the expansion (4.49) of the error variance of the minimum error variance filter satisfies the given bounds.   □

Exercise 4.17 (The soft limiter). Consider the nonlinear "soft limiter" problem

$$dx = -ax\, dt + \sigma\, dw \text{ (for } a > 0), \quad dy = \lambda \arctan\frac{x}{\lambda}\, dt + \rho\, dv.$$

The parameter $2a\lambda/\sigma^2$ characterizes the degree of the saturation: the signal $x(t)$ undergoes very little saturation for $2a\lambda/\sigma^2 \gg 1$, whereas the opposite limit, $2a\lambda/\sigma^2 \ll 1$, represents hard limiting on $x(t)$. Obtain the bounds and approximation LB $= 2.38\rho\sigma$, UB $= 4.12\rho\sigma$, $\mathbb{E}e_0^2 = 3\rho\sigma + o(\rho)$. Conclude that $0.88 \leq \left(\mathbb{E}e^2/\mathbb{E}e_0^2\right)^{1/2} \leq 1.17$ and $0.85 \leq \left(\mathbb{E}e_0^2/\mathbb{E}e^2\right)^{1/2} \leq 1.13$, so that $\left(\mathbb{E}e_0^2\right)^{1/2}$ is quite a good estimator of the minimal error $\left(\mathbb{E}e^2\right)^{1/2}$.   □

Exercise 4.18 (The cubic sensor*). Consider the cubic sensor problem

$$dx = -ax\,dt + \sigma\,dw \text{ (for } a > 0), \quad dy = x^3\,dt + \rho\,dv.$$

What breaks down in the WKB analysis of this case? (see [24] for a simulation study of the problem and Section 4.2).   □

Exercise 4.19 (Filtering of a two-dimensional diffusion [162]). Consider a two-dimensional signal $(x_1(t), x_2(t))$, whose first component is measured in a low-noise channel. This is the case, for example, of FM transmission, as described in (3.4). The signal model is given by

$$dx_i(t) = m^i(x_1, x_2)\,dt + b_{ij}\,dw_j(t) \quad (i = 1, 2), \tag{4.65}$$

where $w_j(t)$ are standard independent Brownian motions, b_{ij} are constants, and $m^i(x_1, x_2)$ (with $i, j = 1, 2$) are smooth functions. The summation convention of summing over repeated indices is used in (4.65). The noisy measurements process $y(t)$ is modeled by the Itô equation

$$dy(t) = h(x_1(t))\,dt + \varepsilon\,d\mu(t), \tag{4.66}$$

where $h(x)$ is a smooth function and $\mu(t)$ is another Brownian motion, independent of $w_j(t)$ $(j = 1, 2)$. Show that the Zakai–Itô equation for an unnormalized conditional joint density function of $x_1(t)$ and $x_2(t)$, given the measurements $y(s)$, $s \leq t$, is given by

$$d\Phi = \left[\left(-m^i\Phi\right)_i + \left(B^{ij}\Phi\right)_{ij}\right] dt + \frac{\Phi h}{\varepsilon^2}\,dy,$$

and its Zakai–Stratonovich version by

$$d\Phi = \left[\left(-m^i\Phi\right)_i + \left(B^{ij}\Phi\right)_{ij} - \frac{\Phi h^2}{\varepsilon^2}\right] dt + \frac{\Phi h}{\varepsilon^2}\,dy, \tag{4.67}$$

where $B^{ij} = \frac{1}{2}b_{ik}b_{jk}$ [162].   □

Exercise 4.20 (The filtering equations for a two-dimensional diffusion). Under the assumption that ε is a small parameter in (4.66), construct a WKB solution to the Zakai–Stratonovich equation (4.67) in the form

$$\Phi(x_1, x_2, t) = \exp\left\{ -\frac{U(x_1, x_2, t, \varepsilon)}{\varepsilon^{\alpha}} \right\}, \qquad (4.68)$$

for some $\alpha > 0$, where $U(x_1, x_2, t, \varepsilon)$ is assumed a regular function of ε. Assume that $U(x_1, x_2, t, \varepsilon)$ has a single maximum at each time t, achieved at a point $x_i = \tilde{x}_i(t)$ $(i = 1, 2)$. Expand $U(x_1, x_2, t, \varepsilon)$ in a Taylor series about $(\tilde{x}_1(t), \tilde{x}_2(t), t, \varepsilon)$. Develop asymptotically optimal filters as in Exercises 4.1–4.16. $\qquad \square$

4.1.4 Applications to Phase and Frequency Tracking

The derivation of phase and frequency trackers from the asymptotic analysis of Zakai's equation is given below in a series of exercises.

Exercise 4.21 (Asymptotically optimal CMSEE filtering of a PM signal). Consider the case of PM transmission of a one-dimensional signal, assuming that either one or both of the time-dependent nonmonotone functions

$$h^1(x(t), t) = \sqrt{A} \sin\left[\omega_0 t + \beta x(t)\right], \quad h^2(x(t), t) = \sqrt{A} \cos\left[\omega_0 t + \beta x(t)\right]$$

are measured in noisy channels with independent noises $\rho \dot{v}_1$ and $\rho \dot{v}_2$ (see (3.3)). Use a linear model for the signal,

$$dx = -mx \, dt + \sigma \, dw \qquad (4.69)$$

and the nonlinear model of the measurements

$$dy_1(t) = \sqrt{A} \sin\left[\omega_0 t + \beta x(t)\right] dt + \rho \, dv_1 \qquad (4.70)$$

$$dy_2(t) = \sqrt{A} \cos\left[\omega_0 t + \beta x(t)\right] dt + \rho \, dv_2.$$

Show that the Zakai–Stratonovich equation for the a posteriori probability density of the signal, given the measurements, is given by

$$p_t = \mathcal{L}p - \frac{\left(\sqrt{A}\right)^2 \sin^2\left(\omega_0 t + \beta x\right)}{\rho^2} p + \frac{\sqrt{A} \sin\left(\omega_0 t + \beta x\right) \dot{y}_1}{2\rho^2} p \qquad (4.71)$$

$$- \frac{\left(\sqrt{A}\right)^2 \cos^2\left(\omega_0 t + \beta x\right)}{\rho^2} p + \frac{\sqrt{A} \cos\left(\omega_0 t + \beta x\right) \dot{y}_2}{2\rho^2} p,$$

where $\mathcal{L}$ is the Fokker–Planck operator corresponding to (4.69). Use (4.70) to simplify (4.71) to

$$p_t = \mathcal{L}p - \frac{A}{\rho^2}p + \frac{A}{2\rho^2}\cos\beta\,(x - x(t))\,p \tag{4.72}$$

$$- \frac{A}{2\rho}\left[\sin\left(\omega_0 t + \beta x\right)\dot{v}_1 + \cos\left(\omega_0 t + \beta x\right)\dot{v}_2\right]p.$$

Use the fact that the expression in brackets is standard white noise to show that (4.72) is the Zakai–Stratonovich equation for the simplified system

$$dx = -mx\,dt + \sigma\,dw,$$
$$d\tilde{y}_1 = \sin\beta x\,dt + \rho\,d\tilde{v}_1, \tag{4.73}$$
$$d\tilde{y}_2 = \cos\beta x\,dt + \rho\,d\tilde{v}_2,$$

where $\dot{\tilde{v}}_1$ and $\dot{\tilde{v}}_2$ are independent white noises. Construct the corresponding block diagrams. □

Exercise 4.22 (WKB analysis of the simplified Zakai–Stratonovich equation). Use the WKB method of Section 4.1 for the analysis of the simplified Zakai–Stratonovich equation (4.72) to obtain the leading-order PM receiver

$$d\tilde{x}_0 = -m\tilde{x}_0\,dt + \frac{\sigma}{\rho}\left[\cos\beta\tilde{x}_0\,dy_1 - \sin\beta\tilde{x}_0\,dy_2\right]. \tag{4.74}$$

Note that the first term in the filter equation can be neglected relative to the second. Draw a block diagram for this filter. □

Exercise 4.23 (The PLL for PM transmission).

(i) Show that if only one measurement is used in (4.73), then the leading-order approximate filter is the first-order phase-locked loop (PLL)

$$d\tilde{x}_0 = -m\tilde{x}_0\,dt + \frac{\sigma}{\rho}\left[dy_1 - \sqrt{A}\sin\beta\tilde{x}_0\,dt\right]. \tag{4.75}$$

Draw a block diagram for the PLL.

(ii) Show that the error $e = x(t) - \tilde{x}_0(t)$ of the filter (4.74) satisfies the equation

$$de = -\left(K\varepsilon e + \sin e\right)dt + \sqrt{2\varepsilon}\,dw_1,$$

where

$$K = \frac{\rho m}{\sigma\sqrt{A}}, \qquad \varepsilon = \frac{\rho\beta\sigma(1 + A)}{2A}.$$

□

Exercise 4.24 (Simplified equations for a high-frequency carrier*). Assume that ω_0 is large and use the method of slow and fast times to derive a simplified model. Use it to design an approximately optimal PLL for tracking the signal [147], [150], [152], [144]. □

Exercise 4.25 (Second order PLL for FM transmission). Consider the case of FM transmission of a two-dimensional signal $(x(t), u(t))$, where $d_f x(t)$ is the frequency and $u(t)$ is the phase (see (3.4) and (3.5)),

$$dx = -mx\, dt + \sigma\, dw, \quad du = d_f x\, dt, \tag{4.76}$$

assuming that either one or both of

$$h^1(u(t), t) = \sqrt{A} \sin\left[\omega_0 t + u(t)\right], \quad h^2(u(t), t) = \sqrt{A} \cos\left[\omega_0 t + u(t)\right] \tag{4.77}$$

are measured in noisy channels with small independent noises.

(i) Use the method of Exercise 4.21 to design a PLL for tracking the signal. Write the filter equations for the leading-order approximations of the phase and frequency estimators $\tilde{u}_0(t)$ and $\tilde{x}_0(t)$ when only $h^1(u(t), t)$ is measured as

$$d\tilde{u}_0 = d_f \tilde{x}_0\, dt + \sqrt{\frac{2\sigma}{\rho\sqrt{A}}}\left[dy - \sqrt{A}\sin\tilde{u}_0\right], \quad d\tilde{x}_0 = \frac{\sigma}{\rho}\left[dy - \sqrt{A}\sin\tilde{u}_0\right].$$

Draw a block diagram of the resulting PLL.
(ii) Do the same when both $h^1(u(t), t)$ and $h^2(u(t), t)$ are measured.
(iii) Show that the equations of the estimation of phase and frequency errors are given by

$$de_u = d_f e_x\, dt - \sqrt{\frac{2\sigma}{\rho}}\sin(u - \tilde{u}_0)\, dt - \sqrt{\frac{2\sigma\rho}{A}}\, dv_1,$$

$$de_x = -me_x\, dt + \sigma\, dw - \frac{\sigma}{\rho}\left[\sqrt{A}\sin(u - \tilde{u}_0)\, dt + \rho\, dv_1\right].$$

Transform the error equations into dimensionless variables

$$t = T\tau, \quad e_x = \varepsilon\xi, \quad e_u = \phi,$$

and choose the constants so that

$$\frac{T\sigma\sqrt{A}}{\rho\varepsilon} = 1, \quad T\sqrt{\frac{2\sigma}{\rho}} = 1.$$

Assume for simplicity that $d_f\sqrt{A} = 1$ and set $\sqrt{\varepsilon} = (2\sigma\rho^3)^{1/4}/\sqrt{A}$ to obtain the simplified error equations

$$d\xi = -[\delta\xi + \sin\phi]\, dt + \sqrt{\varepsilon}\,(dw - dv_1),$$

$$d\phi = \left(\frac{1}{2}\xi - \sin\phi\right) dt - \sqrt{\varepsilon}\,dv_1, \tag{4.78}$$

where $\delta = m\left(\varepsilon A/(2\sigma)^2\right)^{1/3}$.

(iv) Set $\varepsilon = 0$ in (4.78) and investigate the resulting noiseless dynamics in the (ξ,ϕ) plane. Find stable and unstable equilibrium points and the separatrices separating the basins of attraction of the stable ones. Plot the separatrices in this plane (see [18] and Figure 5.4 below). The horizontal axis in Figure 5.4 is the phase estimation error e_1, and the vertical axis is frequency estimation error e_2. The dots are the local attractors at $e_1 = 0, \pm 2\pi, \ldots$, $e_2 = 0$. The bounding curves are the separatrices that converge to the saddle points $e_1 = \pm\pi, \pm 3\pi, \ldots$, $e_2 = 0$. Typical noisy trajectories in the phase plane show escapes from the domains of attraction.   □

4.2 The Cubic Sensor

The small noise analysis of Zakai's equation in Section 4.1 assumes that the measurements are not merely monotone, but that actually $h'(x) > 0$ for all x, or more specifically, that

$$\mathbb{E}\left[\frac{1}{h'(x(t))}\right] < \infty \tag{4.79}$$

(see (4.3)). This implies that the conditional error variance (4.44) is bounded, the approximate filter (4.46) does not blow up, and so on. All this fails if $h'(0) = 0$, for example if $h(x) = x^3$ (the cubic sensor) or any other nonlinearizable strictly monotone measurements function. We consider here the benchmark class $h_n(x) = |x|^n\mathrm{sgn}(x)$ for $n > 1$ (see [146]).

The reason for this failure is the breakdown of the asymptotic expansion of the solution of Zakai's equation (3.25). The balance of terms in Section 4.1.1 fails when the denominators become small. Thus the WKB expansion can be valid only sufficiently far from the zero of $h(x)$. To fix the ideas, we consider the case of the cubic sensor $h(x) = x^3$, so that (4.31) gives

$$P_0(t) = q_{2,0}(t) = \frac{3x_0^2(t)}{\sigma}, \quad q_{3,0}(t) = \frac{6x_0(t)}{\sigma},$$

$$q_{4,0}(t) = \frac{6}{\sigma}q_{k,0}(t) = 0 \text{ for } k \geq 5. \tag{4.80}$$

Therefore the constant-gain filter (4.36) has error variance (4.44),

$$\widehat{e_{x_0}^2} = \frac{\rho\sigma}{3x_0^2(t)} + O\left(\rho^{3/2}\right), \tag{4.81}$$

only if $x_0(t)$ is sufficiently far from 0. To determine this outer region, we note that using (4.80) in (4.19), we see that for $|x_0(t)| = O\left(\rho^{1/4}\right)$, the terms $-(1/\sigma)(\sigma^2 P_0^2 - h'^2)$, $\frac{1}{2}\sigma^2 q_{4,0}$, and $\frac{1}{2}\sigma^2 q_{3,0}^2/P_0$ are all $O(1)$ as $\rho \to 0$. Thus (4.81) is valid for the constant-gain filter (4.36) only for $|x_0(t)| > \rho^{1/4}$, and a different expansion is needed for $|x_0(t)| < \rho^{1/4}$. This region contributes to the error variance more than (4.44), so a different construction of the asymptotic solution to Zakai's equation has to be found in this inner region.

4.2.1 The Inner Region for the Linear Case

In the linear case, $h(x) = gx$ for some constant g. The solution of the filtering problem is Gaussian, and there is apparently no need for separate analysis of an "inner region". Nevertheless, we separate the line into an outer region $|x| > \rho^{1/2}$ and an inner region $|x| < \rho^{1/2}$ and consider the expansion (4.18)–(4.21) of the moments. The leading-order term in the expansion of $q_k(t, \rho)$ is

$$q_{2,0} = \frac{g}{\sigma}, \quad q_{k,0} = 0 \quad \text{for} \quad k \geq 3. \tag{4.82}$$

Disregarding the fact that $q_{k,0}$ is independent of $\tilde{x}(t)$, we scale

$$q_2(t, \rho) = Q_2(t, \rho), \quad q_3(t, \rho) = \rho^{-1/2} Q_3(t, \rho),$$
$$q_k(t, \rho) = \rho^{1-k/2} Q_k(t, \rho), \tag{4.83}$$

and expand the resulting vector of moments $\boldsymbol{Q}(t, \rho)$ in an asymptotic series

$$\boldsymbol{Q} \sim \boldsymbol{Q}_0 + \sum_{j=1}^{\infty} \rho^{j/2} \boldsymbol{Q}_j,$$

with the leading-order equation in the form

$$\dot{\boldsymbol{Q}}_0 = \boldsymbol{f}(\boldsymbol{Q}_0, \tilde{z}) + \boldsymbol{M}(\boldsymbol{Q}_0, \tilde{z}) \dot{\beta}(t), \tag{4.84}$$

where

$$f_k = -k[\sigma^2 Q_{2,0}(t) Q_{k,0}(t) - 3\tilde{z}^2(t) h^{(k-1)}(\tilde{z}(t))]$$

$$-\frac{1}{2}\sum_{i=2}^{k-2}\binom{k}{i}\left[\sigma^2 Q_{i+1,0}(t,\rho) Q_{k+i-1,0}(t,\rho) - h^{(i)}(\tilde{z}(t)) h^{(k-i)}(\tilde{z}(t))\right]$$

$$+\frac{1}{2}\sigma^2 \frac{Q_{k+2,0}(t,\rho) Q_{2,0}(t,\rho) - Q_{3,0}(t,\rho) Q_{k+1,0}(t,\rho)}{Q_{2,0}(t,\rho)} \tag{4.85}$$

and $M(Q_0,\tilde{z})$ is the vector of coefficients of $\dot{\beta}$.

Although the inner region shrinks as $\rho \to 0$, nevertheless, its contribution to the MSEE remains dominant. To construct the conditional pdf in the inner region, we note that the long (scaled) time behavior of the leading equation (4.84) is determined by the stationary point of the drift $f(Q_0,\tilde{z})$ near the stationary point $\tilde{z} = 0$. Thus we have to solve the stationary equation

$$0 = f(\tilde{Q},0) \tag{4.86}$$

for the unknown stationary vector $\tilde{Q}$. Setting $\tilde{z} = 0$ in (4.85), we see that (4.86) has an even solution and that all components $\tilde{Q}_k$ for $k \geq 3$ can be expressed as functions of $\tilde{Q}_2$:

$$\tilde{Q}_{2k+1} = 0, \quad k = 1, 2, 3, \ldots,$$

$$\tilde{Q}_4 = 2\tilde{Q}_2^2 - 2g^2\sigma^{-2}, \quad \tilde{Q}_6 = 16\tilde{Q}_2^3 - 16g^2\tilde{Q}_2\sigma^{-2}, \tag{4.87}$$

$$\vdots$$

We determine $\tilde{Q}_2$ that ensures that the unnormalized conditional pdf $\varphi(x,t)$ (the solution of Zakai's equation) is normalizable and has a unique maximum, and that the stationary point of the drift f is stable. The stretching (4.83) leads to the following WKB form of the conditional pdf $\varphi(x,t)$ in the inner region:

$$\varphi_{\text{inner}}(x,t) \sim \exp\left\{-\hat{\psi}\left(\frac{x - \tilde{x}(t)}{\rho^{1/2}}\right)\right\}, \tag{4.88}$$

where

$$\hat{\psi}(z) = \sum_{k=2}^{\infty} \frac{Q_k(t,\rho)}{k!} z^k. \tag{4.89}$$

Setting

$$z = \frac{x - \tilde{x}(t)}{\rho^{1/2}}, \quad \tilde{z}(t) = \frac{\tilde{x}(t)}{\rho^{1/2}}, \tag{4.90}$$

the substitution of (4.88) in Zakai's equation gives the differential equation

$$\hat{\psi}_{zz} - \hat{\psi}_z^2 - \frac{Q_3}{Q_2}\hat{\psi}_z = -\sigma^{-2}g^2z^2 + Q_2 + \tilde{z}\hat{\psi}_1(z), \quad \hat{\psi}(0) = 0, \qquad (4.91)$$

where $\hat{\psi}_1(z)$ is a linear function. Differentiating (4.91) with respect to z at $z = 0$, and setting $\tilde{z} = 0, Q = \tilde{Q}$ in (4.91), we recover (4.86). Thus, using (4.87), we obtain

$$\tilde{\psi}_{zz} - \tilde{\psi}_z^2 = -\sigma^{-2}z^2 + \tilde{Q}_2, \quad \tilde{\psi}(0) = 0, \qquad (4.92)$$

where $\tilde{\psi}$ is the value of $\hat{\psi}$ at the stationary point. Now we scale σ out of the problem by setting

$$u = \sqrt{\frac{|g|}{\sigma}}\, z, \quad \phi(u) = \sqrt{\frac{\sigma}{|g|}}\, \tilde{\psi}_z, \quad a = \frac{\sigma}{|g|}\, \tilde{Q}_2,$$

to convert (4.92) to the Riccati equation

$$\phi_u - \phi^2 = -u^2 + a, \quad \phi(0) = 0 \quad \text{for} \quad a > 0. \qquad (4.93)$$

The substitution $\phi = -u - uv'/v$ converts the Riccati equation (4.93) into the Hermite equation

$$v'' - 2uv' = -av \qquad (4.94)$$

on the entire line. The first positive eigenvalue $a = 1$ separates solutions of (4.93) that blow up at finite u from those that become negative for some $u > 0$. Because $z = 0$ is the global minimum of $\tilde{\psi}$, the derivative $\phi(u) = \sqrt{\sigma/|g|}\, \tilde{\psi}_z$ cannot change sign for $z > 0$ and we are left with the only possibility $a = 1$ and $\phi(u) = u$ (Figure 4.1 shows the solution of the Riccati equation for linear measurements, $y'(x) - y^2(x) = -x^2 + a$ for $a = 1.0001$ (upper curve), $a = 0.99999$ (lower curve) and $a = 1$ (middle curve)). This, in turn, implies that $\tilde{Q}_2 = |g|/\sigma$, $\tilde{Q}_k = 0$ for $k > 2$, and $\hat{\psi}(z) = |g|z^2/2\sigma$. Thus the solution of Zakai's equation in the inner region is

$$\varphi_{\text{inner}}(x, t) \sim \exp\left\{-\frac{|g|}{2\sigma}\left(\frac{x - \tilde{x}(t)}{\rho^{1/2}}\right)^2\right\}, \qquad (4.95)$$

which is the same as that in the outer region and is identical to that predicted by linear filtering theory (see (3.67)).

Exercise 4.26 (The function $\hat{\psi}_1(z)$). Find $\hat{\psi}_1(z)$. $\qquad\qquad\square$

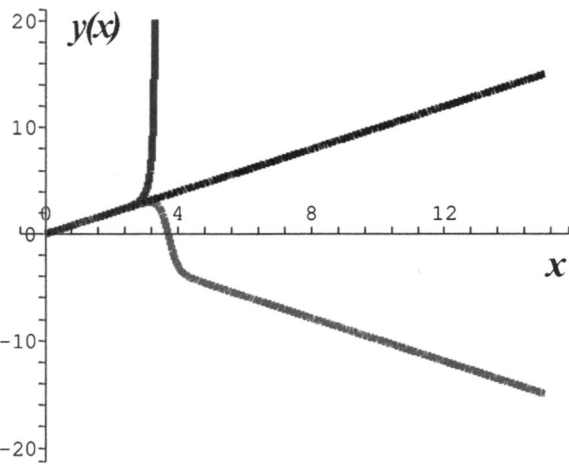

Fig. 4.1 The solution of the Riccati equation $y'(x) - y^2(x) = -x^2 + a$ for $a = 1.0001$ (upper curve), $a = 0.99999$ (lower curve) and $a = 1$ (middle line).

4.2.2 Expansion of the Cubic Sensor in the Inner Region

The expansion of the moments (4.18)–(4.21) for the cubic sensor has to be rebalanced in the inner region $D = \{|x| < \rho^{1/4}\}$. The appropriate scaling of the minimizer $\tilde{x}(t, \rho)$ and the moments in D is

$$\tilde{x}(t, \rho) = \rho^{1/4} \tilde{z}(t, \rho), \quad q_2(t, \rho) = \rho^{1/2} Q_2(t, \rho)$$

$$q_k(t, \rho) = \rho^{1/4} Q_k(t, \rho) \text{ for } k \geq 3, \tag{4.96}$$

in (4.18)–(4.21). Therefore, we obtain for Q_k the differential equations

$$\dot{Q}_2(t, \rho) = \rho^{-1/2} \left[-\sigma^2 Q_2^2(t, \rho) + 9\tilde{z}^4(t, \rho) + \frac{1}{2}\sigma^2 Q_4(t, \rho) - \frac{1}{2}\sigma^2 \frac{Q_3^2(t, \rho)}{Q_2(t, \rho)} \right]$$

$$- 2m'(\tilde{x}(t, \rho)) Q_2(t, \rho) - \rho^{1/4} m''(\tilde{x}(t, \rho)) \frac{Q_3(t, \rho)}{Q_2(t, \rho)}$$

$$+ \rho^{1/2} m^{(3)}(\tilde{x}(t, \rho)) + \rho^{-5/4} \left[3\tilde{z}^2(t, \rho) \frac{Q_3(t, \rho)}{Q_2(t, \rho)} - 6\tilde{z}(t, \rho) \right]$$

$$\times \left[\dot{y}(t) - \rho^{3/4} \tilde{z}^3 \right], \tag{4.97}$$

$$\dot{Q}_3(t, \rho) = \rho^{-1/2} \left[-3\sigma^2 Q_2^2(t, \rho) Q_3^2(t, \rho) + 54\tilde{z}^3(t, \rho) + \frac{1}{2}\sigma^2 Q_5(t, \rho) \right.$$

$$\left. - \frac{1}{2}\sigma^2 \frac{Q_3(t, \rho) Q_4(t, \rho)}{Q_2(t, \rho)} \right]$$

$$- \rho^{1/4} m''(\tilde{x}(t, \rho)) \left[3 Q_2(t, \rho) + \frac{Q_4(t, \rho)}{Q_2(t, \rho)} \right] + \rho^{3/4} m^{(iv)}(\tilde{x}(t, \rho))$$

$$+ \rho^{-5/4} \left[3 \tilde{z}^2(t, \rho) \frac{Q_4(t, \rho)}{Q_2(t, \rho)} - 6 \right] [\dot{y}(t) - \rho^{3/4} \tilde{z}^3] \tag{4.98}$$

and for $k \geq 4$

$$\dot{Q}_k(t, \rho) = \rho^{-1/2} \left\{ - k \left[\sigma^2 Q_2(t, \rho) Q_k(t, \rho) - 3 \tilde{z}^2(t) h^{(k-1)}(\tilde{z}(t)) \right] \right.$$

$$- \frac{1}{2} \sum_{i=2}^{k-2} \binom{k}{i} \left[\sigma^2 Q_{i+1}(t, \rho) Q_{k+i-1}(t, \rho) - h^{(i)}(\tilde{z}(t)) h^{(k-i)}(\tilde{z}(t)) \right]$$

$$\left. + \frac{1}{2} \sigma^2 \frac{Q_{k+2}(t, \rho) Q_2(t, \rho) - Q_3(t, \rho) Q_{k+1}(t, \rho)}{Q_2(t, \rho)} \right\}$$

$$- \rho^{k/4 - 1/2} k m^{(k-1)}(\tilde{x}(t, \rho)) Q_2(t, \rho)$$

$$- \sum_{i=1}^{k-2} \binom{k}{i} \rho^{(i-1)/4} m^{(i)}(\tilde{x}(t, \rho)) Q_{k+2-i}(t, \rho)$$

$$+ \rho \left[m^{(k+1)}(\tilde{x}(t, \rho)) - \rho^{-3/4} m''(\tilde{x}(t, \rho)) \frac{Q_{k+1}(t, \rho)}{Q_2(t, \rho)} \right]$$

$$+ 3 \rho^{-5/4} \tilde{z}^2(t) \frac{Q_{k+1}(t, \rho)}{Q_2(t, \rho)} \left[\dot{y}(t) - \rho^{3/4} \tilde{z}^3 \right], \tag{4.99}$$

where $|\tilde{z}| \leq 1$ and $h(\tilde{z}) = \tilde{z}^3$. Note that in contrast with the situation in the outer region, the leading terms of equations (4.97)–(4.99) remain coupled as $\rho \to 0$, so that there is no truncation rule analogous to that of Section 4.1.2. We rewrite the system (4.97)–(4.99) in vector notation as

$$\dot{Q}(t, \rho) = \rho^{-1/2} f(Q(t, \rho), \tilde{z}(t)) + \rho^{-1/4} M(Q(t, \rho), \tilde{z}(t)) \dot{\beta}(t), \tag{4.100}$$

where the innovation-like process $\dot{\beta}$ is defined as

$$\dot{\beta} = \rho^{-1} [\dot{y}(t) - \rho^{3/4} \tilde{z}^3(t)]. \tag{4.101}$$

We assume, in a self-consistent way, that the process $\dot{\beta}$ is approximately white noise with unit power and a drift term of order $\rho^{-1/4}$ (see below).

As above, we expand $Q(t, \rho)$ in an asymptotic series

$$Q \sim Q_0 + \sum_{j=1}^{\infty} \rho^{j/4} Q_j,$$

with the leading-order equation in the form

$$\dot{Q}_0 = \rho^{-1/2} f(Q_0, \tilde{z}) + \rho^{-1/4} M(Q_0, \tilde{z}) \dot{\beta}(t), \qquad (4.102)$$

where

$$f_k = - k[\sigma^2 Q_{2,0}(t) Q_{k,0}(t) - 3\tilde{z}^2(t) h^{(k-1)}(\tilde{z}(t))]$$

$$- \frac{1}{2} \sum_{i=2}^{k-2} \binom{k}{i} [\sigma^2 Q_{i+1,0}(t, \rho) Q_{k+i-1,0}(t, \rho) - h^{(i)}(\tilde{z}(t)) h^{(k-i)}(\tilde{z}(t))]$$

$$+ \frac{1}{2} \sigma^2 \frac{Q_{k+2,0}(t, \rho) Q_{2,0}(t, \rho) - Q_{3,0}(t, \rho) Q_{k+1,0}(t, \rho)}{Q_{2,0}(t, \rho)} \qquad (4.103)$$

and $M(Q_0, \tilde{z})$ is the vector of coefficients of $\dot{\beta}$. The stochastic forcing term in (4.102) is formally balanced with the drift f in the sense that both give rise to terms of order $\rho^{-1/2}$ in the corresponding Fokker–Planck equation. The leading-order equation (4.102) is not singular in ρ, because ρ can be scaled into t, rendering it independent of ρ.

The asymptotics for $\rho \to 0$ correspond to long (scaled) time asymptotics of the (scaled) equation. Thus in the inner region $|x| < \rho^{1/4}$, the cubic sensor is no longer a low-noise (high-SNR) filtering problem. Although the inner region shrinks as $\rho \to 0$, nevertheless, its contribution to the MSEE remains dominant. To construct the conditional pdf in the inner region, we note that the long (scaled) time behavior of the leading equation (4.102) is determined by the stationary point of the drift $f(Q_0, \tilde{z})$ near the stationary point $\tilde{z} = 0$. Thus we have to solve the stationary equation

$$0 = f(\tilde{Q}, 0) \qquad (4.104)$$

for the unknown stationary vector $\tilde{Q}$. Setting $\tilde{z} = 0$ in (4.103), we see that (4.104) has an even solution and that all components $\tilde{Q}_k$ for $k \geq 3$ can be expressed as functions of $\tilde{Q}_2$:

$$\tilde{Q}_{2k+1} = 0, \quad k = 1, 2, 3, \dots,$$

$$\tilde{Q}_4 = 2\tilde{Q}_2^2, \quad \tilde{Q}_6 = 16\tilde{Q}_2^3, \quad \tilde{Q}_8 = 272\tilde{Q}_2^4 - 720\sigma^{-2}, \qquad (4.105)$$

$$\vdots$$

We determine $\tilde{Q}_2$ that ensures that the unnormalized conditional pdf $\varphi(x, t)$ (the solution of Zakai's equation) is normalizable and has a unique maximum, and that the stationary point of the drift f is stable. The stretching (4.96) leads to the following WKB form of the conditional pdf $\varphi(x, t)$ in the inner region:

$$\varphi_{\text{inner}}(x, t) \sim \exp\left\{-\hat{\psi}\left(\frac{x - \tilde{x}(t)}{\rho^{1/4}}\right)\right\}, \qquad (4.106)$$

where

$$\hat{\psi}(z) = \sum_{k=2}^{\infty} \frac{Q_k(t,\rho)}{k!} z^k. \tag{4.107}$$

Setting

$$z = \frac{x - \tilde{x}(t)}{\rho^{1/4}}, \quad \tilde{z}(t) = \frac{\tilde{x}(t)}{\rho^{1/4}}, \tag{4.108}$$

the substitution of (4.106) in Zakai's equation gives the differential equation

$$\hat{\psi}_{zz} - \hat{\psi}_z^2 - \frac{Q_3}{Q_2}\hat{\psi}_z = -\sigma^{-2}\left[z^6 + 6\tilde{z}z^4 + 15\tilde{z}^2z^4 + 18\tilde{z}^4z^2 - \sigma^2 Q_2\right],$$

$$\hat{\psi}(0) = 0. \tag{4.109}$$

Next, we study the differential equation (4.109) near the stationary point $\tilde{Q}, \tilde{z} = 0, z = 0$. Differentiating (4.109) with respect to z at $z = 0$, and setting $\tilde{z} = 0$, $Q = \tilde{Q}$ in (4.109), we recover (4.104). Thus, using (4.105), we obtain

$$\tilde{\psi}_{zz} - \tilde{\psi}_z^2 = -\sigma^{-2}z^6 + \tilde{Q}_2, \quad \tilde{\psi}(0) = 0, \tag{4.110}$$

where $\tilde{\psi}$ is the value of $\hat{\psi}$ at the stationary point. Now we scale σ out of the problem by setting

$$u = \sigma^{-1/4}z, \quad \phi(u) = \sigma^{1/4}\tilde{\psi}_z, \quad a = \sigma^{1/2}\tilde{Q}_2,$$

to convert (4.110) to the Riccati equation

$$\phi_u - \phi^2 = -u^6 + a, \quad \phi(0) = 0 \quad \text{for} \quad a > 0. \tag{4.111}$$

The Riccati equation is equivalent to the eigenvalue problem for the "super Hermite" equation

$$v''(u) - 3u^3 v'(u) = -\lambda v(u) \tag{4.112}$$

on the entire line. The first positive eigenvalue λ corresponds to the critical value $a_c \approx 1.1448024537$ that separates solutions that blow up at finite u from solutions that become negative for some $u > 0$ (Figure 4.2 shows the solutions of the Riccati equations $y'(x) - y^2(x) = -x^6 + a$, $\phi(0) = 0$ for $a = 1.1448024538$ (upper curve), $a = 1.1448024537$ (lower curve), and the approximation $y(x) = 1.1448024537x + 0.85x^3$ (middle line)). Because $z = 0$ is the global minimum of $\tilde{\psi}$, the derivative $\phi(u) = \sigma^{1/4}\tilde{\psi}_z$ cannot change sign for $z > 0$, and as in Section 4.2.1, we are left with the only possibility $a = a_c$.

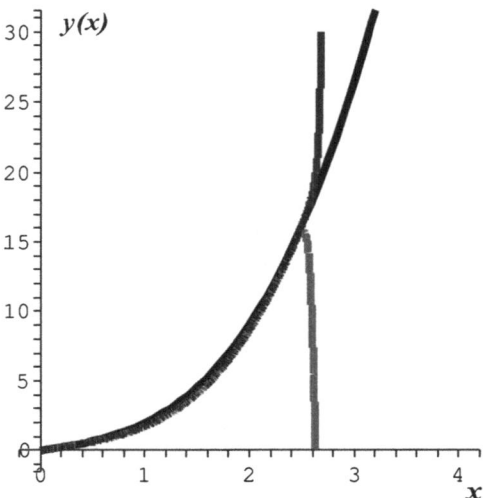

Fig. 4.2 The solutions of the Riccati equations $y'(x) - y^2(x) = -x^6 + a$, $\phi(0) = 0$ for $a = 1.1448024538$ (upper curve), $a = 1.1448024537$ (lower curve), and the approximation $y(x) = 1.1448024537x + 0.85x^3$ (middle curve).

We obtain

$$\tilde{Q}_2 = \sigma^{-1/2} a_c, \tag{4.113}$$

so that $\tilde{Q}_k$ can be now determined from (4.105) and (4.113). Linear stability analysis of the point $\tilde{Q}_k$ involves the estimation of the eigenvalues of the infinite matrix $\{\partial f_i / \partial Q_j\}_{\tilde{Q}}$, which is not easy. However, numerical solutions of the truncated system

$$\dot{\tilde{Q}} = f(\tilde{Q}, 0)$$

indicate that there is a stationary stable equilibrium point near that obtained from (4.105) and (4.113).

We can see from (4.106) that although φ is sharply peaked near $x = \tilde{x}(t)$ in the inner region, it is not in the WKB form, because in general, $Q_k(t, \rho) = O(1)$ as $\rho \to 0$. Consequently, integrals of the form (4.26) are of Laplace type. Therefore, to obtain an expansion in powers of ρ, we cannot truncate the series (4.107) in (4.106), in contrast to the case in Section 4.1.3. To obtain an approximation to (4.26), we use the asymptotic series approximation $\phi(u) \approx a_x u + 0.85u^3$ (see Figure 4.2). This gives the asymptotic representation

$$\tilde{\psi}(z) \approx \frac{a_c}{2} \frac{z^2}{\sigma^{1/2}} + \frac{0.85z^4}{4\sigma}. \tag{4.114}$$

Using (4.106), (4.107) and replacing $\hat{\psi}$ by $\tilde{\psi}$, we define an "average unnormalized conditional pdf" by

$$\tilde{\varphi}_{\text{average}}(x,t) = \exp\left\{-\frac{a_c}{2}\left(\frac{x-\tilde{x}(t)}{(\rho\sigma)^{1/4}}\right)^2 - 0.2125\left(\frac{x-\tilde{x}(t)}{(\rho\sigma)^{1/4}}\right)^4\right\}. \quad (4.115)$$

The approximation $\tilde{\varphi}_{\text{average}}(x,t)$ differs from $\varphi_{\text{inner}}(x,t)$ in that while $Q_k(t,\rho)$ are stochastic processes that depend on $y(t)$, the coefficients $\tilde{Q}_k(t,\rho)$ are the mean values of the system (4.102), linearized about the stable point of the drift f. It follows that the CMSEE and the MSEE obtained by replacing $\varphi_{\text{inner}}(x,t)$ with $\tilde{\varphi}_{\text{average}}(x,t)$ have the same functional dependence on ρ and σ as do those of the optimal filter, to leading-order in ρ.

4.2.3 Asymptotics of the CMSEE of the Cubic Sensor

To obtain an approximation to the central moments of the error of the optimal cubic sensor, we calculate conditional moments by breaking the domain of integration into the inner and outer regions and using the inner and outer expansions there. Then we interpolate the results to obtain a uniform expansion of the filter and its error. First, we define the optimal filter $\hat{x}_0(t)$ and the conditional moments of its error in the outer region by

$$\hat{x}_0(t) = \frac{\int_{-\infty}^{\infty} x\varphi_0(x,t)\,ds}{\int_{-\infty}^{\infty} \varphi_0(x,t)\,ds}, \quad \widehat{e_{x_0}^n}(t) = \frac{\int_{-\infty}^{\infty} (x-\hat{x}_0(t))^n\varphi_0(x,t)\,ds}{\int_{-\infty}^{\infty} \varphi_0(x,t)\,ds},$$

where $\varphi_0(x,t)$ is the outer solution of Zakai's equation, given by

$$\varphi_0(x,t) = \exp\left\{-\frac{1}{\rho}\sum_{k=2}^{\infty}\frac{1}{k!}q_{k,0}(t)(x-\hat{x}(t))^k\right\}, \quad (4.116)$$

with $q_{k,0}(t)$ given, to leading-order in ρ, by (4.31) (higher-order terms are calculated in [84], [85]). According to Section 4.1.2, as long as $|\tilde{x}(t)| > \rho^{1/4}$,

$$\hat{x}_0(t) = \tilde{x}(t) + O(\rho), \quad (4.117)$$

$$\widehat{e_{x_0}^2}(t) = \frac{\rho\sigma}{3x_0^2} + O\left(\rho^{3/2}\right) \quad (4.118)$$

(see (4.44)). Similarly, using the Laplace expansion in the outer region, we obtain the even moments

$$\widehat{e_{x_0}^{2n}}(t) = 1 \cdot 3 \cdot 5 \cdots (2n - 1) \frac{(\rho\sigma)^n}{(3\tilde{x}(t))^n} + O(\rho^{n+1}). \qquad (4.119)$$

Exercise 4.27 (Higher moments). Prove (4.119). $\qquad\qquad\square$

For $|\tilde{x}(t)| < \rho^{1/4}$, we define

$$\hat{x}_{\text{inner}}(t) = \frac{\displaystyle\int_{-\infty}^{\infty} x \varphi_{\text{inner}}(x, t)\, dx}{\displaystyle\int_{-\infty}^{\infty} \varphi_{\text{inner}}(x, t)\, dx},$$

$$\widehat{e^n}_{\text{inner}}(t) = \frac{\displaystyle\int_{-\infty}^{\infty} (x - \hat{x}_{\text{inner}}(t))^n \varphi_{\text{inner}}(x, t)\, dx}{\displaystyle\int_{-\infty}^{\infty} \varphi_{\text{inner}}(x, t)\, dx}, \qquad (4.120)$$

and proceed to calculate the conditional moments. Replacing $\varphi_{\text{inner}}(x, t)$ with $\varphi_{\text{average}}(x, t)$, we obtain the approximations

$$\hat{x}_{\text{inner}}(t) \approx \tilde{x}(t) \qquad (4.121)$$

and

$$\widehat{e^{2n}}_{\text{inner}}(t)$$

$$\approx \frac{\displaystyle\int_{-\infty}^{\infty} (x - \tilde{x}(t))^{2n} \exp\left\{ -\frac{a_c}{2} \left(\frac{x - \tilde{x}(t)}{(\rho\sigma)^{1/4}} \right)^2 - 0.2125 \left(\frac{x - \tilde{x}(t)}{(\rho\sigma)^{1/4}} \right)^4 \right\} dx}{\displaystyle\int_{-\infty}^{\infty} \exp\left\{ -\frac{a_c}{2} \left(\frac{x - \tilde{x}(t)}{(\rho\sigma)^{1/4}} \right)^2 - 0.2125 \left(\frac{x - \tilde{x}(t)}{(\rho\sigma)^{1/4}} \right)^4 \right\} dx}.$$

Hence

$$\widehat{e^{2n}}_{\text{inner}}(t) \approx \frac{(\rho\sigma)^{n/2}}{a_c^n} G_{2n}, \qquad (4.122)$$

where G_n is given by

$$G_{2n} = \frac{\displaystyle\int_{-\infty}^{\infty} x^{2n} \exp\left\{ -\frac{x^2}{2} - \frac{0.2125 x^4}{a_c^2} \right\} dx}{\displaystyle\int_{-\infty}^{\infty} \exp\left\{ -\frac{x^2}{2} - \frac{0.2125 x^4}{a_c^2} \right\} dx}.$$

The values of G_n are

$$2n \quad 2 \quad 4 \quad 6 \quad 8 \quad 10 \quad 12 \; 14 \quad 16 \quad 18 \quad 20$$

$$G_{2n} \; 0.54 \; 0.71 \; 1.38 \; 3.37 \; 9.72 \; 32 \; 116 \; 460 \; 1968 \; 9011$$

An interpolation between (4.119) and (4.122) gives

$$\widehat{e_{\hat{x}}^{2n}} \approx \frac{(\rho\sigma)^n}{(3\tilde{x}^2)^n/(2n-1)!! + (\rho\sigma)^{n/2}a_c^n/G_{2n}}$$

$$\approx \frac{(\rho\sigma)^n}{(3\hat{x}^2)^n/(2n-1)!! + (\rho\sigma)^{n/2}a_c^n/G_{2n}}, \tag{4.123}$$

where the last approximation uses (4.121). Note that the constant values of a_c and G_{2n} are average values of functions that depend on $y(t)$. However, the functional dependence on ρ and σ is correct.

4.2.4 Approximations of the Optimal Cubic Sensor

The Kushner–Itô equation for $\hat{x}$ (see Exercise 3.5) is

$$d\hat{x} = \widehat{m(x)}\,dt + \rho^{-2}\widehat{e_x e_h}[dy - \widehat{h(x)}\,dt], \tag{4.124}$$

where

$$e_x = x - \hat{x}, \quad e_h = h(x) - \widehat{h(x)}.$$

Writing the conditional expectations as moment series, we get

$$\widehat{m(x)} = m(\hat{x}) + \frac{1}{2}m''(\hat{x})\widehat{e_x^2} + \cdots, \tag{4.125}$$

$$\widehat{e_x e_h} = h'(\hat{x})\widehat{e_x^2} + \frac{1}{2}h''(\hat{x})\widehat{e_x^3} + (1/3!)h'''(\hat{x})\widehat{e_x^4} + \cdots, \tag{4.126}$$

$$\widehat{h(x)} = h(\hat{x}) + \frac{1}{2}h''(\hat{x})\widehat{e_x^2} + (1/3!)h'''(\hat{x})\widehat{e_x^3} + \cdots, \tag{4.127}$$

where (4.126) and (4.127) are exact, because the higher derivatives of $h(x) = x^3$ vanish. Thus all terms in (4.126) and (4.127) have to be kept in the inner region. Using $\tilde{\varphi}_{\text{average}}$, we obtain for $\widehat{e_x e_h}$ the interpolation formula

$$\widehat{e_x e_h} \approx \frac{3\hat{x}^2\rho\sigma}{3\hat{x}^2 + (\rho\sigma)^{1/2}a_c/G_2} + \frac{(\rho\sigma)^2}{3\hat{x}^4 + \rho\sigma a_c^2/G_4}. \tag{4.128}$$

Using (4.125)–(4.127) in (4.128), we obtain

$$
d\hat{x} \approx \left[m(\hat{x}) + \frac{1}{2} m''(\hat{x}) \frac{\rho\sigma}{3\hat{x}^2 + (\rho\sigma)^{1/2} a_c/G_2} \right] dt
$$
$$
+ \rho^{-2} \left[\frac{3\hat{x}^2 \rho\sigma}{3\hat{x}^2 + (\rho\sigma)^{1/2} a_c/G_2} + \frac{(\rho\sigma)^2}{3\hat{x}^4 + \rho\sigma a_c^2/G_4} \right]
$$
$$
\times \left[dy - \left(\hat{x}^3 + \frac{3\hat{x}\rho\sigma}{3\hat{x}^2 + (\rho\sigma)^{1/2} a_c/G_2} \right) dt \right]. \tag{4.129}
$$

In the outer region, where $|\hat{x}| > \rho^{1/4}$, the first term on the right-hand side of (4.129) is smaller than the second, as argued in Section 4.1.3. This is also the case in the inner region $|\hat{x}| < \rho^{1/4}$. Indeed,

$$
\rho^{-1}[dy - (\hat{x}^3 + 3\hat{x}\widehat{e_x^2})\, dt] \sim dv + \rho^{-1}[3\hat{x}^2 e_x + 3\hat{x}e_x^2 + e_x^3 - 3\hat{x}\widehat{e_x^2}]\, dt,
$$

and because $e_{x_{\text{inner}}} = O(\rho^{1/4})$, we obtain

$$
\rho^{-1}[dy - (\hat{x}^3 + 3\hat{x}\widehat{e_x^2})\, dt] \sim dv + O\left(\rho^{-1/4}\right), \tag{4.130}
$$

so that $\rho^{-1}(dy - \hat{x}^3\, dt)$ dominates $\rho^{-1} 3\hat{x}\widehat{e_x^2}\, dt$ for $\hat{x}| < \rho^{1/4}$. Thus the resulting approximate cubic sensor is

$$
d\hat{x}^* = \frac{1}{\rho^2} \left[\frac{3x^{*2}\rho\sigma}{3x^{*2} + (\rho\sigma)^{1/2} a_c/G_2} + \frac{(\rho\sigma)^2}{3x^{*4} + \rho\sigma a_c^2/G_4} \right] (dy - x^{*3}\, dt). \tag{4.131}
$$

The CMSEE of x^* is given by

$$
\widehat{(x - x^*)^2} \approx \frac{\rho\sigma}{3x^{*2} + (\rho\sigma)^{1/2} a_c/G_2}. \tag{4.132}
$$

Note that (4.131) reduces to the constant-gain filter when $|x^*| > \rho^{1/4}$. In the inner region $|x^*| < \rho^{1/4}$, (4.131) is again a constant-gain filter, but this time the gain is $(\sigma/\rho)(G_4/a_c^2)$ instead of σ/ρ.

4.2.5 The MSEE of the Optimal Cubic Sensor

To evaluate the MSEE, we take the expectation of (4.123) with $n = 1$. First, we find an approximation to the pdf of $\hat{x}$. In view of (4.125)–(4.127), we know that we can expand all the conditional expectations in (4.124) in terms of $\hat{x}$ and the moments

$\widehat{e_x^{2n}}$ and we can use (4.123) to approximate the even moments. Thus we can write the Fokker–Planck equation for the density $p_{\hat{x}}(x,t)$ as

$$\frac{\partial p_{\hat{x}}(x,t)}{\partial t} = -\frac{\partial}{\partial x}[p_{\hat{x}}(x,t)\widehat{m(x)}] + \frac{\rho^2}{2}\frac{\partial^2}{\partial x^2}[p_{\hat{x}}(x,t)\widehat{e_x e_h}^2], \qquad (4.133)$$

where $\widehat{m(x)}$ and $\widehat{e_x e_h}$ are given by (4.125) and (4.128), respectively. In the steady state, $\partial p_{\hat{x}}(x,t)/\partial t = 0$, and we can integrate the FPE (4.133) to obtain

$$p_{\hat{x}}(x) = C\exp\left\{\frac{\rho^2}{2}\int_0^x \frac{\widehat{m(x)}}{\widehat{e_x e_h}^2}\,dx\right\}, \qquad (4.134)$$

where C is a normalization constant. We assume, for simplicity, that $m(x) = -ax$ for some positive constant a. Then an interpolation formula for the integrand in (4.134) is given by

$$\frac{\widehat{m(x)}}{\widehat{e_x e_h}^2} \approx \frac{-ax}{\left[\dfrac{3x^2\rho\sigma}{3x^2 + (\rho\sigma)^{1/2}a_c/G_2} + \dfrac{(\rho\sigma)^2}{3x^4 + \rho\sigma a_c^2/G_4}\right]^2}. \qquad (4.135)$$

In terms of the scaled variable $\xi = (\rho\sigma)^{-1/2}(G_2/a_c)^{1/2}x$, the denominator in (4.135) is $(\rho\sigma)^2$ times a function of ξ that goes from G_4^2/a_c^4 at $\xi = 0$ to nearly 1 at $\xi = 4$.

To calculate $p_{\hat{x}}(x)$, we note that the probability of the transition region $1 < \xi < 4$ is $O(\rho^{1/2})$, so we replace the denominator in (4.135) by a step function that jumps from $(\rho\sigma)^2 G_4^2/a_c^4$ to $(\rho\sigma)^2$ at $\xi = 1$. We get, therefore,

$$p_{\hat{x}}(x) \approx \begin{cases} C_1\exp\left\{-\dfrac{a_c^4}{\sigma^2 G_4^2}\dfrac{ax^2}{4}\right\} & \text{for } |x| < (\rho\sigma)^{1/2}\left(\dfrac{a_c}{G_2}\right)^{1/2}, \\[4mm] \sqrt{\dfrac{a}{4\pi\sigma^2}}\exp\left\{-\dfrac{ax^2}{4\sigma^2}\right\} & \text{for } |x| > (\rho\sigma)^{1/2}\left(\dfrac{a_c}{G_2}\right)^{1/2}, \end{cases} \qquad (4.136)$$

where C_1 is chosen so that $p_{\hat{x}}(x)$ is normalized. Taking in (4.123) with $(n = 1)$ the expectation of $\widehat{e_x^2}$ with respect to the density (4.136), we obtain from (4.123) the MSEE

$$\widehat{e_x^2} \approx \int_{-\infty}^{\infty} \frac{\rho\sigma}{3x^2 + (\rho\sigma)^{1/2}a_c/G_2}\,p_{\hat{x}}(x)\,dx. \qquad (4.137)$$

The first line in (4.136) contributes $O(\rho\sigma)$ to the integral (4.137), while the second line contributes asymptotically

$$\int_{-\infty}^{\infty} \frac{\rho\sigma}{3x^2 + (\rho\sigma)^{1/2}a_c/G_2} \sqrt{\frac{a}{4\pi\sigma^2}} \exp\left\{-\frac{ax^2}{4\sigma^2}\right\} dx$$

$$\sim \sqrt{\frac{a}{4\pi\sigma^2}} \sqrt{\frac{G_2}{a_c}}(\rho\sigma)^{3/4} \int_{-\infty}^{\infty} \frac{d\xi}{3\xi^2 + 1}$$

$$= \frac{\pi}{\sqrt{3}} \sqrt{\frac{a}{4\pi\sigma^2}} \sqrt{\frac{G_2}{a_c}}(\rho\sigma)^{3/4} = 0.35\sqrt{a}\,\frac{\rho^{3/4}}{\sigma^{1/4}}, \tag{4.138}$$

so that

$$\overline{e_x^2} \approx 0.35\sqrt{a}\,\frac{\rho^{3/4}}{\sigma^{1/4}}. \tag{4.139}$$

The numerical factor 0.35 is obtained from the integration of an interpolation formula. The dependence of the MSEE $\overline{e_x^2}$ on the parameters a, ρ, and σ in (4.138) is, however, exact.

4.2.6 The MSEE of the Constant-Gain Cubic Sensor

The leading-order approximation (4.131) is much more complicated than the constat-gain filter (4.36). To compare the MSEE of the two filters, we approximate first that of (4.36). The joint pdf of $x(t)$ and $x_0(t)$ is the solution of the FPE corresponding to the system

$$dx = m(x)\, dt + \sigma\, dw, \tag{4.140}$$

$$dx_0 = \frac{\sigma}{\rho}(dy - x_0^3)\, dt = \frac{\sigma}{\rho}(x^3 - x_0^3)\, dt + \rho\, dv, \tag{4.141}$$

given by

$$p_t(x, x_0, t) = -[m(x)p\,(x, x_0, t)]_x + \frac{\sigma}{\rho}[(x^3 - x_0^3)p\,(x, x_0, t)]_{x_0}$$

$$+ \frac{\sigma^2}{2}p_{xx}(x, x_0, t) + \frac{\rho^2}{2}p_{x_0x_0}(x, x_0, t).$$

We transform the FPE by the exponential substitution

$$p\,(x, x_0, t) = \exp\left\{-\frac{\theta(x, x_0, t, \rho)}{\rho}\right\} \tag{4.142}$$

to the nonlinear equation

$$-\theta_t = m\theta_x - \rho m' + \frac{\sigma}{\rho}(x^3 - x_0^3)\theta_{x_0} + 3\sigma x_0^2 + \frac{\sigma^2}{2}\left(\frac{\theta_x^2}{\rho} - \theta_{xx}\right)$$

$$+ \frac{\sigma^2}{2}\left(\frac{\theta_{x_0}^2}{\rho} - \theta_{x_0 x_0}\right)$$

and expand

$$\theta(x, x_0, t, \rho) = \theta_0(x, x_0, t) + \rho\theta_1(x, x_0, t) + \cdots.$$

For $x, x_0 = O(1)$, the steady-state equation for the leading term $\theta_0(x, x_0, t)$ satisfies the equation

$$(x^3 - x_0^3)\theta_{0,x_0} + \frac{\sigma}{2}(\theta_{0,x}^2 + \theta_{0,x_0}^2) = 0. \tag{4.143}$$

It can be seen from (4.143) that the minimum of $\theta_0(x, x_0)$ is achieved at $x = x_0$, so that

$$\theta_{0,x}(x_0, x_0) = \theta_{0,x_0}(x_0, x_0) = 0. \tag{4.144}$$

It follows that the value of $\theta_0(x_0, x_0)$ is constant, because $d\theta_0(x_0, x_0)/dx_0 = \theta_{0,x}(x_0, x_0) + \theta_{0,x_0}(x_0, x_0) = 0$. Setting

$$\theta_{ij} = \left.\frac{\partial^{i+j}\theta_0(x, x_0)}{\partial x^i \partial x_0^j}\right|_{x=x_0} \tag{4.145}$$

and differentiating (4.143) twice with respect to x_0 at $x = x_0$ and using (4.144), we obtain

$$-6x_0^2\theta_2 + \sigma\theta_{1,1}^2 + \sigma\theta_{0,2}^2 = 0. \tag{4.146}$$

Similarly, successive differentiation of (4.143) at $x = x_0$ yields

$$-3x_0^2\theta_{1,1} + 3x_0^2\theta_{0,2} + \sigma\theta_{2,0} + \sigma\theta_{1,1}\theta_{0,2} = 0 \tag{4.147}$$

and

$$6x_0^2\theta_{1,1} + \sigma\theta_{2,0}^2 + \sigma\theta_{1,1}^2 = 0. \tag{4.148}$$

Hence

$$\theta_{0,2} = \theta_{2,0} = -\theta_{1,1} = \frac{3x_0^2}{\sigma} \tag{4.149}$$

and

$$\theta_{3,0} = \frac{6x_0}{\sigma}, \quad \theta_{4,0} = \frac{6}{\sigma}. \tag{4.150}$$

It follows that Taylor's expansion of $\theta_0(x, x_0)$ about the diagonal $x = x_0$ is given by

$$\theta_0(x, x_0) - \theta_0(x_0, x_0) = \frac{3x_0^2}{2\sigma}(x - x_0)^2 + \frac{x_0}{\sigma}(x - x_0)^3 + \frac{1}{4\sigma}(x - x_0)^4 + \cdots. \tag{4.151}$$

Scaling $x_0 = (\sigma\rho)^{1/4}\xi, x = (\sigma\rho)^{1/4}\eta$ in the inner region $|x_0| < (\sigma\rho)^{1/4}$, using (4.151) and the approximation

$$e^{-\theta_1\left((\sigma\rho)^{1/4}\xi,(\sigma\rho)^{1/4}\eta\right)} = e^{-\theta_1(0,0)},$$

the stationary MSEE for x_0 in the inner region

$$\mathbb{E}e_{0\,\text{inner}}^2 = \lim_{t \to \infty} \mathbb{E}\left[e_0^2(t) \,|\, y_0^t, |x_0(t)| < (\sigma\rho)^{1/4}\right]$$

can be expressed as

$$\mathbb{E}e_{0\,\text{inner}}^2$$

$$= \frac{\displaystyle\int_{-\infty}^{\infty} \int_{-(\sigma\rho)^{1/4}}^{(\sigma\rho)^{1/4}} (x - x_0)^2 p\,(x, x_0, t)\, dx\, dx_0}{\displaystyle\int_{-\infty}^{\infty} \int_{-(\sigma\rho)^{1/4}}^{(\sigma\rho)^{1/4}} p\,(x, x_0, t)\, dx\, dx_0}$$

$$\approx \frac{\displaystyle(\sigma\rho)^{1/2} \int_{-\infty}^{\infty} \int_{-1}^{1} (\eta - \xi)^2 \exp\left\{-\frac{3}{2}\xi^2(\eta - \xi)^2 - \xi(\eta - \xi)^3 - \frac{1}{4}(\eta - \xi)^4\right\} d\xi\, d\eta}{\displaystyle\int_{-\infty}^{\infty} \int_{-1}^{1} \exp\left\{-\frac{3}{2}\xi^2(\eta - \xi)^2 - \xi(\eta - \xi)^3 - \frac{1}{4}(\eta - \xi)^4\right\} d\xi\, d\eta}$$

$$\approx 0.8(\sigma\rho)^{1/2}.$$

Interpolating with the outer solution (4.118), we obtain

$$\widehat{e_0^2} \approx \frac{\sigma\rho}{3x_0^2 + 1.25(\sigma\rho)^{1/2}},$$

which gives

$$\overline{e_0^2} \approx 0.18(\sigma\rho)^{1/2}. \qquad (4.152)$$

Thus the error of x_0 is of order $O(\rho^{1/2})$ everywhere.

Exercise 4.28 (Asymptotically optimal filter for measurements with a higher order point of inflection). Consider the model

$$dx = m(x)\,dt + \sigma\,dw, \quad dy = h_n(x)\,dt + \rho\,dv,$$

where

$$h_n(x) = \begin{cases} x^n & \text{if } n \text{ is odd,} \\ x^n \operatorname{sgn}(x) & \text{if } n \text{ is even.} \end{cases} \qquad (4.153)$$

Define the outer and the inner regions for the minimizer $\tilde{x}(t,\rho)$ of the a posteriori pdf by $|\tilde{x}| > \rho^{1/(n+1)}$ and $|\tilde{x}| < \rho^{1/(n+1)}$, respectively.

(i) Derive the Riccati equation

$$\phi_u - \phi^2 = -h_n^2 + a_n^{(n)}, \quad \phi(0) = 0, \qquad (4.154)$$

where $a_n^{(n)}$ is the critical value for which the solution of (4.154) is positive for all $u > 0$ and does not blow up.

(ii) Show that the a posteriori unnormalized pdf in the inner region is given by

$$\varphi_{\text{inner}} = \exp\left\{ -\frac{a_c^{(n)}}{2} \frac{(x-\tilde{x})^2}{(\sigma\rho)^{2/(n+1)}} + \frac{|x-\tilde{x}|^{n+1}}{(n+1)\sigma\rho} \right\}. \qquad (4.155)$$

(iii) Derive the interpolation formula for the moments of the error

$$\widehat{|e_x|^k} \approx \frac{\tilde{C}(\sigma\rho)^{k/2}}{(h_n'(\tilde{x}))^{k/2} + \tilde{C}(\sigma\rho)^{k/2 - k/(n+1)}[a_c^{(n)}]^{k/2}[G_k^{(n)}]^{-1}}, \qquad (4.156)$$

where

$$\tilde{C} = \left(\frac{2}{\pi}\right)^{1/2} \Gamma\left(\frac{k+1}{2}\right)$$

and

$$G_k^{(n)} = \frac{\displaystyle\int_{-\infty}^{\infty} |u|^k \exp\left\{ -\frac{1}{2}u^2 - \frac{|u|^{n+1}}{(n+1)[a_c^{(n)}]^{(n+1)/2}} \right\}\,du}{\displaystyle\int_{-\infty}^{\infty} \exp\left\{ -\frac{1}{2}u^2 - \frac{|u|^{n+1}}{(n+1)[a_c^{(n)}]^{(n+1)/2}} \right\}\,du}.$$

(iv) Replace the Taylor expansion of $\widehat{e_x e_{h_n}}$ by

$$\widehat{e_x e_{h_n}} = \mathbb{E}[(x - \tilde{x})(h_n(x) - \widehat{h_n(x)}) \mid y_0^t] = \widehat{x h_n(x)} - \hat{x} \widehat{h_n(x)}.$$

Obtain for $\tilde{x} \sim 0$

$$\widehat{e_x e_{h_n}}_{\text{inner}} - \widehat{x h_n(x)} = \widehat{|x|^{n+1}} \sim \widehat{|e_x|^{(n+1)}}_{\text{inner}}.$$

 (v) Derive the interpolation formula

$$\widehat{e_x e_{h_n}} \approx h_n'(\hat{x}) \widehat{e_x^2 |e_x|^{(n+1)}}.$$

(vi) Derive the leading-order approximation to the optimal filter

$$dx^* = \left[\frac{h_n'(x^*)\sigma}{h_n'(x^*) + (\sigma\rho)^{1-2/(n+1)} a_c^{(n)} [G_2^{(n)}]^{-1}} \right.$$

$$\left. + \frac{\tilde{C}\rho^{(n-1)/2}\sigma^{(n+1)}2}{[h_n'(x^*)]^{(n+1)/2}\tilde{C}(\sigma\rho)^{(n+1)/2-1}[a_c^{(n)}]^{(n+1)/2}[G_4^{(n)}]^{-1}} \right] dJ_0^{(n)},$$

where $dJ_0^{(n)} = \rho^{-1}[dy - h_n(x^*)\,dt]$. $\square$

Exercise 4.29 (The MSEE of the asymptotically optimal filter).

 (i) Use the interpolation formula (4.156) for the conditional moments of the error
to represent the CMSEE as

$$\widehat{e_x^2} \approx \frac{\sigma\rho}{h_n'(x^*) + (\sigma\rho)^{1-2/(n+1)} a_c^{(n)} [G_2^{(n)}]^{-1}}$$

to derive the asymptotic formula for the MSEE for $m = -ax$,

$$\overline{e_x^2} \approx \begin{cases} \rho^{3/(n+1)} \dfrac{\sqrt{a}}{\sigma^{(n-2)/(n+1)}} \dfrac{2}{\sqrt{\pi}} \left(\dfrac{nG_2^{(n)}}{a_c^{(n)}} \right)^{1-1/(n+1)} \dfrac{n-1}{n(n-2)} \\ \qquad\qquad \text{for } n \geq 3, \\ \\ \varepsilon\rho\sqrt{a} \text{ for } \boldsymbol{n} = 2, \end{cases}$$

where $\gamma = O(1)$.

(ii) Find the constant-gain filter and its MSEE. $\square$

4.2.7 Annotations

The path integral approach to nonlinear filtering was proposed in [16], [7], and [49]. The path integral approach to nonlinear smoothing problems presented in Section 3.5.1 is new. Forward–backward equations for the joint fixed-delay filtering-smoothing a posteriori density were given in [165] and [104].

The low-noise analysis of Zakai's equation in Section 4.1 is based on [84], [85]. In problems of phase estimation that lead to loss of lock and cycle slips, an important optimality criterion is maximizing the mean time to lose lock (MTLL) or to exit a given region, which is also a well-known control problem [51], [167], [4], [114]. Approximation methods for finding the various optimal filters have been devised for problems with small noise, including large-deviation and WKB solutions of Zakai's equation, the extended Kalman filter [71], [85], [125], [66], and others. The extended Kalman filter and WKB approximations produce explicit suboptimal finite-dimensional filters, which in case of phase estimation are the well-known phase trackers, such as the phase-locked loop (PLL), delay-locked loop (DLL), angle-tracking loops, and so on [147]. The MSEE in these phase trackers is asymptotically optimal [84], [85], [66].

The cubic sensor was considered in [25], [24], and [5]; rigorous results, though not practical realizations, were given in [148]; a proof that no finite-dimensional realization exists was given in [63], [64]; probability densities were discussed in [122]; and geometrical methods were proposed in [111], [20], [21]. Suboptimal solutions were proposed in [27], and real-time realizations in [62], [7]. The early history of the algebraic approach to the cubic sensor problem is described in [163] as follows: "*In the late 1970s, Brockett and Clark* [23] *Brockett* [22] *and Mitter* [116] *proposed the idea of using estimation algebras to construct a finite-dimensional nonlinear filter. In 1983 Brockett proposed to classify all finite dimensional estimation algebras.*" More recent developments are also mentioned.

Chapter 5
Loss of Lock in Phase Trackers

The trackers of phase, frequency, angle, range, and other parameters are notorious for their tendency to lose their lock on the tracked signal. Phase is usually defined mod 2π, so a noise-induced jump of 2π in the tracked phase, a so-called cycle slip, causes only a short-lived disturbance (see Figure 5.3). If cycle slips occur frequently, as is the case in PLLs for tracking FM signals, the signal is lost altogether and a sharp degradation in the tracking-loop performance ensues. In range or angle tracking (radar), once the tracking error exceeds a certain threshold, the lock detector indicates that the target is lost and has to be reacquired. In certain synchronization systems losses of lock are catastrophic and have to be made rare. Therefore the mean time between losses of lock (MTLL) is an important performance criterion for trackers. In this chapter the phenomenon of loss of lock is investigated in one- and two-dimensional trackers and an asymptotic method is developed for the calculation of the MTLL as a function of the tracking-loop parameters. This method can be generalized in a straightforward manner to higher-order phase trackers.

5.1 Loss of Lock in a First-Order PLL for PM

It was shown in Exercise 4.23 that the phase estimation error $e = x(t) - \tilde{x}_0(t)$ of the phase tracker

$$d\tilde{x}_0 = -m\tilde{x}_0 \, dt + \frac{\sigma}{\rho} [\cos \beta \tilde{x}_0 \, dy_1 - \sin \beta \tilde{x}_0 \, dy_2] \tag{5.1}$$

satisfies the equation

$$de = -(K\varepsilon e + \sin e) \, dt + \sqrt{2\varepsilon} \, dw_1, \tag{5.2}$$

Z. Schuss, *Nonlinear Filtering and Optimal Phase Tracking*, Applied Mathematical Sciences 180, DOI 10.1007/978-1-4614-0487-3_5, © Springer Science+Business Media, LLC 2012

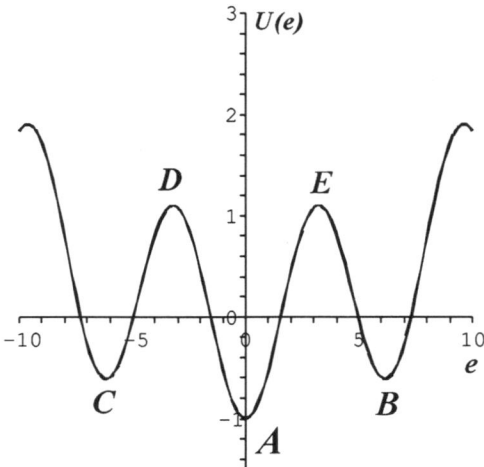

Fig. 5.1 The potential $U(e) = 0.01e^2 - \cos e$. The local minima at A, B, C are stable equilibria of the noiseless dynamics $\dot{e} = -U'(e)$, and the local maxima D, E are unstable equilibria.

which can be interpreted as the equation of motion of an overdamped Brownian particle in the potential field

$$U(e) = \frac{K\varepsilon e^2}{2} - \cos e. \tag{5.3}$$

Figure 5.1 shows the potential $U(e) = 0.01e^2 - \cos e$. The local minima at A, B, C are stable equilibria of the noiseless dynamics $\dot{e} = -U'(e)$, and the local maxima D, E are unstable equilibria.

Thus the error equation (5.2) can be written as

$$de = -U'(e)\,dt + \sqrt{2\varepsilon}\,dw_1. \tag{5.4}$$

When the noise is small, the error $e(t)$ stays for long periods of time near the stable equilibrium point A, where $e = 0$. The noise, however, regardless how small, will eventually drive the error over the potential barrier at D or E, and $e(t)$ will end up near another locally stable equilibrium at B or C. It will then spend a long period of time in the potential well near B or C and will then be pushed by the noise over a potential barrier either back into the well near A or into the next potential well. The tracker $x_0(t)$ is said to be locked on the true phase $x(t)$ as long as the tracking error stays near A. A noise-induced transition into a neighboring well is called *loss of lock* or *cycle slip*. The escape from a well, when it occurs, is quite rapid in the sense that the mean last passage time from A to B is about the same as the relaxation time of the noiseless system $\dot{e} = -U'(e)$, which is much shorter than the mean first exit time of $e(t)$ from the well at A (see [137, Section 7.5.1, Example 7.2]).

The mean time between consecutive losses of lock (MTLL) is an important performance and design parameter of trackers. The MTLL of the tracker (5.1) is the mean time for $e(t)$ to go from A to B or C. This mean time is twice the mean time to reach either D or E, because only 50% of the trajectories that reach the top of the barrier, at E, say, reach B before they reach A (see [137, Example 10.2]). Thus the MTLL is twice the MFPT from A to D or to E.

In the notation $D = (e_D, U(e_D))$, $E = (e_E, U(e_E))$, the MFPT is the mean first exit time $\bar{\tau}$ of the process $e(t)$ from the interval (e_D, e_E), and the MTLL is $2\bar{\tau}$. To calculate the MFPT, we define the random exit time $\tau = \min\{t \mid u(t) = e_E, e_D\}$ and its mean $u(e) = \mathbb{E}[\tau \mid e(0) = e]$. According to the Andronov–Vitt–Pontryagin theorem (Theorem 1.5.2), the function $u(e)$ is the solution of the boundary value problem

$$\varepsilon u''(e) - U'(e)u'(e) = -1 \quad \text{for } e_D < e < e_E, \quad u(e_D) = u(e_E) = 0. \quad (5.5)$$

The symmetry $U(-e) = U(e)$ implies the symmetry $u(-e) = u(e)$; hence $u'(0) = 0$. It is enough, therefore, to solve the Andronov–Vitt–Pontryagin equation (5.5) in the interval $0 < e < e_E$ with the boundary conditions $u(0) = u(e_E) = 0$. The solution is given by

$$u(e) = \frac{1}{\varepsilon} \int_e^{e_E} \int_0^y \exp\left\{\frac{U(y) - U(s)}{\varepsilon}\right\} ds\,dy. \quad (5.6)$$

Exercise 5.1 (The mean exit time in one dimension). Find the explicit solution of (5.5) for a general asymmetric potential forming a well and plot its graph. □

5.1.1 Small Noise: a Singular Perturbation Problem

When the noise intensity ε is small relative to the potential barrier height (for (5.3) this condition is $\varepsilon/2 \ll 1$), equation (5.5) becomes a singular perturbation problem, because in the limit $\varepsilon \to 0$, the degree of the equation drops from 2 to 1. The solution of a first-order equation cannot possibly satisfy two boundary conditions, so the solution has to have a singularity at this value of ε. Although an asymptotic evaluation of the integral in the explicit solution (5.6) reveals the nature of the singularity of the solution in the limit $\varepsilon \to 0$, a qualitative analysis is needed that can be generalized to higher-dimensional problems, for which no explicit solutions are known.

To find the dependence of the solution on ε near the singular value $\varepsilon = 0$, we assume first that the solution, which we now denote by $u_\varepsilon(e)$, has a series expansion in powers of ε,

$$u_\varepsilon(e) = u_0(e) + \varepsilon u_1(e) + \varepsilon^2 u_2(e) + \cdots. \quad (5.7)$$

Using the regular power series (5.7) in (5.5) and comparing like powers of ε, we obtain at the leading-order

$$-U_0'(e)u_0'(e) = -1, \tag{5.8}$$

where $U_0(e) = \sin e$, so (5.8) can be written as

$$\frac{d u_0(e(t))}{dt} = -1, \tag{5.9}$$

where $e(t)$ is the solution of

$$\frac{de}{dt} = -U_0'(e) = -\sin e. \tag{5.10}$$

Because $U_0(e)$ has the single attractor $e = 0$ in the domain $D = (e_D, e_E)$, we must have $e(t) \to 0$ as $t \to \infty$. It follows that $u_0(e(t)) = u_0(e(0)) - t \to -\infty$ as $t \to \infty$. Thus, on the one hand, $u_0(e(t)) \to u_0(0)$, and on the other, $u_0(e(t)) \to -\infty$ as $t \to \infty$. It follows that there is no regular expansion (5.7) and that $u_\varepsilon(0) \to \infty$ as $\varepsilon \to 0$.

5.1.2   Boundary Layers and Matched Asymptotics

The fact that $u_\varepsilon(0) \to \infty$ as $\varepsilon \to 0$ but $u_\varepsilon(e_D) = u_\varepsilon(e_E) = 0$ indicates that the solution $u_\varepsilon(e)$ develops a singularity everywhere in the domain. A further insight into the structure of the solution is gained by scaling

$$v_\varepsilon(e) = \frac{u_\varepsilon(e)}{u_\varepsilon(0)}, \tag{5.11}$$

so that $v_\varepsilon(0) = 1$ and

$$\varepsilon v_\varepsilon''(e) - U'(e)v_\varepsilon'(e) = \frac{1}{u_\varepsilon(0)} \sim 0 \quad \text{for} \quad \varepsilon \ll 1. \tag{5.12}$$

Expanding

$$v_\varepsilon(e) = v_0(e) + \varepsilon v_1(e) + \varepsilon^2 v_2(e) + \cdots, \tag{5.13}$$

we find that

$$U_0'(e)v_0'(e) = 0, \tag{5.14}$$

so that $v_\varepsilon(0) = 1$ implies that $v_\varepsilon(e) = 1$, which fails to satisfy the boundary conditions $v_\varepsilon(e_D) = v_\varepsilon(e_E) = 0$. We have to conclude that the regular expansion, which assumes that $\varepsilon v_0''(e) \ll U'(e)v_0'(e)$ and can hold away from the boundary, cannot be valid near the boundary. The series (5.13) is called the *outer expansion* or the *outer solution*.

The local behavior of $v_\varepsilon(e)$ near the boundary can be determined by stretching the neighborhood of the boundary e_D with the *boundary layer variable*

$$z = \frac{e - e_D}{\sqrt{\varepsilon}}, \tag{5.15}$$

which maps the domain D and its boundary into the closed interval $[0, (e_E - e_D) /\sqrt{\varepsilon}] \sim \mathbb{R}^+$. Setting $V_\varepsilon(z) = v_\varepsilon(e)$ and expanding everything in sight in powers of $\sqrt{\varepsilon}$, we find that the leading term in the expansion, $V_0(z)$, satisfies the boundary value problem

$$V_D''(z) - U_0''(e_D)zV_D'(z) = 0 \text{ for } z > 0, \tag{5.16}$$

$$V_D(0) = 0, \tag{5.17}$$

$$V_D(z) \to 1 \text{ as } z \to \infty. \tag{5.18}$$

The function $V_D(z)$ is called the *inner solution* or *boundary layer solution*. Equation (5.16) is called a *boundary layer equation*, (5.17) is the original boundary condition, and (5.18) is the *matching condition* that connects the inner solution to the outer solution $v_0(e) = v_\varepsilon(0) = 1$ at the leading-order. Boundary layer equations and matching conditions can be written at all orders.

Writing $-U''(e_D) = \omega_D^2 > 0$, we obtain the boundary layer solution in the form

$$V_D(z) = \text{erf}\left(\frac{z\omega_D}{\sqrt{2\varepsilon}}\right),$$

that is, near e_D, we have

$$v_D(e) = \text{erf}\left(\frac{(e - e_D)\omega_D}{\sqrt{2\varepsilon}}\right). \tag{5.19}$$

Similar analysis near e_E gives the inner solution

$$v_E(e) = \text{erf}\left(\frac{(e_E - e)\omega_E}{\sqrt{2\varepsilon}}\right). \tag{5.20}$$

Because both boundary layer functions $v_D(e)$ and $v_E(e)$ match to the outer solution $v_0(e) = 1$, in the sense that

$$\lim_{\varepsilon \to 0} v_D(e) = \lim_{\varepsilon \to 0} v_E(e) = v_0(e) = 1 \tag{5.21}$$

for every $e_D < e < e_E$, a uniform approximation, valid in D and on its boundary, can be obtained by setting

$$v_{\mathrm{unif}}(e) = v_D(e)v_E(e) \tag{5.22}$$

or

$$v_{\mathrm{unif}}(e) = v_D(e) + v_E(e) - v_0(e). \tag{5.23}$$

The method of constructing the uniform approximations (5.22) and (5.23) is called the *method of matched asymptotics* [87], [12]. It consists in constructing first the outer solution of the singular perturbation problem in the form of a regular expansion. The outer solution, which in general does not satisfy all boundary conditions, can be an approximation to the solution only sufficiently far from the boundary. To correct the outer solution for the missing boundary conditions, a boundary layer solution is constructed by introducing a stretched variable near the boundary, which rebalances the equation locally. The solution of the boundary layer equation, also called the inner solution, is required to satisfy the given boundary condition and to match the outer solution in the matching region, that is, where the original variable and the stretched variable are equal for all sufficiently small values of the singular perturbation parameter ε.

In the case at hand, the matched uniform asymptotic expansion (5.23) fails to determine all constants. Specifically, the value of $u_\varepsilon(0)$ in the scaling (5.11) is still unknown, so an additional criterion is needed for its determination. We find the additional criterion by rewriting (5.5) as

$$\left[\exp\left\{ -\frac{U(e)}{\varepsilon} \right\} u_\varepsilon'(e) \right]' = -\frac{1}{\varepsilon} \exp\left\{ -\frac{U(e)}{\varepsilon} \right\} \tag{5.24}$$

and integrating over the interval D. We get

$$\exp\left\{ -\frac{U(e_E)}{\varepsilon} \right\} u_\varepsilon'(e_E) - \exp\left\{ -\frac{U(e_D)}{\varepsilon} \right\} u_\varepsilon'(e_D)$$

$$= -\frac{1}{\varepsilon} \int_{e_D}^{e_E} \exp\left\{ -\frac{U(e)}{\varepsilon} \right\} de.$$

Using (5.19), (5.20), and (5.22) or (5.23), we find that

$$v_{\mathrm{unif}}'(e_D) = \frac{\omega_D}{\sqrt{\pi \varepsilon/2}}, \quad v_{\mathrm{unif}}'(e_E) = -\frac{\omega_E}{\sqrt{\pi \varepsilon/2}},$$

and the Laplace expansion of integrals gives

$$\int_{e_D}^{e_E} \exp\left\{ -\frac{U(e)}{\varepsilon} \right\} de = \sqrt{\frac{2\pi \varepsilon}{\omega_A^2}} \exp\left\{ -\frac{U_0(0)}{\varepsilon} \right\} \left(1 + O(\sqrt{\varepsilon}) \right),$$

where $\omega_A^2 = U''(0)$. It follows that

$$u_\varepsilon(0) \approx \frac{\dfrac{\pi}{\omega_A}\exp\left\{-\dfrac{U_0(0)}{\varepsilon}\right\}}{\omega_D\exp\left\{-\dfrac{U_0(e_D)}{\varepsilon}\right\} + \omega_E\exp\left\{-\dfrac{U_0(e_E)}{\varepsilon}\right\}}. \tag{5.25}$$

For the symmetric potential $U_0(e) = -\cos e$ and $-e_D = e_E = \pi$, equation (5.25) gives

$$u_\varepsilon(0) \approx \frac{\pi}{2\omega_A\omega_E}\exp\left\{\frac{U_0(e_E) - U_0(0)}{\varepsilon}\right\} = \frac{\pi e^{2/\varepsilon}}{2}. \tag{5.26}$$

For an asymmetric potential, if $U(e_E) < U(e_D)$, equation (5.25) gives

$$u_\varepsilon(0) \approx \frac{\pi}{\omega_A\omega_E}\exp\left\{\frac{U_0(e_E) - U_0(0)}{\varepsilon}\right\}. \tag{5.27}$$

Equation (5.27) is *Kramers' formula* [94], [137, Exercise 6.4 and Section 10.3.2]. Because $u_\varepsilon(0)$ is the MFPT to the boundary, the MTLL for the potential at hand is

$$\mathrm{MTLL} = 2u_\varepsilon(0) \approx \pi\exp\left\{\frac{2}{\varepsilon}\right\}. \tag{5.28}$$

Exercise 5.2 (The next term). Find $v_1(e)$ in the expansion (5.13) and the next-order correction to the leading-order approximation $u_\varepsilon(0)$ in (5.25).   □

Exercise 5.3 (Asymptotically optimal smoothers of PM signals*).

 (i) Find asymptotically optimal fixed-interval and fixed-delay smoothers for PM transmission.
 (ii) Is the MTLL of a smoother of PM transmission in a low-noise channel longer than that of a filter?   □

   Figure 5.2 shows the exact normalized solution $u_\varepsilon(e)/e(0)$ given in (5.6) (solid line), the uniform approximation (5.22) (dashed line), and the approximation (5.23) (dotted line) for $\varepsilon = 0.1$.

5.2   Loss of Lock in a Second-Order PLL

It was shown in Section 4.1.2 (see Exercise 4.25) that the scaled PLL equations for FM transmission of Brownian motion ($m = 0$ in (4.76), $\tilde{x}_1$ is the scaled u and $\tilde{x}_2$ is the scaled x), when both $h^1(u(t), t)$ and $h^2(u(t), t)$ are measured, are given by the dimensionless equations

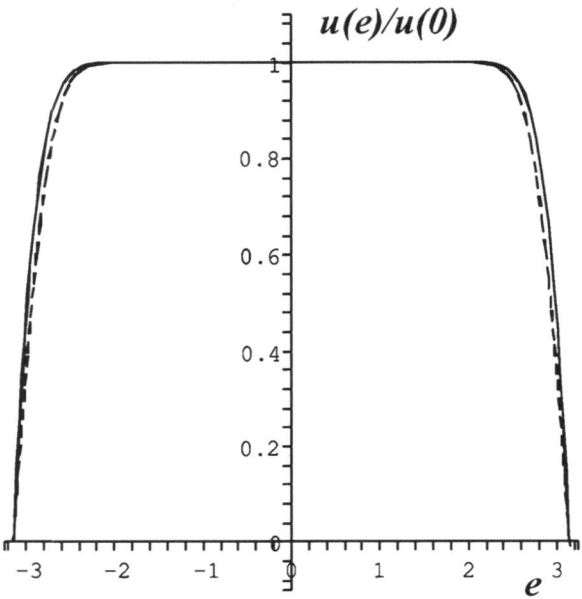

Fig. 5.2 The exact normalized solution (5.6) (solid line), the uniform approximation (5.22) (dashed line), and the approximation (5.23) (dotted line) for $\varepsilon = 0.1$.

$$d \begin{bmatrix} \tilde{x}_1 \\ \tilde{x}_2 \end{bmatrix} = \begin{bmatrix} 0 & 1 \\ 0 & 0 \end{bmatrix} \begin{bmatrix} \tilde{x}_1 \\ \tilde{x}_2 \end{bmatrix} dt + \sqrt{\varepsilon} \begin{bmatrix} 0 \\ 1 \end{bmatrix} dw,$$

$$d \begin{bmatrix} \tilde{y}_1 \\ \tilde{y}_2 \end{bmatrix} = \begin{bmatrix} \sin \tilde{x}_1 \\ \cos \tilde{x}_1 \end{bmatrix} dt + \sqrt{\varepsilon} \, d \begin{bmatrix} v_1 \\ v_2 \end{bmatrix}, \tag{5.29}$$

where w, v_1, and v_2 are standard Brownian motions. The phase and frequency estimation errors, e_1 and e_2, respectively, have the dynamics

$$d \begin{bmatrix} e_1 \\ e_2 \end{bmatrix} = \begin{bmatrix} e_2 - \sin e_1 \\ -\sin e_1 \end{bmatrix} dt + \sqrt{\varepsilon} \begin{bmatrix} 1 & 0 \\ 1 & -1 \end{bmatrix} \begin{bmatrix} dv \\ dw \end{bmatrix}, \tag{5.30}$$

where $v(t)$ is a standard Brownian motion independent of $w(t)$. To examine the loss of lock in the second-order PLL (5.29), we consider the case of small noise, $\varepsilon \ll 1$. We proceed as in Section 5.1. First, we examine the noiseless error dynamics (5.30). Linearizing the noiseless system (5.30) near its critical points $e_1 = 0, e_2 = n\pi$, where $n = 0, \pm 1, \pm 2, \ldots$, we find that the critical points corresponding to even n are attractors, while the ones corresponding to odd n are saddle points. Thus the (e_1, e_2) plane is partitioned into domains of attraction of the stable equilibria at $(0, 2n\pi)$, which are separated by the trajectories that converge to the saddle

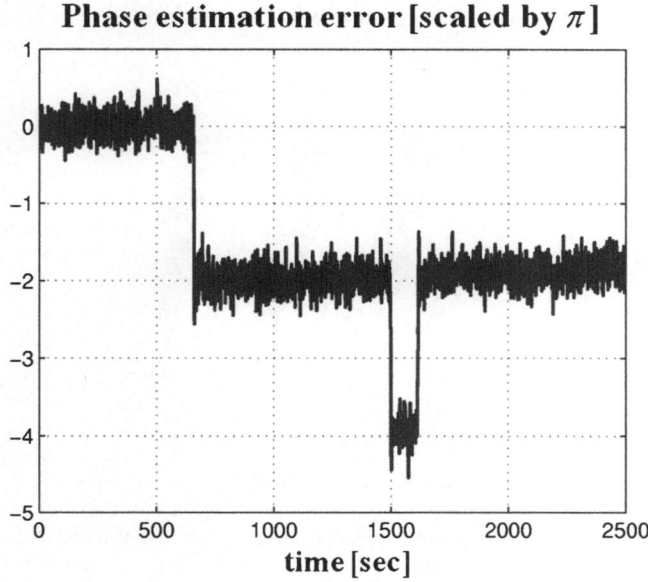

Fig. 5.3 A typical trajectory of the phase estimation error.

points (the bounding trajectories in Figure 5.4). This partition of the phase plane is analogous to the partition of the e-axis in Figure 5.1 into domains of attraction of the stable equilibria of the potential $U(e)$.

Simulated noisy error trajectories of (5.30) of the phase tracker (5.29) are shown in in Figure 5.4. When a noisy error trajectory crosses a bounding separatrix it continues into another domain of attraction, so a typical phase estimation error trajectory looks like that in Figure 5.3. The frequency estimation error, which looks like the derivative of the phase error, has sharp peaks, called *FM clicks*, which are distinctly audible in FM radio receivers. Figure 5.4 shows also trajectories that wander across many separatrices, forming bunches of phase slips, and last longer than a single phase slip.

Losses of lock are rare events if the noises are weak. As the noise increases, the frequency error spends longer and longer periods of time wandering in the tails of the separatrices, far from the locked state $e_1 = 0$, and the performance of the tracker deteriorates. This happens when the SNR falls below a certain threshold (the dimensionless noise intensity ε crosses a certain threshold), beyond which the PLL becomes useless [152], [144], [18], [136], [147].

Exercise 5.4 (Loss of lock in second-order PLL).

 (i) Derive (5.29) and (5.30).
(ii) Plot the domains of attraction of (5.30), shown in Figure 5.4.
(iii) Run simulations of (5.29) and (5.30) (see Figures 5.3 and 5.4) and compare the error trajectories created by each system.   □

Error trajectories in phase plane

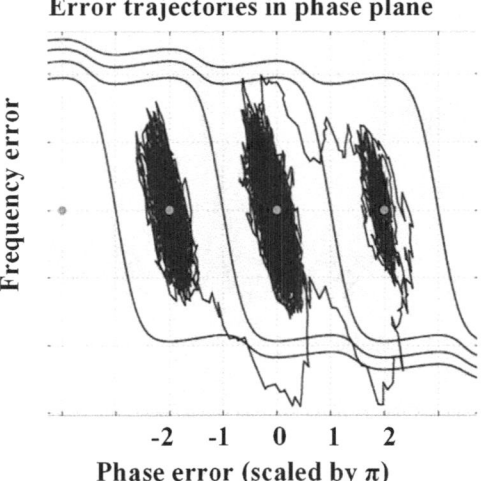

Phase error (scaled by π)

Fig. 5.4 The horizonal axis is phase estimation error e_1 and the vertical axis is frequency estimation error e_2. The dots are the local attractors at $e_1 = 0, \pm 2\pi, \ldots, e_2 = 0$. The bounding curves are the separatrices that converge to the saddle points $e_1 = \pm\pi, \pm 3\pi, \ldots, e_2 = 0$. Typical noisy error trajectories in the phase plane show escapes from the domains of attraction.

Exercise 5.5 (The threshold in the second-order PLL).

(i) Use the simulations of Exercise 5.4 to calculate the MTLL. Show that up to the pre-exponential factor, the MTLL is $\tau \propto \exp\{0.78525/\varepsilon\}$ for frequency estimation and $\tau \propto \exp\{2/\varepsilon\}$ for phase estimation of a Brownian motion in the model of Exercise 4.23.

(ii) Plot the MTLL vs SNR $= 1/\varepsilon$ to examine the threshold in the two PLLs.   □

5.2.1   The Phase Plane of the Noiseless Error Equations

To study the MTLL for $\varepsilon \ll 1$, we study first the noiseless dynamics of the error $x = (e_1, e_2)^T$,

$$\frac{d}{dt}\begin{bmatrix} e_1 \\ e_2 \end{bmatrix} = \begin{bmatrix} e_2 - \sin e_1 \\ -\sin e_1 \end{bmatrix}, \tag{5.31}$$

which we write as

$$\dot{x} = a(x) = \begin{bmatrix} x_2 - \sin x_1 \\ -\sin x_1 \end{bmatrix}. \tag{5.32}$$

The system has a stable critical point at the origin, and its domain of attraction is denoted by D. The matrix A of the linearized system about the origin,

$$\dot{x} = Ax, \qquad (5.33)$$

is given by

$$A = \left\{ \frac{\partial a^i(0)}{\partial x^j} \right\}_{i,j=1}^{2} = \begin{bmatrix} -1 & -1 \\ 1 & 0 \end{bmatrix}, \qquad (5.34)$$

and its eigenvalues, $\lambda_{\pm} = \frac{1}{2}(-1 \pm i\sqrt{3})$, have negative real parts. Thus the trajectories of the noiseless system (5.32) that start in D are attracted to the origin.

To identify the boundary ∂D, we note that the noiseless dynamics (5.32) has saddle points at $x_1 = \pm\pi$, $x_2 = 0$, and the matrix A_1 of the linearized system there,

$$\dot{\xi} = A_1 \xi, \qquad (5.35)$$

where $x_1 = 2(n+1)\pi + \xi_1$, $x_2 = \xi_2$, is given for $n = 0, \pm1, \pm2 \ldots$ by

$$A_1 = \left\{ \frac{\partial a^i(\pm\pi, 0)}{\partial x^j} \right\}_{i,j=1}^{2} = \begin{bmatrix} 1 & 1 \\ 1 & 0 \end{bmatrix}. \qquad (5.36)$$

Its eigenvalues are $\lambda_{+,-} = \frac{1}{2}(-1 \pm \sqrt{5})$, so the all trajectories of (5.32) are repelled from the saddle points, except the stable trajectories that enter the saddle point $(0, \pm\pi)$ in the direction of the eigenvector

$$\tau_1 = \frac{-1}{\sqrt{1+\lambda_-^2}} \begin{bmatrix} \lambda_- \\ 1 \end{bmatrix}. \qquad (5.37)$$

Therefore, on the segment of the separatrix near the saddle point, where the solution of linearized system (5.35) is asymptotically close to the solution of (5.32),

$$\begin{bmatrix} x_1(t) \\ x_2(t) \end{bmatrix} = \begin{bmatrix} 2(n+1)\pi \\ 0 \end{bmatrix} - \delta\tau_1 e^{\lambda_-(t-t_1)} \text{ for } t > t_1, \qquad (5.38)$$

where $[2(n+1)\pi, 0]^T - \delta\tau_1$ is assumed to be a point on the separatrix, and the arc length from the saddle point $[2(n+1)\pi, 0]^T$ to $[x_1(t), x_2(t)]^T$ is $s = \delta e^{\lambda_-(t-t_1)}$.

To construct ∂D numerically, we can integrate the ODE

$$\frac{dx_2}{dx_1} = \frac{-\sin x_1}{x_2 - \sin x_1} \qquad (5.39)$$

with the initial point $x_2(x_1) = \lambda_-(x_1 \mp \pi)$ for sufficiently small $|x_2|$. The integration produces the separatrices shown in Figure 5.4.

The local behavior of the noiseless error dynamics (5.32) near the separatrices (see Figure 5.4) is determined by the drift vector $a(x)$, which is tangent to the boundary, so its normal component vanishes there. We expand $a(x)$ near the separatrix in a Taylor series in powers of the distance to the boundary. At each point $x \in D$ near the boundary we denote its orthogonal projection on the boundary by x' and the unit outer normal and unit tangent there by $n(x')$ and $\tau(x')$, respectively. We choose $\tau(0) = \tau_1$. We define the signed distance to the boundary

$$\rho(x) = -|x - x'| \quad \text{for} \quad x \in D, \quad \rho(x) = |x - x'| \quad \text{for} \quad x \notin D. \tag{5.40}$$

The boundary corresponds to $\rho(x) = 0$, and the unit outer normal at x' is $n(x') = \nabla\rho|_{\rho=0}$. The unit outer normal at ∂D can also be expressed in terms of the drift, which is tangential to the boundary,

$$n(x') = \frac{-1}{|a(x')|} \begin{bmatrix} \sin x_1' \\ x_2' - \sin x_1' \end{bmatrix}. \tag{5.41}$$

The signed arc length $s(x)$ is measured from the saddle point to x' on the separatrix through the saddle point $(0, \pi)$. Choosing the eigenvector τ_1 as the positive direction on the separatrix, the tangent $\tau(s)$ is defined as a continuous function of s for $-\infty < s < \infty$. The transformation $x \to (\rho, s)$, where $\rho = \rho(x)$, $s = s(x)$, maps a finite strip near a connected component of the boundary onto the strip $|\rho| < \rho_0$, $-S < s < S$ for some $S, \rho_0 > 0$. The transformation is given by $x = x' + \rho n(x)$, where the projection x' is a function of s. We write $(n(x), \tau(x)) = (n(s), \tau(s))$.

Because $a(x') \cdot n(x') = 0$, a Taylor expansion of the normal component of the drift in powers of ρ in the strip $|\rho| < \rho_0$ is

$$a(x) \cdot n(x') = \sum_{i,j=1}^{2} \frac{\partial a_i(x')}{\partial x_j} n_i(x') n_j(x') \rho + O(\rho^2)$$

$$= \frac{\sin x_1'(x_2'(1 - \cos x_1') - \sin x_1')}{\sin^2 x_1' + (x_2' - \sin x_1')^2} \rho + O(\rho^2).$$

Setting

$$\frac{\sin x_1'(x_2'(1 - \cos x_1') - \sin x_1')}{\sin^2 x_1' + (x_2' - \sin x_1')^2} = a^0(s), \tag{5.42}$$

we find that

$$a^0(0) = \frac{-\lambda_-(2 + \lambda_-)}{\lambda_-^2 + (1 + \lambda_-)^2} > 0,$$

which implies that $a^0(s) > 0$ for all s, because the function is continuous and does not vanish. Therefore

$$\dot{\rho} = \nabla\rho \cdot \dot{x} = a^0(s)\rho + O(\rho^2)$$

implies that $\rho(t) \approx \rho(0)e^{a^0(s)t}$, which decreases for every $\rho(0) < 0$ as t increases. This means that the trajectories of the noiseless dynamics inside D are repelled from the boundary.

The tangential component of the drift is the speed of motion on ∂D toward the saddle point, that is,

$$B(s) = a(x') \cdot \tau(x')$$

$$= -\operatorname{sgn}(x_2')|a(x')| = -\operatorname{sgn}(x_2')\sqrt{(x_2' - \sin x_1')^2 + \sin^2 x_1'}. \tag{5.43}$$

Near the saddle point $(0, \pi)$ the speed is given by

$$B(s) \approx -x_2'\sqrt{(1 + \lambda_-)^2 + \lambda_-^2}, \tag{5.44}$$

so it changes sign at the saddle point; it is thus a *stable critical point* of the noiseless error dynamics (5.32) on the boundary ∂D. The local structure of the drift near ∂D is therefore

$$a(x) = \{\rho a^0(s)n(s) + B(s)\tau(s)\}\{1 + o(1)\}, \tag{5.45}$$

which in local coordinates is

$$a(\rho, s) = a^0(s)\rho\nabla\rho + B(s)\nabla s + o(\rho). \tag{5.46}$$

5.3 The MFPT in Planar Systems

The error dynamics (5.30) exhibit the general properties of higher-dimensional loss of lock problems. We have

$$b(x) = \begin{bmatrix} 1 & 0 \\ 1 & -1 \end{bmatrix}, \quad \sigma(x) = \frac{1}{2}\begin{bmatrix} 1 & 1 \\ 1 & 2 \end{bmatrix}, \tag{5.47}$$

so (5.30) has the autonomous form

$$dx = a(x)\, dt + \sqrt{\varepsilon}\, b(x)\, dw(t), \quad x(0) = x, \tag{5.48}$$

where the noiseless dynamics

$$dx = a(x)\,dt, \quad x(0) = x \tag{5.49}$$

has a stable attractor at the origin (see Figure 5.4). The MFPT $u_\varepsilon(x) = \mathbb{E}[\tau_D \mid x(0) = x]$ from a point $x \in D$ (the domain of attraction of the origin) to the boundary ∂D is the solution of the boundary value problem (1.145), (1.146) for the Andronov–Vitt–Pontryagin equation

$$\mathcal{L}_\varepsilon^* u_\varepsilon(x) = -1 \ \text{ for } x \in D, \quad u_\varepsilon(x) = 0 \ \text{ for } x \in \partial D, \tag{5.50}$$

where the backward Kolmogorov operator for (5.48) is given by

$$\mathcal{L}_\varepsilon^* u_\varepsilon(x) = \sum_{i,j=1}^{2} \varepsilon \sigma^{i,j}(x) \frac{\partial^2 u_\varepsilon(x)}{\partial x^i \partial x^j} + \sum_{i=1}^{2} a^i(x) \frac{\partial u_\varepsilon(x)}{\partial x^i}. \tag{5.51}$$

The similarity of the noise-induced escape of the two-dimensional error $(e_1(t), e_2(t))$ from the domain of attraction D of the stable equilibrium point at the origin (the dot in Figure 5.4) to that of the one-dimensional problem discussed in Sections 5.1.1 and 5.1.2 is apparent. In both cases the noiseless dynamics is stable, so the MFPT to the boundary becomes infinite in the limit $\varepsilon \to 0$. As in Section 5.1.2, the outer solution fails to satisfy the boundary condition, which in the planar case is given not merely at two points, but rather on an entire curve (the separatrix in Figure 5.4). The matched asymptotics method described in Section 5.1.2 has to be extended to a much more complicated geometry of the two-dimensional case. The scaling (5.11) and the outer expansion (5.13) are generalized in a straightforward manner. However, the boundary layer analysis requires geometric considerations. First, the stretched boundary layer variable (5.15) has to be chosen in a manner that reflects the singularity of the solution. Specifically, due to the homogeneous boundary condition, the solution does not change along the boundary, so no boundary layer should be expected in the direction tangent to ∂D. Therefore the boundary layer variable should be the stretched distance to the boundary, in the direction of the normal. The boundary layer function should satisfy a boundary layer equation with boundary and matching conditions similar to (5.16)–(5.18). An important difference in the evaluation of the MFPT in the two-dimensional case is that the boundary value problem (5.50) cannot be written in the form (5.24), unless it is self-adjoint in the sense that there exists a function $U(x_1, x_2)$ such that

$$a(x) = -\sigma \nabla U(x)$$

for some function $U(x)$, which in the original variables takes the form

$$\begin{bmatrix} e_2 - \sin e_1 \\ -\sin e_1 \end{bmatrix} = -\frac{1}{2} \begin{bmatrix} 1 & 1 \\ 1 & 2 \end{bmatrix} \begin{bmatrix} \partial U(e_1, e_2)/\partial e_1 \\ \partial U(e_1, e_2)/\partial e_2 \end{bmatrix}. \tag{5.52}$$

This is not in the case at hand (see Exercise 5.6 below). Therefore a different criterion for the determination of the missing constant $u_\varepsilon(0, 0)$ in the matched asymptotic expansion has to be found.

Exercise 5.6 (The boundary value problem (5.50) is not self-adjoint). Why is (5.52) impossible? [136], [137, Exercise 10.16]. $\qquad\square$

The calculation of the MFPT is given in the following theorem.

Theorem 5.3.1 (The asymptotics of the MFPT). *The asymptotic approximation to the MFPT for small ε is given by*

$$\bar{\tau}(x) = K(\varepsilon)\exp\left\{\frac{\hat{\Psi}}{\varepsilon}\right\}(1 + o(1)), \tag{5.53}$$

where $K(\varepsilon)$ has an asymptotic series expansion in powers of ε, and $\hat{\Psi}$ is the minimum on the boundary ∂D of the domain of attraction D of the stable equilibrium point x_0 of the nonzero solution of the eikonal equation $\Psi(x)$,

$$\sum_{i,j=1}^{d} \sigma^{i,j}(x)\frac{\partial\Psi(x)}{\partial x^i}\frac{\partial\Psi(x)}{\partial x^j} + \sum_{i=1}^{d} a^i(x)\frac{\partial\Psi(x)}{\partial x^i} = 0,$$

$$\Psi(x_0) = 0. \tag{5.54}$$

$\qquad\square$

Note that (5.54) defines $\Psi(x)$ up to an additive constant, so if the condition at x_0 is changed to any other value, then $\hat{\Psi}$ is redefined as $\hat{\Psi} = \min_{x \in \partial D} \Psi(x) - \Psi(x_0)$, so that (5.53) remains unchanged. The proof of Theorem 5.3.1 is divided into several steps.

5.3.1 The Boundary Layer Structure of $u_\varepsilon(x)$

First, we note that $u_\varepsilon(x) \to \infty$ as $\varepsilon \to 0$, because due to the stability of the attractor at the origin, all trajectories of the noiseless dynamics (5.49) never leave D. Setting

$$C_\varepsilon = \sup_{x \in D} u_\varepsilon(x), \quad U_\varepsilon(x) = \frac{u_\varepsilon(x)}{C_\varepsilon}, \tag{5.55}$$

we obtain for all $x \in D$,

$$\sum_{i,j=1}^{2} \varepsilon\sigma^{i,j}(x)\frac{\partial^2 U_\varepsilon(x)}{\partial x^i \partial x^j} + \sum_{i=1}^{2} a^i(x)\frac{\partial U_\varepsilon(x)}{\partial x^i} = -\frac{1}{C_\varepsilon} = o(1) \text{ as } \varepsilon \to 0$$

$$U_\varepsilon(x) = 0 \text{ for } x \in \partial D. \tag{5.56}$$

The outer expansion of $U_\varepsilon(x)$,

$$U_\varepsilon(x) \sim U^0(x) + \varepsilon U^1(x) + \cdots,$$

gives

$$\sum_{i=1}^{2} a^i(x) \frac{\partial U^0(x)}{\partial x^i} = 0, \tag{5.57}$$

which can be written as

$$\frac{dU^0(x(t))}{dt} = 0 \tag{5.58}$$

along the trajectories of the noiseless system (5.49). This implies that $U^0(x(t))$ is constant on the noiseless trajectories, which all converge to the origin. Thus $U^0(x)$ is constant throughout D. The normalization (5.55) implies that $U^0(x) = 1$ for all $x \in D$. We note, however, that $U^0(x)$ fails to satisfy the boundary condition (5.56), and the higher-order corrections $U^i(x)$ cannot remedy this failure.

The reason for this failure is the expansion (5.57), which can be valid only under the assumption that the first term in (5.56) is smaller than the second one. Apparently, this assumption fails near the boundary, where both terms become of the same order of magnitude. To resolve the structure of the solution $U_\varepsilon(x)$ in this boundary layer zone, we change to local variables (ρ, s) (see Section 5.2.1, from (5.40)) and write

$$U_\varepsilon(x) = v_\varepsilon(s, \rho) \tag{5.59}$$

Now we introduce the stretched variable $\xi = \rho/\sqrt{\varepsilon}$ and the boundary layer function $v_\varepsilon(s, \rho) = V_\varepsilon(\xi, s)$. Using the local structure (5.46) and expanding all functions in powers of $\varepsilon^{1/2}$, we transform the boundary value problem (5.56) to

$$\sigma^0(s) \frac{\partial^2 V_\varepsilon(\xi, s)}{\partial \xi^2} + a^0(s)\xi \frac{\partial V_\varepsilon(\xi, s)}{\partial \xi} + B(s) \frac{\partial V_\varepsilon(\xi, s)}{\partial s} = 0, \tag{5.60}$$

to leading-order in $\sqrt{\varepsilon}$, with the boundary and matching conditions

$$V_\varepsilon(0, s) = 0, \quad \lim_{\xi \to -\infty} V_\varepsilon(\xi, s) = \lim_{\rho \to 0} v^0(\rho, s) = 1, \tag{5.61}$$

where

$$\sigma^0(s) = \sum_{i,j=1}^{2} \sigma^{i,j}(0, s)\rho_i \rho_j > 0, \quad v^0(\rho, s) = U^0(x) = 1. \tag{5.62}$$

The solution of the boundary value problem (5.60), (5.61) is given by

$$V_\varepsilon(\xi, s) = -\sqrt{\frac{2}{\pi}} \int_0^{\gamma(s)\xi} e^{-z^2/2} \, dz, \tag{5.63}$$

where $\gamma(s)$ is the solution of Bernoulli's equation

$$B(s)\gamma'(s) + a^0(s)\gamma(s) - \sigma^0(s)\gamma^3(s) = 0, \quad \gamma(0) = \sqrt{\frac{a^0(0)}{\sigma^0(0)}}. \tag{5.64}$$

The substitution $\beta(s) = \gamma^{-2}(s)$ converts (5.64) into the linear equation

$$\beta'(s) - \frac{2a^0(s)}{B(s)}\beta(s) = -\frac{2\sigma^0(s)}{B(s)}, \quad \beta(0) = \frac{\sigma^0(0)}{a^0(0)}. \tag{5.65}$$

Because $B(0) = 0$, we construct the solution of (5.65) in the form $\beta(s) = \beta(0) + \beta_1(s)$, where $\beta_1(s)$ satisfies the linear equation

$$\beta_1'(s) - \frac{2a^0(s)}{B(s)}\beta(s) = f(s), \quad \beta_1(0) = 0, \tag{5.66}$$

where

$$f(s) = 2\frac{a^0(0)\sigma^0(s) - a^0(s)\sigma^0(0)}{a^0(s)B(s)}. \tag{5.67}$$

Because both numerator and denominator in (5.67) vanish linearly as $s \to 0$, the limit $f(0)$ is finite. The solution (5.66) is given by

$$\beta_1(s) = \int_0^s f(s') \exp\left\{\int_{s'}^s \frac{2a^0(s'')}{B(s'')} \, ds''\right\} \, ds'. \tag{5.68}$$

All integrals in (5.68) are finite, because $a^0(s) > 0$ and $B(s) < 0$ for $s > 0$. It follows that $\gamma(s)$ in (5.63) is a positive function.

Exercise 5.7 (Integration of the Bernoulli equation). Integrate the Bernoulli equation (5.64) numerically for the case of a second-order PLL and plot the graph of $\gamma(s)$ along the boundary. $\qquad \square$

In view of (5.63), the uniform leading-order approximation to $U_\varepsilon(x)$ is

$$U_\varepsilon(x) = v_\varepsilon(s, \rho) \sim -\sqrt{\frac{2}{\pi}} \int_0^{\rho\gamma(s)/\sqrt{\varepsilon}} e^{-z^2/2} \, dz \tag{5.69}$$

(see (5.59)). Consequently, the uniform leading-order approximation to $u_\varepsilon(x)$ is $u_\varepsilon(x) = C_\varepsilon v_\varepsilon(\rho, s)$, and C_ε is a yet undetermined constant. To determine C_ε, we need to construct a normalized asymptotic approximation to the solution of the stationary Fokker–Planck equation

$$
\mathcal{L}_\varepsilon p_\varepsilon(x) = \sum_{i,j=1}^{2} \varepsilon \frac{\partial^2}{\partial x^i \partial x^j} \left[\sigma^{i,j}(x) p_\varepsilon(x) \right] - \sum_{i=1}^{2} \frac{\partial}{\partial x^i} \left[a^i(x) p_\varepsilon(x) \right]
$$

$$
= 0 \text{ for } x \in D. \tag{5.70}
$$

Note that no boundary conditions are imposed on $p_\varepsilon(x)$.

The following lemma is proved by applying Green's identity.

Lemma 5.3.1 (The Lagrange identity). *If $p_\varepsilon(x)$ is a solution of the Fokker–Planck equation (5.70) and $u_\varepsilon(x)$ is a sufficiently regular function in D that satisfies the boundary condition (5.50), then*

$$
\int_D p_\varepsilon(x) \mathcal{L}_\varepsilon^* u_\varepsilon(x) \, dx = \oint_{\partial D} p_\varepsilon(x) \varepsilon \sum_{i,j} \sigma^{ij}(x) \frac{\partial u_\varepsilon(x)}{\partial x^j} n^i(x) \, ds_x. \tag{5.71}
$$

To proceed with the proof of Theorem 5.3.1, we multiply both sides of (5.50) by the solution $p_\varepsilon(x)$, using the Lagrange identity (5.71), the boundary layer expansion (5.69), and (5.55). We obtain

$$
-\int_D p_\varepsilon(x) \, dx \sim -C_\varepsilon \sqrt{\frac{2\varepsilon}{\pi}} \oint_{\partial D} p_\varepsilon(x) \sum_{i,j} \sigma^{ij}(x) n^i(x) \frac{\partial \rho(x)}{\partial x^j} \gamma(s) \, ds; \tag{5.72}
$$

hence

$$
C_\varepsilon \sim \frac{\displaystyle\int_D p_\varepsilon(x) \, dx}{\displaystyle\sqrt{\frac{2\varepsilon}{\pi}} \oint_{\partial D} p_\varepsilon(x) \sum_{i,j} \sigma^{ij}(x) n^i(x) n^j(x) \gamma(s) \, ds}. \tag{5.73}
$$

In view of (5.55) and (5.69), it suffices to show that (5.73) implies (5.53).

5.3.2 Asymptotic Solution of the Stationary FPE

We construct the asymptotic solution to (5.70), as in Section 4.1, by seeking a solution in the WKB form

$$
p_\varepsilon(x) = K_\varepsilon(x) \exp\left\{ -\frac{\Psi(x)}{\varepsilon} \right\}, \tag{5.74}
$$

where $K_\varepsilon(x)$ has an asymptotic series expansion in powers of ε,

$$K_\varepsilon(x) = K_0(x) + \varepsilon K_1(x) + \cdots, \tag{5.75}$$

with $K_0(x)$, $K_1(x)$, ... regular functions in D and on its boundary and $\Psi(x)$ is a regular function. Substituting (5.74) in the FPE (5.70) and comparing like powers of ε, we find at the leading-order $O(\varepsilon^{-1})$ that the eikonal function $\Psi(x)$ has to satisfy the eikonal equation (5.54) and $K_\varepsilon(x)$ has to satisfy the transport equation

$$\varepsilon \sum_{i,j=1}^{2} \frac{\partial^2 \sigma^{i,j}(x) K_\varepsilon(x)}{\partial x^i \partial x^j} - \sum_{i=1}^{2} \left(2 \sum_{j=1}^{2} \sigma^{i,j}(x) \frac{\partial \Psi(x)}{\partial x^j} + a^i(x) \right) \frac{\partial K_\varepsilon(x)}{\partial x^i}$$

$$- \sum_{i=1}^{2} \left(\frac{\partial a^i(x)}{\partial x^i} + \sum_{j=1}^{2} \left(\sigma^{i,j}(x) \frac{\partial^2 \Psi(x)}{\partial x^i \partial x^j} + 2 \frac{\partial \sigma^{i,j}(x)}{\partial x^j} \frac{\partial \Psi(x)}{\partial x^j} \right) \right) K_\varepsilon(x) = 0. \tag{5.76}$$

The expansion (5.75) implies that the transport equation for $K_0(x)$ reduces to

$$\sum_{i=1}^{2} \left(2 \sum_{j=1}^{2} \sigma^{i,j}(x) \frac{\partial \Psi(x)}{\partial x^j} + a^i(x) \right) \frac{\partial K_0(x)}{\partial x^i}$$

$$= - \sum_{i=1}^{2} \left(\frac{a^i(x)}{\partial x^i} + \sum_{j=1}^{2} \left(\sigma^{i,j}(x) \frac{\partial^2 \Psi(x)}{\partial x^i \partial x^j} + 2 \frac{\partial \sigma^{i,j}(x)}{\partial x^j} \frac{\partial \Psi(x)}{\partial x^j} \right) \right) K_0(x). \tag{5.77}$$

5.3.3 The Eikonal Equation

The eikonal function can be constructed by solving the eikonal equation (5.54) by the method of characteristics [29], [143]. In this method a first-order partial differential equation of the form

$$F(x, \Psi, p) = 0, \tag{5.78}$$

with $p = \nabla \Psi(x)$, is converted into the system of ordinary differential equations

$$\frac{dx}{dt} = \nabla_p F,$$

$$\frac{dp}{dt} = - \left(\frac{\partial F}{\partial \Psi} p + \nabla_x F \right),$$

$$\frac{d\Psi}{dt} = p \cdot \nabla_p F. \tag{5.79}$$

The function $\Psi(x)$ is defined by the third equation at each point x of the trajectory of the first equation. There is a neighborhood of the initial conditions (see below) that is covered by these trajectories.

In the case at hand, the function $F(x, \Psi, p)$ in the eikonal equation (5.54) has the form

$$F(x, \Psi, p) = \sum_{i,j=1}^{2} \sigma^{i,j}(x) p^i p^j + \sum_{i=1}^{2} a^i(x) p^i$$

$$= \frac{1}{2} p_1^2 + p_1 p_2 + p_2^2 + (x_2 - \sin x_1) p_1 - \sin x_1 p_2,$$

so that the characteristic equations (5.79) are

$$\frac{dx}{dt} = 2\sigma(x) p + a(x) = \begin{bmatrix} 1 & 1 \\ 1 & 2 \end{bmatrix} \begin{bmatrix} p_1 \\ p_2 \end{bmatrix} + \begin{bmatrix} x_2 - \sin x_1 \\ -\sin x_1 \end{bmatrix}, \tag{5.80}$$

$$\frac{dp}{dt} = -\nabla_x p^T \sigma(x) p - \nabla_x a^T(x) p = \begin{bmatrix} (p_1 + p_2) \cos x_1 \\ -p_1 \end{bmatrix}, \tag{5.81}$$

$$\frac{d\Psi}{dt} = p^T \sigma(x) p = \frac{1}{2} p_1^2 + p_1 p_2 + p_2^2. \tag{5.82}$$

First, we observe that the trajectories of the autonomous system (5.80), (5.81), which begin near the attractor $x = p = 0$ in the (x, p) space, diverge. To see this, we linearize the system (5.80), (5.81) around this point and obtain

$$\frac{dx(t)}{dt} = 2\sigma(0) p(t) + A x(t),$$

$$\frac{dp(t)}{dt} = -A p(t),$$

where A is defined in (5.34). It follows that $p(t) = e^{-At} p(0)$, and hence

$$x(t) = e^{At} x_0 + 2 \int_0^t e^{A(t-u)} \sigma(0) e^{-Au} p(0) \, du.$$

For any $(x(0), p(0)) \neq (0,0)$ both $x(t)$ and $p(t)$ diverge as $s \to \infty$, because the eigenvalues of $-A$ have positive real parts.

To integrate the characteristic equations (5.80), (5.81), initial conditions can be imposed near the unstable critical point $(0,0)$ by constructing $\Psi(x)$ in the form of a power series. The truncation of the power series near the attractor provides an approximation to $\Psi(x)$ and to $p = \nabla \Psi(x)$, whose error can be made arbitrarily small. Expanding $\Psi(x)$, $a(x)$, and $\sigma(x)$ in Taylor series about the origin, we find from the eikonal equation (5.54) that $\nabla\Psi(0) = 0$, so that the power series expansion of $\Psi(x)$ begins as a quadratic form

$$\Psi(x) = \frac{1}{2}x^T Q x + o\left(|x|^2\right).$$ (5.83)

Substituting (5.83) into the eikonal equation (5.54) with the linearized drift $a(x) \approx A x$ near the origin, we find (use Maple or Mathematica) that

$$Q = \begin{pmatrix} 1.2 & -0.8 \\ -0.8 & 1.2 \end{pmatrix}, \quad \Psi(x) \approx 0.6x_1^2 - 0.8x_1x_2 + 0.6x_2^2.$$ (5.84)

The matrix Q is also the solution of the Riccati equation

$$2Q\sigma(0)Q + QA + A^T Q = 0.$$ (5.85)

Note that Q is the matrix of the second partial derivatives of $\Psi(x)$ at the critical point $x = 0$ (the so called *Hessian matrix*). Obviously, the first term in the power series expansion of $p = \nabla\Psi(x)$ is given by

$$p = Qx + O\left(|x|^2\right) \approx \begin{bmatrix} 1.2x_1 - 0.8x_2 \\ -0.8x_1 + 1.2x_2 \end{bmatrix}.$$ (5.86)

In deriving (5.85), use is made of the facts that Q and σ are symmetric matrices and that a quadratic form vanishes identically if and only if it is defined by an antisymmetric matrix. The solution of (5.85) is a positive definite matrix [136, Exercise 7.5.2], [57].

Exercise 5.8 (Square root of a positive definite symmetric matrix*). Show that a positive definite symmetric matrix has a positive definite symmetric square root. □

Exercise 5.9 (The Riccati equation). Reduce the Riccati equation (5.85) to

$$AY + Y^T A^T = -I$$ (5.87)

by the substitutions $X = Q\sqrt{\sigma}$, where X is the solution of

$$2XX^T + XA + A^T X^T = 0$$

and $X = -\frac{1}{2}Y^{-1}$. Show that the solution of (5.87) is a symmetric matrix given by

$$Y = \int_0^\infty e^{At} e^{A^T t}\, dt$$

and show that the integral converges. □

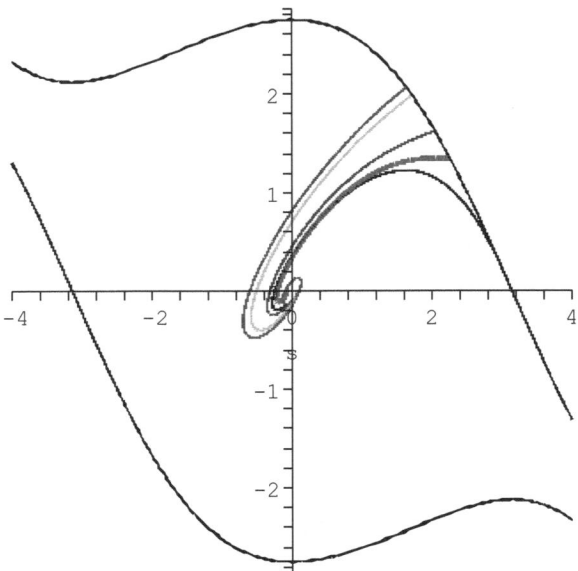

Fig. 5.5 The lock domain D and characteristics that hit the separatrix.

Choosing for the initial surface for the system (5.80)–(5.82) the contour

$$\frac{1}{2} x^T Q x = \delta, \tag{5.88}$$

for some small positive δ, and using the approximate initial values $\Psi(x) = \delta$ and (5.86) at each point of the surface, we can integrate the system (5.80)–(5.82) analytically or numerically. Once the domain D is covered with characteristics, the approximate value of $\Psi(x)$ can be determined at each point $x \in D$ as the value of the solution $\Psi(t)$ of (5.82) at s such that the solution of (5.80) satisfies

$$x(t) = x. \tag{5.89}$$

The initial condition on the surface (5.88) determines the unique trajectory of the system (5.80)–(5.82) that satisfies (5.89) for some s. It can be found numerically by the method of shooting. Figure 5.6 shows the lock domain D and characteristics that hits the separatrix. The lowest characteristic hits the saddle point $(\pi, 0)$. The initial conditions are given on the ellipse (5.88), $0.6x_1^2 - 0.8x_1x_2 + 0.6x_2^2 = 0.06$. The initial values are $x_1(0) = -0.08215, x_2(0) = -0.1344583556$, $p_1(0) = 0.0089866845$, $p_2(0) = -0.0956300267$. The characteristic above it hits at $(2.2500, 1.3384)$, the next ones at $(2.0000, 1.6239)$ and at $(1.7250, 1.9809)$, and the top one at $(1.6250, 2.0522)$. Figure 5.6 shows the values of $\Psi(\cdot)$ along the characteristics of Figure 5.5. The endpoints of the characteristic curves are on the separatrix, at arc lengths s and values $\Psi(0) = 0.78525$ (the

Fig. 5.6 The values of $\Psi(s)$ as a function of arc length along the characteristics of Figure 5.5. The endpoints are on the separatrix.

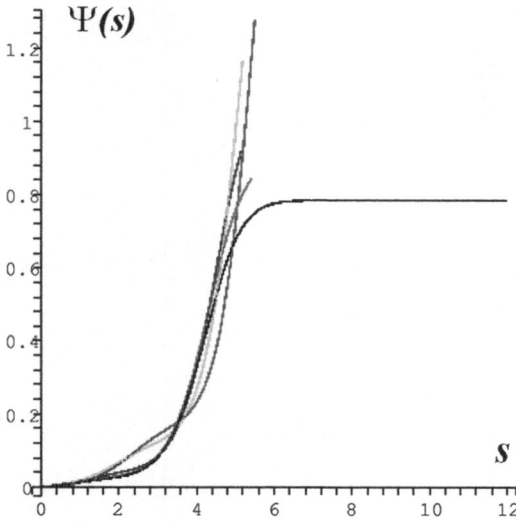

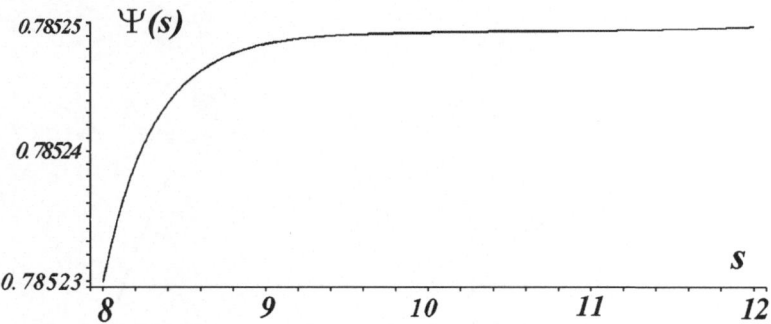

Fig. 5.7 Blowup of the graph of $\Psi(t)$ near the saddle point $(\pi, 0)$. The value is $\Psi(\pi, 0) \approx 0.78525$.

bottom characteristic), $\Psi(1.66) = 0.85$ (the one above it), $\Psi(2.052) = 0.9205$ (the next one), $\Psi(2.45) = 1.1611$ (the one above it), and $\Psi(2.6) = 1.2814$ (the top characteristic). Figure 5.8 shows the graph of $\Psi(s) \approx P_{12}(s)$ vs arc length s on the separatrix. The points of Figure 5.6 are marked with circles. Figure 5.8 shows an interpolation (with Maple) of the data points in Figure 5.6 by the 12th-order polynomial

$$P_{12}(s) = 0.78525 + 10^{-2} \times (0.33s - 0.36s^2 - 0.5s^3 - 0.3s^4 + 0.42s^6 + 0.6s^7$$

$$+ 0.37s^8 - 0.23s^9 - 0.6s^{10} + 0.41s^{11} - 0.07s^{12}). \tag{5.90}$$

Fig. 5.8 Graph of
$\Psi(s) \approx P_{12}(s)$ vs arc length
s on the separatrix. The
points of Figure 5.6 are
marked with circles.

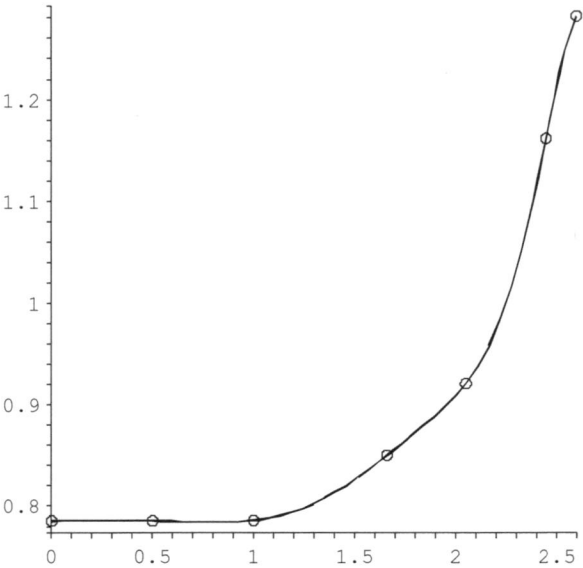

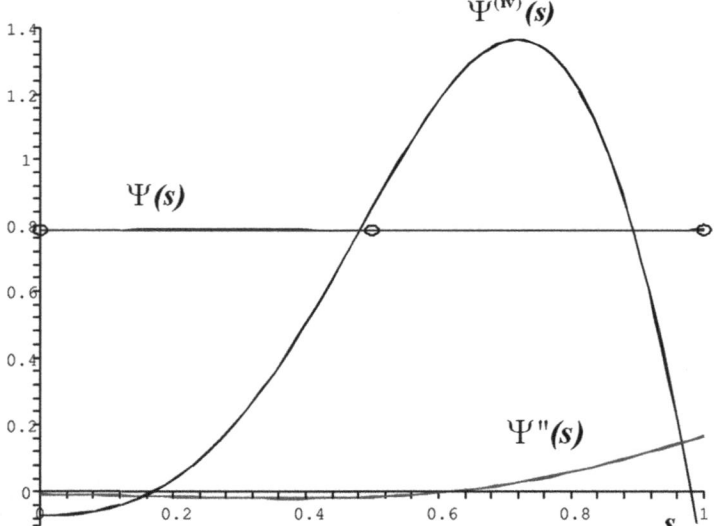

Fig. 5.9 The graphs of $\Psi(s)$ (flat line segment), $\Psi''(s)$ (flat curve near the axis), and $\Psi^{(iv)}(s)$ near the saddle point $s = 0$.

Fig. 5.10 The partial
derivatives $p_1(t) = \Psi_{x_1}(t)$
and $p_2(t) = \Psi_{x_2}(t)$ along the
characteristic in Figure 5.5.

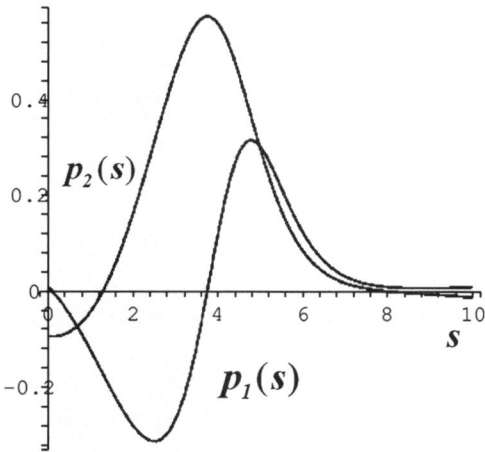

5.3.4   The Eikonal on the Separatrix

The eikonal equation in local coordinates (ρ, s) on ∂D can be written as

$$\sum_{i,j=1}^{2} \sigma^{i,j}(0,s) \frac{\partial \Psi(0,s)}{\partial x^i} \frac{\partial \Psi(0,s)}{\partial x^j} + B(t) \frac{\partial \Psi(0,s)}{\partial s} = 0. \qquad (5.91)$$

It follows that $\Psi(0,s)$ is minimal on ∂D at the saddle point $s = 0$. Changing the partial derivatives to local variables, we can write (5.91) as

$$\sigma^0(t)\Psi_\rho^2(0,s) + 2 \sum_{i,j=1}^{2} \sigma^{i,j}(0,s)\rho_i s_j \Psi_\rho(0,s)\Psi_s(0,s)$$

$$+ \sum_{i,j=1}^{2} \sigma^{i,j}(0,s)s_i s_j \Psi_s^2(0,s) + B(t)\frac{\partial \Psi(0,s)}{\partial s} = 0, \qquad (5.92)$$

where $\sigma^0(t)$ is given in (5.62). If $\Psi(0,s)$ is constant on a segment of the separatrix near the saddle point (see Figures 5.8 and 5.10, which shows the partial derivatives $p_1(t) = \Psi_{x_1}(t)$ and $p_2(t) = \Psi_{x_2}(t)$ along the characteristic in Figure 5.5), then the local expansion of $\Psi(\rho, s)$ about the separatrix in this segment is

$$\Psi(\rho,s) = \hat{\Psi} + \frac{\rho^2}{2}\Psi_{\rho\rho}(0,s) + O(\rho^3). \qquad (5.93)$$

Setting $\phi(t) = \Psi_{\rho\rho}(0,s)$, the eikonal equation (5.92) on the segment of the separatrix can be written as

$$\rho^2 \left[\sigma^0(s)\phi^2(s) + B(s)\phi'(s)\right] + O\left(\rho^3\right) = 0.$$

It follows that

$$\sigma^0(s)\phi^2(s) + B(s)\phi'(s) = 0,$$

so

$$\phi(s) = \left[\phi^{-1}(s_0) + \int_{s_0}^{s} \frac{\sigma^0(u)}{B(u)}\,du\right]^{-1}, \tag{5.94}$$

where s_0 is the arc length to a point on the separatrix. Using the approximate values on the segment

$$\sigma^0(s) = \sigma^0(0), \quad B(s) \approx -\beta\,s, \tag{5.95}$$

where β is a positive constant (see Exercise 5.10 below), we obtain

$$\phi(s) \sim \frac{1}{\phi^{-1}(s_0) + \dfrac{\sigma^0(0)}{\beta} \log \dfrac{s}{s_0}} \quad \text{for } 0 < s < s_0. \tag{5.96}$$

Finally, it follows from (5.93) and (5.96) that

$$\Psi(\rho, s) = \hat{\Psi} + \frac{1}{2} \frac{\rho^2}{\phi^{-1}(s_0) + \dfrac{\sigma^0(0)}{\beta} \log \dfrac{s}{s_0}} + O(\rho^3) \quad \text{for } 0 < s < s_0. \tag{5.97}$$

The value of $\phi(s_0)$ is negative near the saddle point, so that $\Psi_{\rho\rho}(0, s) < 0$ on the segment and $\Psi_{\rho\rho}(0, 0) = 0$.

Exercise 5.10 (The constants). Prove (5.95) with

$$\sigma^0(0) = \frac{\lambda_-^2 - \lambda_- + 2}{2(1 + \lambda_-^2)}, \quad \beta \approx \frac{-\lambda_- \sqrt{(1 + \lambda_-)^2 + \lambda_-^2}}{\sqrt{1 + \lambda_-^2}}. \tag{5.98}$$

$\square$

5.3.5   The Transport Equation

Recall that $K_\varepsilon(x)$ satisfies the transport equation (5.76). First, we note that $K_\varepsilon(x)$ cannot have an internal layer at the global attractor point $\mathbf{0}$ in D. This is due to the fact that stretching $x = \sqrt{\varepsilon}\,\xi$ and taking the limit $\varepsilon \to 0$ (5.76) converts the transport equation into

$$\sum_{i,j=1}^{2} \frac{\partial^2 \sigma^{i,j}(0) K_0(\xi)}{\partial \xi^i \partial \xi^j} - (2AQ + A)\xi \cdot \nabla_{\xi} K_0(\xi) - \text{tr}\,(A + \sigma(0)Q)\,K_0(\xi) = 0,$$

whose bounded solution is $K_0(\xi) = \text{const}$, because $\text{tr}\,(A + \sigma(0)Q) = 0$. The last equality follows from the Riccati equation (5.85) (left multiply by Q^{-1} and take the trace). Because the characteristics diverge, the initial value (at $s = 0$) on each characteristic is given at $x = 0$ as $K_0(0) = \text{const}$, which we can choose as const$= 1$.

Exercise 5.11 (The potential case). Show that if the diffusion matrix σ is constant and $a(x) = -\sigma \nabla \phi(x)$ for some function $\phi(x)$, then $\Psi(x) = \phi(x)$ and the WKB solution of the homogenous Fokker–Planck equation (5.70) is given by $p_\varepsilon(x) = e^{-\Psi(x)/\varepsilon}$, that is, the solution of the transport equation (5.76) is $K_0 = \text{const}$. $\square$

The transport equation has to be integrated numerically, together with the characteristic equations (5.80), (5.81). To evaluate the partial derivatives $\partial^2 \Psi(x)/\partial x^i \partial x^j$ along the characteristics, we use (5.83), (5.86), and set

$$\left. \frac{\partial^2 \Psi(x)}{\partial x^i \partial x^j} \right|_{x=0} = Q^{i,j}$$

on the initial ellipsoid (5.88). The differential equations for $\partial^2 \Psi(x)/\partial x^i \partial x^j$ along the characteristics are derived by differentiating the characteristic equations (5.80), (5.81) with respect to the initial values $x(0)$ on the initial ellipse. Writing

$$x_j(t) = \frac{\partial x(t)}{\partial x_0^j}, \quad p_j(t) = \frac{\partial p(t)}{\partial x_0^j}, \quad Q^{i,j}(t) = \frac{\partial^2 \Psi(x(t))}{\partial x^i \partial x^j}, \qquad (5.99)$$

we get the identity $p_j(t) = Q(t)x_j(t)$. Thus the matrix $P(t)$, whose columns are the vectors $p_j(t)$, and the matrix $X(t)$, whose columns are the vectors $x_j(t)$, are related by $P(t) = Q(t)X(t)$, or

$$Q(t) = P(t)X^{-1}(t). \qquad (5.100)$$

The initial conditions are

$$x_j^i(0) = \delta_{i,j}, \qquad (5.101)$$

$$p_j^i(0) = \left. \frac{\partial^2 \Psi(x)}{\partial x^i \partial x^j} \right|_{x=0} = Q^{i,j}(0) = Q^{i,j}, \qquad (5.102)$$

and the dynamics is

$$\frac{dx_j(t)}{dt} = \sum_{k=1}^{2} \left\{ 2\frac{\partial \sigma(x(t))}{\partial x^k} p(t) + 2\sigma(x(t))p_k(t) + \frac{\partial a(x(t))}{\partial x^k} \right\} x_k^j(t), \quad (5.103)$$

$$\frac{dp_j(t)}{dt} = -\sum_{k=1}^{2} \left[\nabla_x p^T(t) \frac{\partial \sigma(x(t))}{\partial x^k} p(t) + 2\nabla_x p_k^T(t)\sigma(x(t))p(t) \right.$$

$$\left. + \nabla_x a^T(x(t))p_k(t) + \frac{\partial}{\partial x^k} \nabla_x a^T(x(t))p(t) \right] x_k^j(t).$$

$$(5.104)$$

Exercise 5.12 (The system (5.103), (5.104) for the second-order PLL).

(i) Show that the system (5.103), (5.104) for the PLL model is

$$\frac{d}{dt}\frac{\partial x_1}{\partial x_1^0} = \frac{\partial p_1}{\partial x_1^0} + \frac{\partial p_2}{\partial x_1^0} + \frac{\partial x_2}{\partial x_1^0} - \cos x_1 \frac{\partial x_1}{\partial x_1^0}, \quad (5.105)$$

$$\frac{d}{dt}\frac{\partial x_1}{\partial x_2^0} = \frac{\partial p_1}{\partial x_2^0} + \frac{\partial p_2}{\partial x_2^0} + \frac{\partial x_2}{\partial x_2^0} - \cos x_1 \frac{\partial x_1}{\partial x_2^0},$$

$$\frac{d}{dt}\frac{\partial x_2}{\partial x_1^0} = \frac{\partial p_1}{\partial x_1^0} + 2\frac{\partial p_2}{\partial x_1^0} - \cos x_1 \frac{\partial x_1}{\partial x_1^0},$$

$$\frac{d}{dt}\frac{\partial x_2}{\partial x_2^0} = \frac{\partial p_1}{\partial x_2^0} + 2\frac{\partial p_2}{\partial x_2^0} - \cos x_1 \frac{\partial x_1}{\partial x_2^0},$$

$$\frac{d}{dt}\frac{\partial p_1}{\partial x_1^0} = \left(\frac{\partial p_1}{\partial x_1^0} + \frac{\partial p_2}{\partial x_1^0} \right) \cos x_1 - (p_1 + p_2) \sin x_1 \frac{\partial x_1}{\partial x_1^0}, \quad (5.106)$$

$$\frac{d}{dt}\frac{\partial p_1}{\partial x_2^0} = \left(\frac{\partial p_1}{\partial x_2^0} + \frac{\partial p_2}{\partial x_2^0} \right) \cos x_1 - (p_1 + p_2) \sin x_1 \frac{\partial x_1}{\partial x_2^0},$$

$$\frac{d}{dt}\frac{\partial p_2}{\partial x_1^0} = -\frac{\partial p_1}{\partial x_1^0}, \quad \frac{d}{dt}\frac{\partial p_2}{\partial x_2^0} = -\frac{\partial p_1}{\partial x_2^0}.$$

Then equations (5.99), $\partial x_i / \partial x_j^0 = x_{i,j}$, and $\partial p_i / \partial x_j^0 = p_{i,j}$ give

$$\dot{x}_{1,1} = p_{1,1} + p_{2,1} + x_{2,1} - \cos x_1 x_{1,1}, \quad (5.107)$$

$$\dot{x}_{1,2} = p_{1,2} + p_{2,2} + x_{2,2} - \cos x_1 x_{1,2},$$

$$\dot{x}_{2,1} = p_{1,1} + 2p_{2,1} - \cos x_1 x_{1,1},$$

$$\dot{x}_{2,2} = p_{1,1} + 2p_{2,2} - \cos x_1 x_{1,2},$$

$$\dot{p}_{1,1} = (p_{1,1} + p_{2,1}) \cos x_1 - (p_1 + p_2) \sin x_1 x_{1,1}, \quad (5.108)$$

$$\dot{p}_{1,2} = (p_{1,2} + p_{2,2}) \cos x_1 - (p_1 + p_2) \sin x_1 x_{1,2},$$

$$\dot{p}_{2,1} = -p_{1,1}, \quad \dot{p}_{2,2} = -p_{1,2}.$$

(ii) Show that the transport equation (5.77) can be written on the characteristics $x(t)$ as

$$\frac{dK_0(x(t))}{dt} = -K_0(x(t)) \tag{5.109}$$

$$\times \sum_{i=1}^{2} \left(\frac{a^i(x(t))}{\partial x^i} + \sum_{j=1}^{2} \left(\sigma^{i,j}(x(t)) Q^{i,j}(t) + 2 \frac{\partial \sigma^{i,j}(x(t))}{\partial x^j} p_j(t) \right) \right).$$

(iii) Show that as $t \to \infty$, the characteristic that hits the saddle point coalesces with the separatrix on a segment near the saddle point.

(iv) Show that because $\sum_{i=1}^{2} \partial a^i(x(t))/\partial x^i = -\cos x_1$, along this segment

$$\sum_{i=1}^{2} \frac{a^i(x(t))}{\partial x^i} = \to 1, \quad \frac{\partial \sigma^{i,j}(x(t))}{\partial x^j} p_j(t) y = 0, \quad \sum_{i=1}^{2} \sum_{j=1}^{2} \sigma^{i,j}(x(t)) Q^{i,j}(t) \to 0,$$

which implies that the transport equation near the saddle point can be written as

$$\frac{dK_0(x(t))}{dt} = (-1 + o(1)) K_0(x(t)) \text{ as } t \to \infty.$$

(v) Conclude that $K_0(x(t)) = K_0(x(t_1)) e^{-(t-t_1)(1+o(1))} \to 0$ as $t \to \infty$, where $x(t_1)$ is a point on the segment of the separatrix near the saddle point.

(vi) To express $K_0(x(t))$ on the segment of the separatrix in terms of arc length s from the saddle point, recall that $s = \delta e^{\lambda_-(t-t_1)}$ (see (5.38)); hence

$$K_0(s) = K_0(s_1) \left(\frac{s}{\delta} \right)^{-(1+o(1))/\lambda_-} \longrightarrow 0 \text{ as } s \to 0, \tag{5.110}$$

because $\lambda_- < 0$. Figure 5.5 shows that $\delta = 1$ can be assumed. $\square$

In summary, the numerical integration of the eikonal and the transport equations consists in integrating numerically the differential equations (5.80)–(5.82), (5.103)–(5.109) with initial values $x(0)$ that cover the ellipse (5.88), with $p(0)$ and $\Psi(x(0))$ given by $p(0) = Q(0)x(0)$ and $\Psi(x(0)) = \delta$, and the initial values (5.101), (5.102), and $K_0(x(0)) = 1$. The matrix $Q(t)$ has to be evaluated from (5.100) at each step of the integration.

Exercise 5.13 (The characteristics for the second-order PLL).

(i) Use the fact that $\sigma(x)$ in the case of the error dynamics (5.30) of second-order PLL is a constant matrix to simplify the characteristic equations (5.80), (5.81), (5.103), (5.104).

(ii) Write the transport equation (5.109) in the form

$$\frac{dK_0(\boldsymbol{x}(t))}{dt} = -\left[\nabla \cdot \boldsymbol{a}(\boldsymbol{x}(t)) + \text{tr}\left(\sigma\left(\boldsymbol{x}(t)\right)\boldsymbol{Q}(t)\right)\right] K_0(\boldsymbol{x}(t)). \qquad (5.111)$$

(iii) Integrate the characteristic equations (5.80)–(5.82) together with (5.103), (5.104) (that is, with (5.107), (5.108)) and calculate $\boldsymbol{Q}(t)$ from (5.100).

(iv) Integrate the transport equation (5.109)) and plot $K_0(t)$ on ∂D.   □

5.3.6 Proof of Theorem 5.3.1

To conclude the proof Theorem 5.3.1, we have to show that (5.73) implies (5.53). To do so, we use the WKB solution (5.74) in (5.73) and evaluate the integrals asymptotically for small ε by the Laplace method. The main contribution to the numerator comes from the minimum of $\Psi(\boldsymbol{x})$ in D at $\boldsymbol{x} = \boldsymbol{0}$. The value of the integral is given by

$$\int_D p_\varepsilon(\boldsymbol{x})\,d\boldsymbol{x} = \int_D K_\varepsilon(\boldsymbol{x})e^{-\Psi(\boldsymbol{x})/\varepsilon}\,d\boldsymbol{x} = \frac{2\pi\varepsilon K_0(\boldsymbol{0})e^{-\Psi(\boldsymbol{0})/\varepsilon}}{\mathcal{H}(\boldsymbol{0})}(1 + O(\varepsilon)), \tag{5.112}$$

where $\mathcal{H}(\boldsymbol{0})$ is the determinant of the Hessian matrix of $\Psi(\boldsymbol{0})$ and is equal to the determinant of $\boldsymbol{Q}$ (see (5.84)). For the second-order PLL, we have $K_0(\boldsymbol{0}) = 1$, $\Psi(\boldsymbol{0}) = 0$, and $\det\boldsymbol{Q} = 0.8$. It follows that the value of the integral is $2.5\pi\varepsilon + O\left(\varepsilon^2\right)$.

Using the notation (5.62), the initial value $\gamma(0)$ given in (5.64), and the WKB solution (5.74), the integral in the denominator of (5.73) is evaluated by the Laplace method on ∂D as

$$\sqrt{\frac{2\varepsilon}{\pi}} \oint_{\partial D} p_\varepsilon(\boldsymbol{x}(s)) \sum_{i,j} \sigma^{ij}(\boldsymbol{x}(s))n^i(s)n^j(s)\gamma(s)\,ds$$

$$= \sqrt{\frac{2\varepsilon}{\pi}}\sqrt{2\pi\varepsilon}K_0(0)e^{-\Psi(0)/\varepsilon}\sigma^0(0)\sqrt{\frac{a^0(0)}{\sigma^0(0)}}(1 + O(\sqrt{\varepsilon}))$$

$$= 2\varepsilon K_0(0)\sqrt{\frac{a^0(0)\sigma^0(0)}{\Psi''(0)}}e^{-\Psi(0)/\varepsilon}. \tag{5.113}$$

The approximation (5.113) is valid if $K_0(0) \neq 0$ and $\Psi''(0) > 0$.

If $\Psi''(0) > 0$, but $K_0(s)$ on the boundary vanishes at the saddle point $s = 0$ as $K_0(s) = K_0 s^{2k}$, then the value of the integral is

$$
\sqrt{\frac{2\varepsilon}{\pi}} \oint_{\partial D} p_\varepsilon(\mathbf{x}(s)) \sum_{i,j} \sigma^{ij}(\mathbf{x}(s)) n^i(s) n^j(s) \gamma(s) \, ds
$$

$$
= K_0 \left(\frac{\varepsilon}{\Psi''(0)} \right)^{k+1/2} 2^{k+1} \Gamma(k+1) \sqrt{a^0(0)\sigma^0(0)} \, e^{-\Psi(0)/\varepsilon}(1 + o(1)),
$$

$$(5.114)$$

where $\Gamma(\cdot)$ is Euler's gamma function. Figure 5.9 shows the graphs of $\Psi(s)$ (flat line segment), $\Psi''(s)$ (flat curve near the axis), and $\Psi^{(iv)}(s)$ near the saddle point $s = 0$ for the characteristics of Figure 5.5.

If $\Psi(s) = \Psi(0) = \hat{\Psi}$ on a finite interval $0 \le s \le s_0$ (see, e.g., Figures 5.8 and 5.9), then

$$
\sqrt{\frac{2\varepsilon}{\pi}} \oint_{\partial D} p_\varepsilon(\mathbf{x}(s)) \sum_{i,j} \sigma^{ij}(\mathbf{x}(s)) n^i(s) n^j(s) \gamma(s) \, ds
$$

$$
= \sqrt{\frac{2\varepsilon}{\pi}} e^{-\hat{\Psi}/\varepsilon} \int_0^{s_0} K_0(s)\sigma^0(s)\gamma(s)(1 + o(1)) \, ds \tag{5.115}
$$

(see (5.110)).

In each of these cases, (5.53) follows by using (5.112) and (5.113) (or (5.114), or (5.115)) in (5.73). $\qquad\square$

Exercise 5.14 (Flat $\Psi(x)$). Evaluate the integral (5.114) for the cases that $\Psi^{(k)}(0) = 0$ for $k = 1, 2, \ldots, 2m - 1$ and $\Psi^{(2m)}(0) > 0$. $\qquad\square$

5.3.7 Survival Probability and Exit Density

The loss of lock problem in trackers is equivalent to the classical problem of escape of a multidimensional diffusion process from the domain of attraction of an attractor [137, Chapter 10], [138], [106], [118], [134]. We consider first a general multidimensional system

$$
d\mathbf{x} = \mathbf{a}(\mathbf{x}, t) \, dt + \sqrt{\varepsilon} \mathbf{b}(\mathbf{x}, t) \, d\mathbf{w}, \tag{5.116}
$$

in a domain $D \subset \mathbb{R}^d$, whose trajectories are terminated when they hit the boundary ∂D for the first time. The transition probability density function $p_\varepsilon(\mathbf{x}, t \mid \mathbf{y}, s)$ of the process $x(t)$ satisfies the Fokker–Planck equation with respect to the forward variables (x, t),

$$\frac{\partial p_\varepsilon(x, t \mid y, s)}{\partial t} = \mathcal{L}_\varepsilon p_\varepsilon(x, t \mid y, s)$$

$$= \varepsilon \sum_{i,j} \frac{\partial^2 \sigma^{ij}(x) p_\varepsilon(x, t \mid y, s)}{\partial x^i \partial x^i} - \sum_i \frac{\partial a^i(x) p_\varepsilon(x, t \mid y, s)}{\partial x^i} \quad \text{for } x, y \in D$$

$$(5.117)$$

with the initial and absorbing boundary conditions

$$p_\varepsilon(x, t \mid y, s) \to \delta(x - y) \text{ as } t \downarrow s \tag{5.118}$$

$$p_\varepsilon(x, t \mid y, s) = 0 \text{ for } x \in \partial D, \ y \in D. \tag{5.119}$$

Definition 5.3.1 (The survival probability). *The survival probability $S(t \mid x, s)$ of trajectories of (5.116) in D at time t that started at time $s < t$ at a point $x \in D$ is the conditional probability that the first passage time τ to the boundary ∂D of the domain does not exceed t,*

$$S(t \mid x, s) = \Pr\{\tau > t \mid x, s\} = \int_D p_\varepsilon(y, t \mid x, s) \, dy. \tag{5.120}$$

Obviously, the MFPT to the boundary after time s is

$$\mathbb{E}[\tau \mid y, s] = \int_s^\infty S(t \mid y, s) \, dt = \int_s^\infty \int_D p_\varepsilon(x, t \mid y, s) \, dx \, dt. \tag{5.121}$$

Because $p_\varepsilon(x, t \mid y, s)$ is also the solution of the BKE (1.169) with respect to the backward variable (y, s) and satisfies the terminal condition $p_\varepsilon(x, t \mid y, s) \to \delta$ $(x - y)$ as $s \uparrow t$, differentiation of (5.121) with respect to s shows that $\mathbb{E}[\tau \mid y, s]$ is the solution of boundary value problem (1.145), (1.146) for the Andronov–Vitt–Pontryagin equation (see Theorem 1.5.2).

For the autonomous system (5.48) the transition density function is time homogeneous, $p_\varepsilon(x, t \mid y, s) = p_\varepsilon(x, t - s \mid y, 0)$, so we can set $s = 0$ in (5.121) and write

$$\mathbb{E}[\tau \mid y] = \int_0^\infty S(t \mid y) \, dt = \int_D p_\varepsilon(x \mid y) \, dx, \tag{5.122}$$

where

$$p_\varepsilon(x \mid y) = \int_0^\infty p_\varepsilon(x, t \mid y, 0) \, dt. \tag{5.123}$$

The function $p_\varepsilon(x \mid y)$ satisfies the stationary FPE

$$\sum_{i,j=1}^{2} \varepsilon \frac{\partial^2 \left[\sigma^{i,j}(x) p_\varepsilon(x \mid y)\right]}{\partial x^i \partial x^j} - \sum_{i=1}^{2} \frac{\partial \left[a^i(x) p_\varepsilon(x \mid y)\right]}{\partial x^i} = -\delta(x - y) \quad (5.124)$$

with the boundary condition

$$p_\varepsilon(x \mid y)|_{x \in \partial D, y \in D} = 0, \quad (5.125)$$

where

$$\sigma(x) = \frac{1}{2} b(x) b^T(x).$$

The function $p_\varepsilon(x \mid y)$ has two probabilistic interpretations. It is the conditional probability density of the time a trajectory $x(t)$ spends at x prior to absorption in the boundary ∂D, given that it started at $y \in D$. It is also the stationary pdf of the trajectories of (5.48) with a unit source of trajectories placed at $y \in D$. Accordingly, the integral in (5.122) can be interpreted at the stationary population of trajectories in D. Because integration of the FPE (5.124) over D gives

$$\oint_{\partial D} J(x \mid y) \cdot n(x) \, dS_x = 1, \quad (5.126)$$

where $n(x)$ is the unit outer normal at ∂D and the probability flux density of absorbed trajectories in the boundary is

$$J_i(x \mid y) = a^i(x) p_\varepsilon(x \mid y) - \sum_{j=1}^{2} \varepsilon \frac{\partial^2}{\partial x^j} \left[\sigma^{i,j}(x) p_\varepsilon(x \mid y)\right], \quad (5.127)$$

we can write (5.124) as

$$\mathbb{E}[\tau \mid y] = \frac{N(y)}{F(y)}, \quad (5.128)$$

where

$$N(y) = \int_D p_\varepsilon(x \mid y) \, dx \quad (5.129)$$

is the stationary population of trajectories emitted by the source at $y \in D$ and

$$F(y) = \oint_{\partial D} J(x \mid y) \cdot n(x) \, dS_x \quad (5.130)$$

is their total absorption flux on the boundary. If the source is not concentrated at a point, but rather distributed with a given density, both numerator and denominator in (5.128) have to be averaged with respect to this density. Thus, in order to calculate the MTLL in the PLL (5.29) it suffices to solve the boundary value problem (5.124), (5.125) and use the solution in (5.128). Note that if the source strength in (5.124) is changed by a constant factor, (5.128) remains unchanged, because the factor cancels in the numerator and denominator.

In the autonomous case, the solution of the initial and boundary value problem (5.117)–(5.119) can be found by separation of variables in the form

$$p_\varepsilon(x,t \mid y,s) = \sum_{i=0}^{\infty} e^{-\lambda_n(t-s)} \varphi_n(x) \bar{\psi}_n(y), \qquad (5.131)$$

where $\lambda_n, \varphi_n(x), \psi_n(y)$ are the eigenvalues and eigenfunctions of the boundary value problems for the Fokker–Planck and backward Kolmogorov equations,

$$\mathcal{L}_\varepsilon \varphi_n(x) = -\lambda_n \varphi_n(x) \text{ for } x \in D, \quad \varphi_n(x) = 0 \text{ for } x \in \partial D,$$

$$\mathcal{L}_\varepsilon^* \psi_n(y) = -\bar{\lambda}_n \psi_n(y) \text{ for } y \in D, \quad \psi_n(y) = \text{ for } y \in \partial D,$$

respectively. The eigenfunctions are bi-orthogonal, that is,

$$\int_D \varphi_m(x) \bar{\psi}_n(x) \, dx = \delta_{mn}, \qquad (5.132)$$

and in particular, it follows from the analysis above that

$$\lim_{\varepsilon \to 0} \int_D \varphi_0(x) \, dx = 1, \quad \lim_{\varepsilon \to 0} \psi_0(y) = 1 \text{ for } y \in D.$$

The principal eigenvalue λ_0 is positivem and the eigenvalues are ordered so that $\lambda_0 < \mathfrak{Re}\,\lambda_1 < \mathfrak{Re}\,\lambda_2 < \cdots \to \infty$. If the initial density is $p_0(y)$, then the pdf is given by

$$p_\varepsilon(x,t) = \int_D p_\varepsilon(x,t \mid y,0) p_0(y) \, dy = \sum_{i=0}^{\infty} a_n e^{-\lambda_n(t)} \varphi_n(x),$$

where the Fourier coefficients of $p_0(y)$ are

$$a_n = \int_D p_0(y) \bar{\psi}_n(y) \, dy. \qquad (5.133)$$

It follows that the MFPT is given by

$$\mathbb{E}\tau = \sum_{i=0}^{\infty} \frac{a_n}{\lambda_n}.$$

The normalization condition (5.133) and Section 5.3.1 show that $a_0 \sim 1$, $\psi_0(y) \sim U_\varepsilon(y)$ for $y \in D \cup \partial D$, and that $\mathbb{E}\tau \sim \lambda_0^{-1}$ for $\varepsilon \ll 1$. Because $0 < \lambda_0 \ll \mathfrak{Re}\,\lambda_1$ for $\varepsilon \ll 1$, the survival probability in the autonomous case is given asymptotically by

$$S(x, t \mid y) \sim \varphi_0(x)\psi_0(y)e^{-\lambda_0 t} \quad \text{for } x, y \in D, \varepsilon \ll 1 \qquad (5.134)$$

for t such that $\left| e^{-\lambda_1 t} \right| \ll e^{-\lambda_0 t}$.

5.3.8   The Singularity of the FPE as $\varepsilon \to 0$

The construction of the solution $p_\varepsilon(x \mid y)$ of the boundary value problem (5.124), (5.125) is similar to that of (5.70), as presented in Sections 5.3.2–5.3.5. The difference between the two cases is the boundary condition (5.125), which gives rise to a boundary layer in the solution of (5.124). Such a boundary layer is absent in the solution of (5.70).

The form of $p_\varepsilon(x \mid y)$ is now

$$p_\varepsilon(x \mid y) = K_\varepsilon(x \mid y)\exp\left\{ -\frac{\Psi(x)}{\varepsilon} \right\}, \qquad (5.135)$$

where $K_\varepsilon(x \mid y)$ is the solution of the transport equation. The eikonal function $\Psi(x)$ is the same as in Sections 5.3.2 and 5.3.3. The boundary condition (5.125) implies the boundary condition

$$K_\varepsilon(x \mid y) = 0 \quad \text{for } x \in \partial D, \ y \in D. \qquad (5.136)$$

The function $K_\varepsilon(x \mid y)$ has a regular outer expansion in powers of ε for $x, y \in D$, but when $\varepsilon \to 0$, it develops a boundary layer. Therefore we construct a uniform approximation to $K_\varepsilon(x \mid y)$, valid for $y \in D$ and for $x \in D \cup \partial D$ and for all $\varepsilon \ll 1$,

$$K_\varepsilon(x \mid y) = [K_0(x \mid y) + \varepsilon K_1(x \mid y) + \cdots]q_\varepsilon(x \mid y), \qquad (5.137)$$

where $K_0(x \mid y)$, $K_1(x \mid y)$, ... are regular functions in D and on its boundary and are independent of ε, and $q_\varepsilon(x \mid y)$ is a boundary layer function. The functions $K_j(x \mid y)$ $(j = 0, 1, \ldots)$ satisfy first-order partial differential equations and therefore cannot satisfy the boundary condition (5.136). The boundary layer function $q_\varepsilon(x \mid y)$ satisfies the boundary condition

$$q_\varepsilon(x \mid y) = 0 \text{ for } x \in \partial D, \ y \in D, \qquad (5.138)$$

the matching condition

$$\lim_{\varepsilon \to 0} q_\varepsilon(x \mid y) = 1 \text{ for all } x, y \in D, \ x \neq y, \qquad (5.139)$$

and the smoothness condition

$$\lim_{\varepsilon \to 0} \frac{\partial^i q_\varepsilon(x \mid y)}{\partial (x^j)^i} = 0, \quad \text{for all } x, y \in D, \ x \neq y, \ i \geq 1, \ 1 \leq j \leq 2. \quad (5.140)$$

Note that the delta function $\delta(x - y)$ in the FPE (5.124) can be multiplied by any constant C_ε, which may depend on ε, without changing (5.128). The role of $\delta(x - y)$ is to ensure that the solution does not vanish identically, so that $q_\varepsilon(x \mid y)$ converges to 1 in (5.140) rather than to 0. The eikonal equation for $\Psi(x)$ and the transport equation for $K_0(x \mid y)$ remain the same as in Sections 5.3.2–5.3.5.

5.3.9 The Boundary Layer Equation

The boundary layer function $q_\varepsilon(x \mid y)$ satisfies the boundary layer equation

$$\sum_{i=1}^{2} \left\{ \sum_{j=1}^{2} \sigma^{ij}(x) \left[\varepsilon \frac{\partial^2 q_\varepsilon(x \mid y)}{\partial x^i \partial x^j} - 2 \frac{\partial \Psi(x)}{\partial x^j} \right] - a^i(x) \right\} \frac{\partial q_\varepsilon(x \mid y)}{\partial x^i} = 0. \quad (5.141)$$

The boundary layer equation (5.141) and the boundary, matching, and smoothness conditions (5.138)–(5.140) are similar to those of Section 5.3.1. There is, however, an important difference: while the drift in (5.56) is the vector $a(x)$, the drift in (5.141) is $\tilde{a}(x) = -[2\sigma(x)\nabla\Psi(x) + a(x)]$. Both are the same in the potential case, that is, when $a(x) = -\sigma(x)\nabla\Psi(x)$, but not otherwise. Therefore $\tilde{a}(x)$ is not tangent to the boundary in general, and it does not have the decomposition (5.45) (or (5.46)). Linearization near the origin, as in Sections 5.3.2–5.3.5, shows that the origin is a global attractor in D for the system

$$\dot{x} = \tilde{a}(x) \quad (5.142)$$

and that the saddle points $(\pm\pi, 0)$ of (5.32) (see Figure 5.4) are also saddle points of (5.142).

Exercise 5.15 (The system (5.142)).

(i) Calculate the drift $\tilde{a}(x)$ explicitly for the error equations (5.30) of the second-order PLL (5.29).

(ii) Prove that the system (5.142) has the above-mentioned properties.   □

If, however, $\Psi(s)$ is constant along the boundary near the saddle point, as shown in Figure 5.7, which shows a blowup of the graph of $\Psi(t)$ near the saddle point $(\pi, 0)$ (the limiting value is $\Psi(\pi, 0) \approx 0.78525$), then $\nabla\Psi$ is parallel to the normal there and the eikonal equation implies that $\tilde{a}(x)$ has an expansion analogous to (5.45) (or (5.46)),

$$\tilde{a}(x) = \left\{\rho\tilde{a}^0(s)n(s) + \tilde{B}(s)\tau(s)\right\}\{1 + o(1)\},\tag{5.143}$$

which in local coordinates is

$$\tilde{a}(\rho, s) = \tilde{a}^0(s)\rho\nabla\rho + \tilde{B}(s)\nabla s + o(\rho).\tag{5.144}$$

The MFPT can be calculated from the population/flux formula (5.128) using the WKB solution (5.135) to calculate $N(y)$ (see (5.129)) and the boundary layer (see (5.69))

$$q_\varepsilon(x \mid y) \sim -\sqrt{\frac{2}{\pi}} \int_0^{\rho\tilde{\gamma}(s)/\sqrt{\varepsilon}} e^{-z^2/2}\, dz\tag{5.145}$$

to calculate $F(y)$ (see (5.130)). The results of Theorem 5.3.1 can be recovered from this expansion as well.

5.3.10 The Exit Density

According to Theorem 2.5.3, the (normalized) absorption flux density of trajectories on the separatrix, is calculated from the solution $p_\varepsilon(x \mid y)$ of the boundary value problem (5.124), (5.125). In the case at hand, $\Psi(s) = \Psi(0) = \hat{\Psi}$ on a finite interval $0 \leq s \leq s_0$ (see, e.g., Figures 5.8 and 5.9), so

$$\begin{aligned}
p_\varepsilon(s)\, ds &= \Pr\{x(\tau) \in x(s) + ds \mid x(0) = y\}\\
&= \frac{J(x(s) \mid y) \cdot n(x(s))}{F(y)}\, ds\\
&= \sqrt{\frac{2\varepsilon}{\pi}}\frac{p_\varepsilon(x(s) \mid y))}{F(y)}\sum_{i,j}\sigma^{ij}(x(s))n^i(s)n^j(s)\tilde{\gamma}(s)\, ds\\
&= \frac{e^{-\Psi(s)/\varepsilon}K_0(s)\sigma^0(s)\hat{\gamma}(s)(1 + o(1))\, ds}{\displaystyle\int_{\partial D} e^{-\Psi(s)/\varepsilon}K_0(s)\sigma^0(s)\hat{\gamma}(s)(1 + o(1))\, ds}\\
&\approx \frac{e^{-\Psi(s)/\varepsilon}(s)^{-1)/\lambda_-}\, ds}{\displaystyle\int_{\partial D} e^{-\Psi(s)/\varepsilon}K_0(s)\sigma^0(s)\hat{\gamma}(s)\, ds}\tag{5.146}
\end{aligned}$$

(see (5.110)), where $\sigma^0(s)$ and $\tilde{\gamma}(s)$ have been approximated by their values at $s = 0$. Figure 5.11 shows the exit density $p_\varepsilon(s)$ (5.146) on the upper branch of

Fig. 5.11 The exit density (5.146) on the upper branch of separatrix for $\varepsilon = 0.005, 0.05, 0.1, 0.3$ (from top down at the origin).

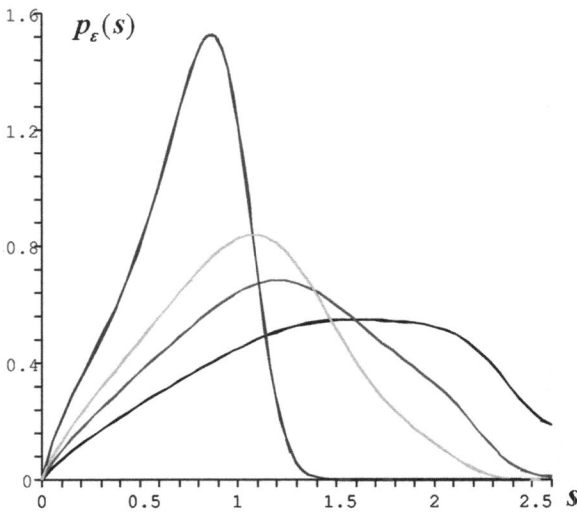

Fig. 5.12 The exit density (lower at $\varepsilon = 0$, upper at $\varepsilon = 0.3$) is maximal at $s_m \approx 0.8 + 1.3\sqrt{\varepsilon}$ (upper at $\varepsilon = 0$, lower at $\varepsilon = 0.3$).

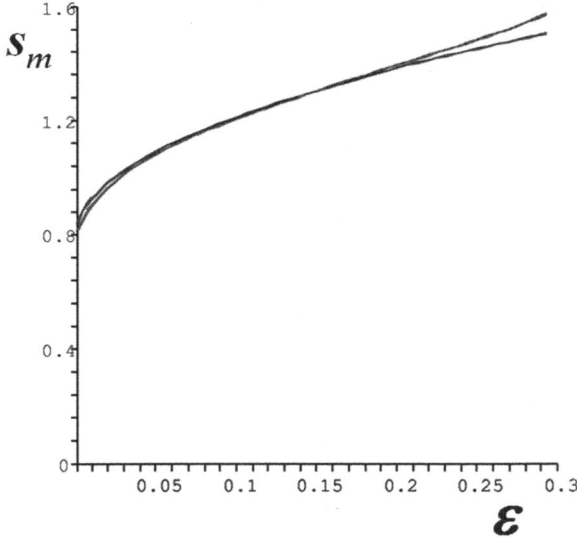

the separatrix for $\varepsilon = 0.005, , 0.05, 0.1$, and 0.3 (from the top down at the origin). Figure 5.12 shows the point of maximal exit probability on the upper branch of the separatrix. The exit density (lower at $\varepsilon = 0$, upper at $\varepsilon = 0.3$) is maximal at $s_m \approx 0.8 + 1.3\sqrt{\varepsilon}$ (upper at $\varepsilon = 0$, lower at $\varepsilon = 0.3$) at arc length s_m from the saddle point [136], [18].

Chapter 6
Loss of Lock in Radar and Synchronization

In this model, $y_1(t)$ is the received signal, $y_2(t)$ is the local replica of the transmitted signal, and $n_1(t)$ is the (stationary) noise signal, statistically independent from the signal $s(t)$; τ denotes the (unknown) time delay between the satellite and the received signal

Tracking range and angle is mathematically similar to tracking phase or frequency in a PLL or to maintaining fine synchronization in a delay-locked loop (DLL). There is, however, an important difference. While the purpose of the designer of both the transmitting and receiving ends of a communication system is to improve the SNR and keep the phase tracker locked on the signal, the designer of a radar system is often faced with jamming, whose purpose is to break the tracker's lock on the target. Thus the designers of trackers and those of countermeasures work at cross purposes. Both need an efficient performance index to assess their designs. This is not the case, however, in synchronizing pseudo-noise (PN) codes in cellular telephony and other applications. Here the enemy may be the Doppler shift between mobile receivers and transmitters, clock drift, and any number of other noisy kinds of interference. For example, the line-of-sight communication between the high-frequency ground or satellite antennas of cellular telephony requires the system to stay synchronized for months, lest the entire network collapses and has to reacquire all signals. There is no feasible computer simulation or laboratory test that can confirm that a given design can maintain synchronization for that long under the given load and interference conditions. The most efficient way to assess the durability of a synchronization design is to obtain an analytical approximation to its MTLL, or another performance index, from a reliable mathematical model.

Experience shows that maneuvering may enhance the efficiency of jamming with noise against certain tracking radars. This effect is caused, to a large extent, by inherent properties of the tracking loop. The choice of the loop's time constant is always a compromise between the requirement of tracking highly maneuverable targets on the one hand, and the improvement of the SNR at the loop output, on the other. This situation is well illustrated by a first-order loop with a large time constant that stays locked on a stationary target even at relatively high levels of jamming.

Z. Schuss, *Nonlinear Filtering and Optimal Phase Tracking*, Applied
Mathematical Sciences 180, DOI 10.1007/978-1-4614-0487-3_6,
© Springer Science+Business Media, LLC 2012

However, if the target is maneuvering, the efficiency of the loop drops drastically, because it loses lock in relatively short time. The mean time to lose lock is therefore an important performance measure of a tracking system.

6.1   How is Range Tracked? The Tracking Loop

A range tracker estimates the time delay t_0 of the radar pulse reflected from the target. The range is estimated as

$$r = \frac{ct_0}{2},\tag{6.1}$$

where c is the speed of light propagation. The radar pulse envelope $s(t)$ is usually chosen to be a positive function in an interval $0 < t < \tau$ and the range gate envelope $s_1(t)$ is chosen to be orthogonal to $s(t)$ in the interval $[0, \tau]$, e.g., $s_1(t) = \dot{s}(t)$ (matched filter). Some choices of $s(t)$ (top) and $s_1(t)$ (bottom) are shown below in Figures 6.1 (a rectangular pulse envelope and the corresponding range gate), 6.2 (a trapezoidal pulse envelope and its range gate [10]), and 6.3 (a Gaussian pulse envelope $s(t)$ and its range gate $s_1(t) = \dot{s}(t)$ [150]).

The reflected pulse $s(t - t_0)$ is convoluted at the discriminator with the delayed range gate envelope $s_1(t - \hat{t}_0)$, where $\hat{t}_0$ is the loop estimate of the delay t_0. Figure 6.7

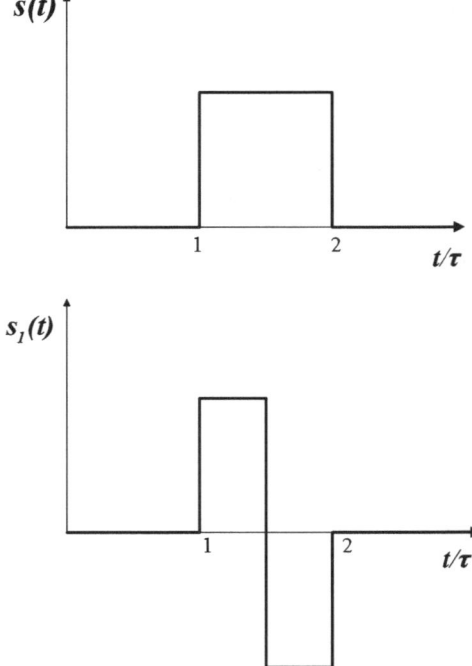

Fig. 6.1 Rectangular pulse envelope $s(t)$ (top) and the corresponding range gate $s_1(t)$ (bottom).

Fig. 6.2 Trapezoidal pulse envelope $s(t)$ (top) and its range gate $s_1(t)$ [10].

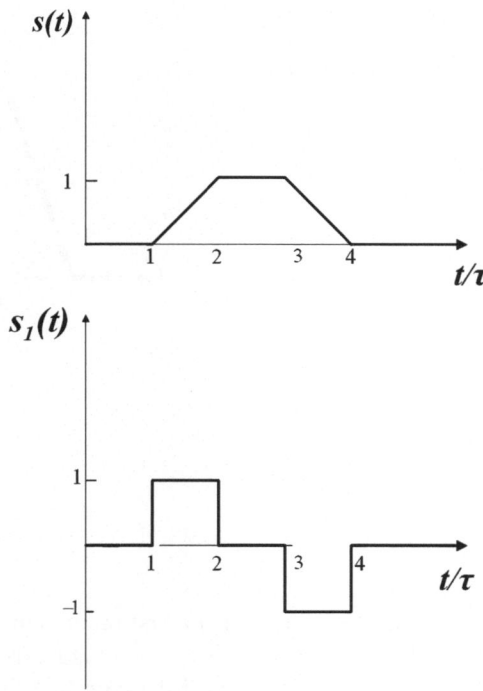

Fig. 6.3 A Gaussian pulse envelope $s(t)$ and its range gate $s_1(t) = \dot{s}(t)$ (matched filter) [150].

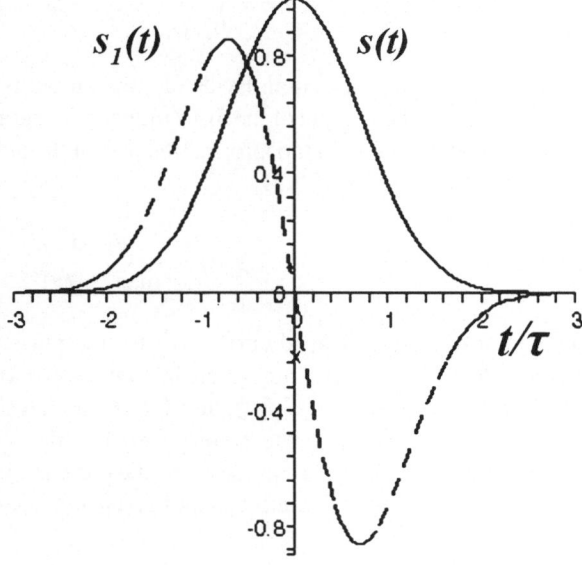

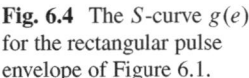

Fig. 6.4 The S-curve $g(e)$
for the rectangular pulse
envelope of Figure 6.1.

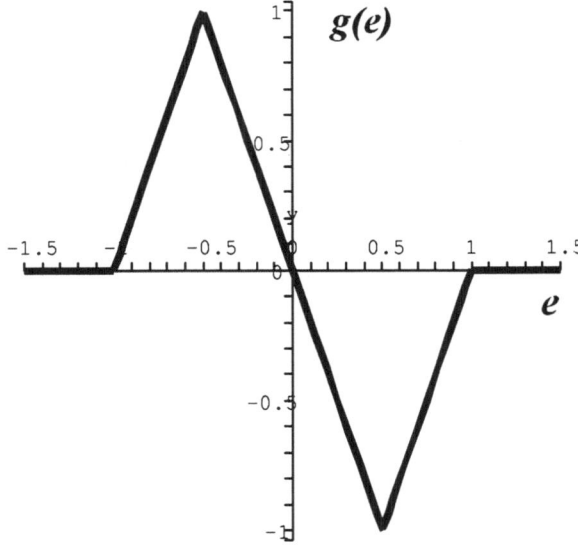

shows the block diagram of a first-order loop (that has no loop filter that follows the
range discriminator, i.e., $K_1 = 0$ in Figure 6.8). AGC is the automatic gain control.
A loop is of second-order if the loop filter has a single pole ($K_1 \neq 0$). Thus the
delayed pulse envelope $s(t - t_0)$ emerges from the discriminator in the form

$$g_1(t_0 - \hat{t}_0) = \int_{\hat{t}_0 - \tau/2}^{\hat{t}_0 - \tau/2} \alpha s(t - t_0) s_1(t - \hat{t}_0) \, dt, \qquad (6.2)$$

where τ is the pulse duration. Evidently, the output is a function of the estimation
error $e = t_0 - \hat{t}_0$. The output of the discriminator is a sum of the incoming information
and noise, where $g_1(e)$ is the information part of the noisy output $\tilde{y}$ (see Figure 6.8).
We normalize $g_1(e)$ by setting

$$g(e) = \frac{g_1(e)}{\max_e g_1(e)}. \qquad (6.3)$$

The graphs of the S-shaped curve $g(e)$ for the pulse and range gate envelopes of
Figures 6.1, 6.2, and 6.3 are given in Figures 6.4 (rectangular pulse envelope),
6.5 (trapezoidal pulse envelope), and 6.6 (Gaussian pulse envelope), respectively.
We refer henceforward to the graph of $g(e)$ as the S-curve. These graphs can be
characterized mainly by two parameters, the pulse length τ and $g'(0)$. Typical values
for $g'(0)$, for example, for pulse shapes given above, are

$$-\frac{6}{\tau} \leq g'(0) \leq -\frac{2}{\tau}.$$

Fig. 6.5 The S-curve $g(e)$
for the rectangular pulse
envelope of Figure 6.2.

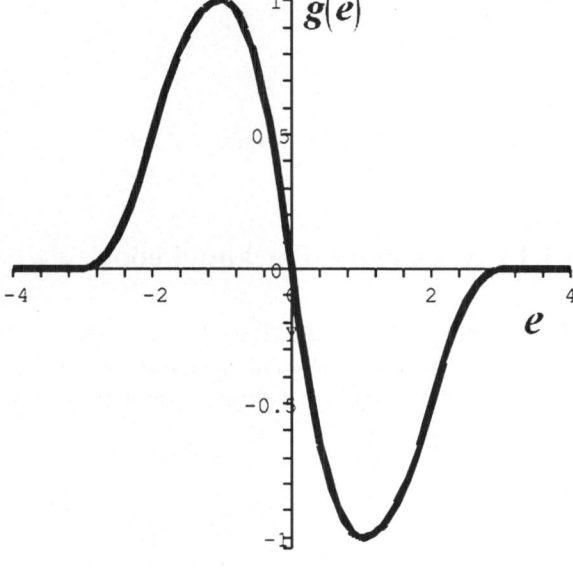

Fig. 6.6 The S-curve $g(e)$
for the Gaussian pulse
envelope of Figure 6.3.

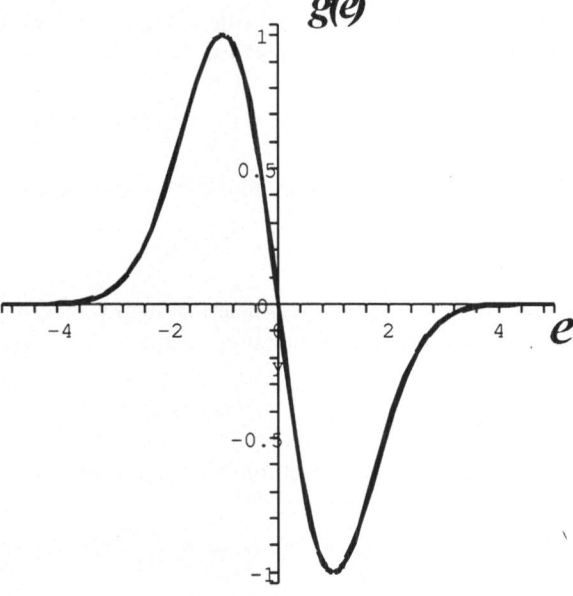

We choose for simplicity the piecewise linear S-curve of Figure 6.4 with $g'(0) = -3/\tau$. The loop estimate $\hat{r}$ of the range r (see (6.1)) is defined by

$$G(r - \hat{r}) = -g(e), \tag{6.4}$$

so that

$$G'(0) = \frac{3}{r_0}, \tag{6.5}$$

where $r_0 = c\tau$ is the pulse length in meters.

6.1.1 Noise in the Tracking Loop

The jamming noise in the loop 6.7 is assumed white with two-sided spectral height J. The error standard deviation per pulse in the split gate tracker of Figure 6.2, with envelope energy S and with time bandwidth product $Br \approx 1.4$, is given by (see [10])

$$\frac{\sigma_r}{r} = \frac{\sigma_r}{r_0} \approx \frac{1}{4}\sqrt{\frac{J}{S}}. \tag{6.6}$$

The pulse repetition rate PRF determines the repetition time $\Delta t = 1/\text{PRF}$. Numbering the pulses $i = [t/\Delta t]$ (the integral part), we can express the discriminator output $\tilde{t} y_i$ for pulse i in terms of the range as

$$\tilde{y}_i = \frac{r_0}{3} G(e_i) + n_i, \tag{6.7}$$

where $e_i = r - r_i$ and n_i are i.i.d. zero-mean Gaussian variables with variance

$$\overline{n_i^2} = \sigma_r^2. \tag{6.8}$$

The S-curve in Figure 6.1 is linear for small values of e, that is, for $(r_0/3)\, G(e) \approx e$, so in the first-order loop the range estimator $\hat{r}_i$ satisfies

$$\hat{r}_{i+1} = \hat{r}_i + \Delta t K \tilde{y}_i, \tag{6.9}$$

where K is the loop gain. It follows from (6.7) that

$$\hat{r}_{i+1} = \hat{r}_i + \Delta t \frac{K r_0}{3} G(e_i) + \Delta t K n_i. \tag{6.10}$$

Equation (6.10) is the Euler scheme (2.3), whose solution is well approximated by that of the Itô stochastic differential equation

$$d\hat{r}(t) = \frac{K r_0}{3} G(e(t))\, dt + \sqrt{\Delta t}\, K\, dW(t), \tag{6.11}$$

if PRF is high. Normalizing

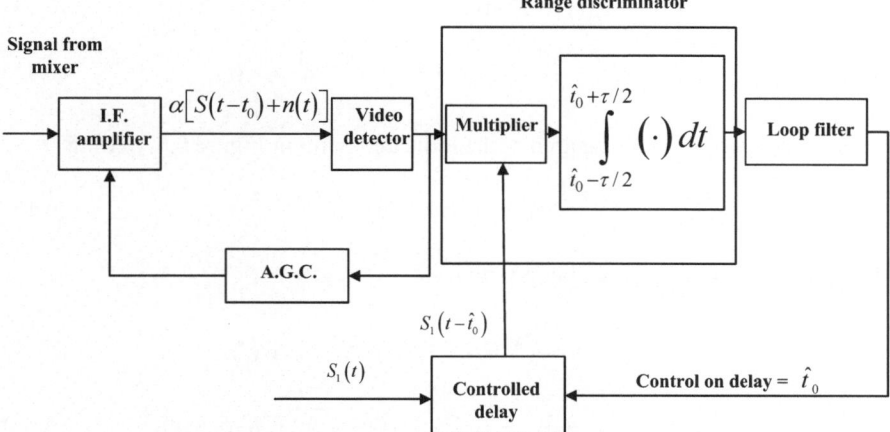

Fig. 6.7 The range tracking loop. AGC is the automatic gain control.

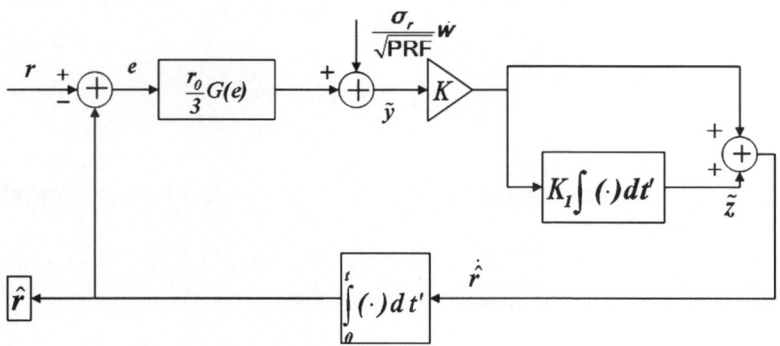

Fig. 6.8 Equivalent block diagram of a second-order loop.

$$W(t) = \sigma_r w(t),$$

we find that $\mathbb{E}w(t) = 0$ and $\mathbb{E}w^2(t) = t$ and write (6.11) as

$$\dot{\hat{r}}(t) = \frac{Kr_0}{3} G(e(t)) + \sqrt{\Delta t}\, K\sigma_r \dot{w}(t). \tag{6.12}$$

Subtracting (6.12) from $\dot{r}(t)$, we obtain the error equation

$$\dot{e}(t) = -\frac{Kr_0}{3} G(e(t)) + \dot{r}(t) - \sqrt{\Delta t}\, K\sigma_r \dot{w}(t). \tag{6.13}$$

The second-order loop in Figure 6.7 (with a loop filter) is equivalent to the block diagram in Figure 6.8 with $K_1 \neq 0$. Proceeding as above, we obtain a system of two equations that describe the loop,

$$\dot{\hat{r}}(t) = \hat{z}(t) + K\tilde{y}(t) \tag{6.14}$$

$$\dot{\hat{z}}(t) = \frac{KK_1r_0}{3} G(e(t)) + \sqrt{\Delta t}\, KK_1\sigma_r\dot{w}(t), \tag{6.15}$$

where $\tilde{y}(t)$ and $\hat{z}(t)$ are measured at the indicated point in Figure 6.8. Setting

$$z(t) = \hat{z}(t) - \dot{r}(t), \tag{6.16}$$

the system (6.14), (6.15) can be rewritten as

$$\dot{e}(t) = -z(t) - \frac{Kr_0}{3} G(e(t)) - \sqrt{\Delta t}\, K\sigma_r\dot{w}(t), \tag{6.17}$$

$$\dot{z}(t) = -\ddot{r}(t) + \frac{KK_1r_0}{3} G(e(t)) + \sqrt{\Delta t}\, KK_1\sigma_r\dot{w}(t). \tag{6.18}$$

Obviously, the system (6.17), (6.18) reduces to (6.13) if $K_1 = 0$.

The transfer function between $r(t)$ and $\hat{r}(t)$ in the linearized loop of Figure 6.7 is given by [10]

$$L(s) = \frac{\hat{R}(s)}{R(s)} = \frac{2\omega_n\zeta s + \omega_n^2}{s^2 + 2\omega_n\zeta s + \omega_n^2},$$

where $R(s)$ and $\hat{R}(s)$ are the Laplace transforms of $r(t)$ and $\hat{r}(t)$, respectively, ω_n is the natural frequency

$$\omega_n = \sqrt{KK_1}\ \text{rad/sec},$$

and the damping factor ζ is

$$\zeta = \frac{1}{2}\sqrt{\frac{K}{K_1}}.$$

In order to convert to dimensionless variables, we introduce

$$x = \frac{e}{r_0}, \quad \tilde{t} = \frac{Kt}{3}, \quad y = \frac{z}{K_1r_0}, \tag{6.19}$$

$$H(x) = G(e), \quad \beta = \frac{3K_1}{K} = \frac{3}{4\zeta^2} v_{max} = \frac{Kr_0}{3},$$

$$a_{max} = \frac{KK_1r_0}{3}, \quad u = \frac{\dot{r}}{v_{max}}, \quad a = \frac{\ddot{r}}{a_{max}},$$

$$\varepsilon = \frac{3}{2} \frac{\Delta t K\sigma_r^2}{r_0^2}.$$

Note that the numbers 3 in (6.5) and 4 in (6.6) are chosen as typical values and may vary for different pulse shapes $s(t)$, different gate shapes $s_1(t)$, and different bandwidths B.

Employing (6.19) in (6.13), we obtain

$$\dot{x} = u - H(x) - \sqrt{2\varepsilon}\,\dot{\tilde{w}}, \tag{6.20}$$

where $\tilde{w}(\tilde{t}) = \sqrt{K/3}\,w(t)$ and $(\,\dot{}\,) = d(\,)/d\tilde{t}$. The process $\tilde{w}(\tilde{t})$ is a standard Brownian motion [137, Exercise 2.4]. Similarly, we obtain from (6.17)–(6.19)

$$\dot{x} = -\beta y - H(x) - \sqrt{2\varepsilon}\,\dot{\tilde{w}}, \quad \dot{y} = -a + H(x) + \sqrt{2\varepsilon}\,\dot{\tilde{w}}. \tag{6.21}$$

Turning next to the automatic gain control (AGC), characteristics, we denote the signal and jamming energies by S and J, respectively, and note that for an ideal AGC

$$J_{\text{out}} + S_{\text{out}} = \text{const} \tag{6.22}$$

and

$$\frac{S_{\text{out}}}{J_{\text{out}}} = \frac{S_{\text{in}}}{J_{\text{in}}}. \tag{6.23}$$

Thus the open loop gain K in the presence of noise (jamming) is given by

$$K = \frac{K_0}{\sqrt{1 + \dfrac{J}{S}}}, \tag{6.24}$$

where K_0 is the gain in the absence of noise.

The detector, under high noise conditions, is no longer linear; hence we have for $S/J \ll 1$ [100], [32]

$$\left(\frac{S}{J}\right)_{\text{video}} \approx \frac{2\left(\dfrac{S}{J}\right)_{\text{IF}}}{1 + 2\left(\dfrac{S}{J}\right)_{\text{IF}}}, \tag{6.25}$$

so that

$$\left(\frac{J}{S}\right)_{\text{video}} \approx \frac{1}{2}\left(\frac{J}{S}\right)_{\text{IF}}^2 + \left(\frac{J}{S}\right)_{\text{IF}}. \tag{6.26}$$

Because J/S in (6.6) and (6.24) is, in fact, $(J/S)_{\text{video}}$, equations (6.19) and (6.24) with (6.6) can be used to express σ_r and K in terms of $(J/S)_{\text{IF}}$. First, we note that because

$$0.707 \leq \frac{1 + \frac{1}{2}\left(\dfrac{J}{S}\right)_{\text{IF}}}{\sqrt{1 + \left(\dfrac{J}{S}\right)_{\text{IF}} + \frac{1}{4}\left(\dfrac{J}{S}\right)_{\text{IF}}^{2}}} \leq 1,$$

the noise parameter ε in (6.21) is essentially linear in $(J/S)_{\text{IF}}$.

Next, to understand the effects of K, $\dot{r}$, and $\ddot{r}$ on the time to loss of lock, we describe the process of losing lock in a first-order loop in terms of a mechanical analogy with Kramers' activated escape problem (see [137, Sections 10.1.5 and 10.2.9]). Equation (6.20) can be interpreted as the equation of motion of an overdamped particle (neglecting acceleration) in a potential well, given by the potential

$$U(x) = -ux + \int_{0}^{x} H(x)\,dx, \tag{6.27}$$

and forced by a random force $\sqrt{2\varepsilon}\,\dot{w}(t)$. The locked state of the loop corresponds to the stable equilibrium state of the particle at the bottom of the well, at the point $x = x_0$, where $U'(x_0) = 0$ and $U''(x_0) > 0$. The motion of the particle consists of a deterministic motion (drift) due to the potential force, which tends to drive the particle toward its stable equilibrium state, and random fluctuations. Thus, if the noise intensity ε is relatively low, the particle spends a long period of time fluctuating about the stable equilibrium state at x_0. Due to the noisy driving force, it can make a large excursion at a random time and escape the well. This is a rare event relative to the loop time constants.

Figure 6.10 shows the graph of the "potential" $U(x) = \int_{-\infty}^{x} H(x)\,dx - ux$ for "bias" $u = 0$ (solid), $u = 0.5$ (dash), and $u = 1$ (dot). The barrier height for $u = 0$ is $\Delta U = U(1) - U(0) = 1/2$. For $u = 0.5$ the height is $\Delta U = U(x_1) - U(x_0) = 0.146$. For $u = 1$ there is no barrier. It can be seen from the figure that the height of the potential barrier to be overcome in order to escape (to lose lock) decreases as the normalized velocity u increases. For $u = 1$ the height of the barrier is zero, and escape occurs immediately, even without noise. This is clear also from simple engineering intuition, because at velocity $r_0 K/3 = v_{\max}$, the error in the loop is maximal.

It is shown below (see Sections 6.1.2 and 6.2) that the MTLL depends exponentially on the quotient

$$\frac{\Delta U}{\varepsilon} = \frac{U_{\max} - U_{\min}}{\varepsilon}, \tag{6.28}$$

where $U_{\max} = U(x_1)$, $U_{\min} = U(x_0)$, and ΔU is the height of the potential barrier. Equation (6.28) thus clarifies the influence of each parameter on the mean time to loss of lock. A similar dependence appears in second-order loops, but the structure of the function U is more complicated (see Section 6.2).

Fig. 6.9 Graph of the "force" $u - H(x)$ for "bias" $u = 0$ (solid), $u = 0.5$ (dash), and $u = 1$ (dot).

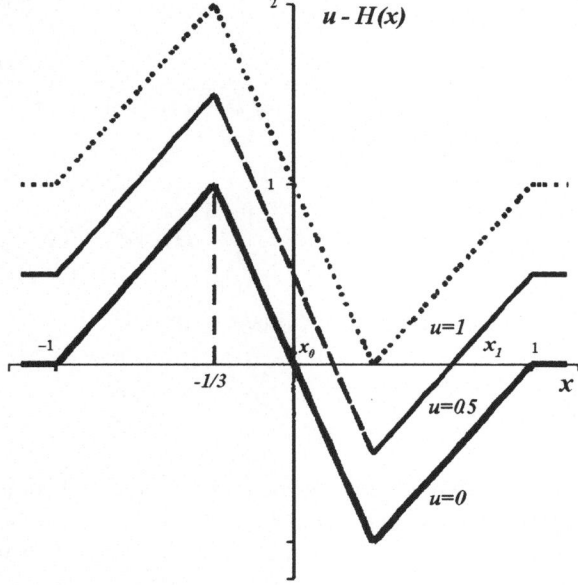

Fig. 6.10 Graph of the "potential" $U(x) = \int_{-\infty}^{x} H(x)\,dx - ux$ for $u = 0$ (solid), $u = 0.5$ (dash), and $u = 1$ (dot). The barrier height for $u = 0$ is $\Delta U = U(1) - U(0) = 1/2$. For $u = 0.5$ the height is $\Delta U = U(x_1) - U(x_0) = 0.146$. For $u = 1$ there is no barrier.

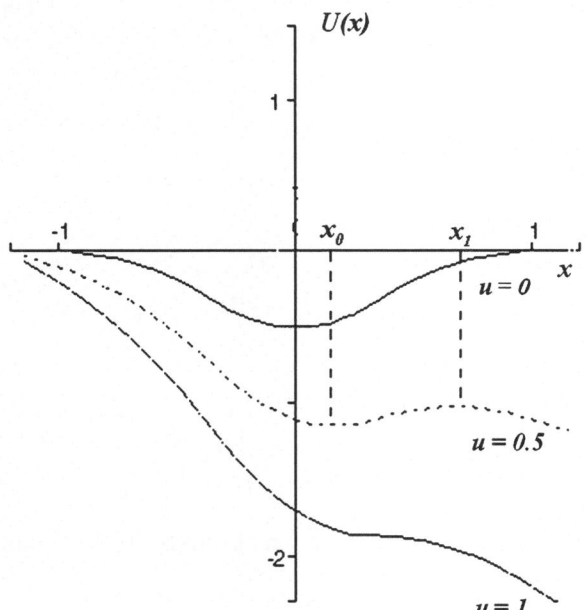

6.1.2   Loss of Lock in First-Order Tracking Loops

Setting $K_1 = 0$ in the block diagram in Figure 6.8 reduces the second-order loop to first-order, for which the calculation of the scaled mean time to loss of lock, $\bar{\tilde{t}}$, has been carried out in Section 5.1.1 (see also [137, Section 10.1.5]). It is given by

$$\tilde{\tilde{t}} = \frac{C\pi}{\sqrt{H'(x_0)H'(x_1)}}\exp\left\{\frac{\Delta U}{\varepsilon}\right\}, \tag{6.29}$$

where

$$\Delta U = \int_{x_0}^{x_1} H(x)\,dx - (x_1 - x_0)u. \tag{6.30}$$

The unscaled mean escape time $\bar{t}$ is given by

$$\bar{t} = \frac{3\tilde{\tilde{t}}}{K} \tag{6.31}$$

with $\tilde{\tilde{t}}$ given by (6.29). The factor C is equal to 1 for $u = 0$, that is, for a stationary target. For a moving target with $u \gg \varepsilon$ it is given by $C = 2$. In the range $0 < u < \delta$, where $\delta \gg \varepsilon$ (e.g., $\delta = 5\varepsilon$), C varies continuously from 1 to 2.

As mentioned in Section 6.1, we confine our attention to the case of a piecewise linear approximation of $G(e)$. Figure 6.9 shows graphs of the "force" $u - H(x)$ for values of the "bias" $u = 0$ (solid), $u = 0.5$ (dash), and $u = 1$ (dot). In this case,

$$\Delta U = \frac{1}{2}\left(1 - \frac{3\dot{r}}{Kr_0}\right)^2. \tag{6.32}$$

For $\Delta U > \varepsilon$, that is, for

$$\sqrt{\frac{3K_0}{16\text{PRF}}\left(\frac{J}{S}\right)_{\text{IF}}} + \frac{3\dot{r}}{K_0 r_0}\xi < 1,$$

where

$$\xi = \sqrt{1 + \left(\frac{J}{S}\right)_{\text{IF}} + \frac{1}{2}\left(\frac{J}{S}\right)_{\text{IF}}^2}, \tag{6.33}$$

Equations (6.6), the definition of ε in (6.19), the definition (6.24) of K with AGC in (6.32), give

$$\bar{t} \approx \frac{C\pi\sqrt{2}}{K_0}\xi\exp\left[\frac{\frac{16}{3}\left(1 - \frac{3\dot{r}}{K_0 r_0}\xi\right)^2\xi\text{PRF}}{(\xi^2 - 1)K_0}\right]. \tag{6.34}$$

Fig. 6.11 A plot of $T = \bar{t}$ vs $(S/J)_{\text{IF}}$ ((6.35)) for a first-order loop without AGC for $K_0 = 30$ rad/sec (solid) and $K_0 = 15$ rad/sec (dashed). Here $v_{\max} = 450$ m/sec.

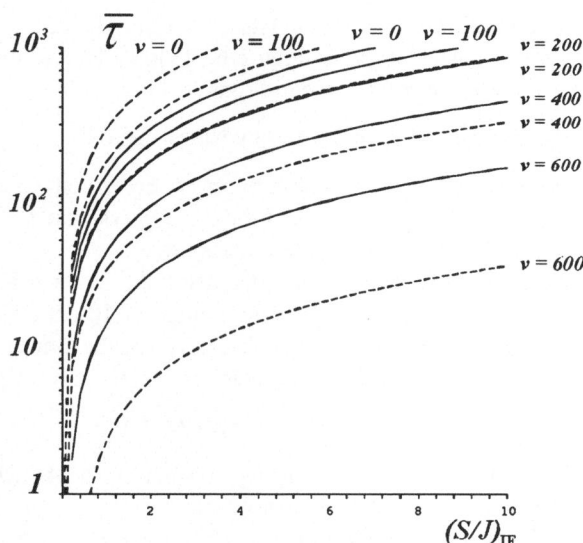

In the absence of AGC, equation (6.34) is replaced by

$$\bar{t} \approx \frac{C\pi\sqrt{2}}{K_0}\exp\left[\frac{\frac{16}{3}\left(1 - \frac{3\dot{r}}{K_0 r_0}\right)^2 \text{PRF}}{K_0 \left(\dfrac{J}{S}\right)_{\text{IF}}}\right]. \tag{6.35}$$

Example 6.1 (Tracking a moving target with a first-order loop). Consider a pulse of length $r_0 = 90$ m, PRF=800 pulses/sec, and $v_{\max}|_{J=0} = K_0 r_0/3 = 900$ m/sec, that is, $K_0 = 30$ rad/sec, or the loop bandwidth of approximately 5 Hz. Figure 6.11 shows a logarithmic plot of $T = \bar{t}$ vs $(S/J)_{\text{IF}}$ for $0.1 < (S/J)_{\text{IF}} < 1000$, according to equation (6.35), for a first-order loop without AGC for $K_0 = 30$ rad/sec (solid) and $K_0 = 15$ rad/sec (dashed). The curves correspond to various values of $v = \dot{r} > 0$ ($\ddot{r} = 0$). Here $v_{\max} = 450$ m/sec. In this case we also neglect the effect of the detector by neglecting the quadratic term in (6.26). The time $T = \bar{t}$ is plotted also for $K_0 = 15$ rad/sec. $\qquad\square$

The mean square estimation error calculated by the linearization of the equations about the stable equilibrium point x_0 is given by

$$\frac{\overline{e_r^2}}{r_0^2} = \frac{K_0}{32\text{PRF}}\left(\frac{J}{S}\right)_{IF}, \tag{6.36}$$

which seems to indicate that a small K_0 should be preferred in order to increase accuracy. This agrees with the trend seen in Figure 6.11 for $v = \dot{r} = 0$. However, as the target's velocity increases, the performance of the slower loop deteriorates, due to the nonlinearity in $G(e)$. Thus, for example, at $v = \dot{r} = 400$ m/sec the slower loop (the dashed curve in Figure 6.11) loses lock at much lower jamming than the faster loop (the solid curve in Figure 6.11).

Exercise 6.1 (First order loop without AGC).

(i) Show that the level of jamming is equivalent to a velocity of the target for a fixed loss of lock time. Plot $(J/S)_{\text{IF}}$ vs $\dot{r}$ at $\bar{t} = 10$ sec. Choose the value $\bar{t} = 10$ sec to keep the graphs apart.

(ii) Show, in particular, that increasing the target velocity from 450 m/sec to 750 m/sec is equivalent to an increment of 10 dB of jamming noise for the faster loop ($K_0 = 30$ rad/sec). Show that for the slower loop ($K_0 = 15$ rad/sec) the same effect is achieved if velocity increases from 300m/sec to 400 m/sec. □

Exercise 6.2 (First order loop with AGC).

(i) Plot $\bar{t}$ vs $(J/S)_{\text{IF}}$ according to equation (6.34), which incorporates the effects of the AGC and the detector.

(ii) Note that linearization and (6.27) lead to (6.36), as above. Argue that this procedure may be justified only under a high SNR assumption. □

Exercise 6.3 (The effect of AGC in a first-order loop).

(i) Compare the graphs of Exercises 6.1 and 6.2 to show that for $u = 0$, the AGC increases the mean time $\bar{t}$.

(ii) Interpret this fact in terms of (6.24) by arguing that the open loop gain K decreases as a result of the AGC reaction to noise, so the loop becomes slower. Thus it integrates the noise over a longer period of time and therefore suppresses the noise to a higher degree. However, as the velocity increases, a drastic degradation in the MTLL $\bar{t}$ occurs. □

Exercise 6.4 (AGC and detector).

(i) Plot $(J/S)_{\text{IF}}$ vs $\dot{r}$ ($\ddot{r} = 0$) at $\bar{t} = 10$ sec for a loop with AGC and detector for $K_0 = 30$ rad/sec and $K_0 = 15$ rad/sec and compare with the graph of Figure 6.2.

(ii) Note that even at low velocities the graphs fall off drastically. Conclude that a target moving at a relatively low velocity achieves loss of lock corresponding to a relatively high power of jamming.

(iii) Show that for $K_0 = 30$ rad/sec an increase in velocity from 25 m/sec to 180 m/sec has the effect of 10 dB of jamming power. Show that for $K_0 = 15$ rad/sec this effect is even more pronounced. □

6.2 Loss of Lock in Second-Order Range Trackers

Loss of lock in a second-order range-tracking loop is similar to that in a second-order PLL, as described in Section 5.2. The calculation of the MTLL here is

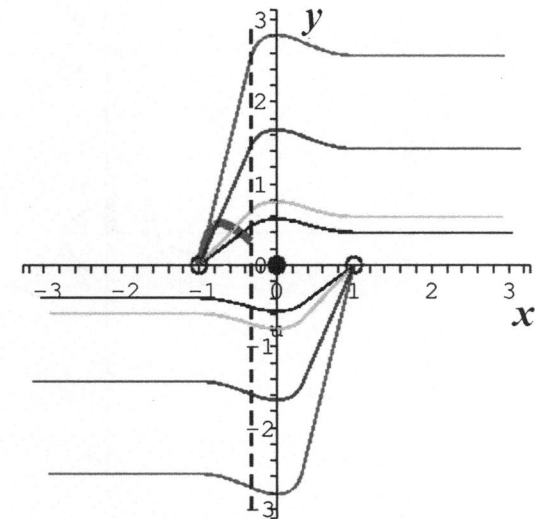

Fig. 6.12 The lock domain D and its boundary ∂D for $a = 0$. The boundaries of the domains corresponding to $\beta = 0.5, 1, 3$, and 5 consist of the negative x-axis and the upper curves, in descending order, and their mirror images with respect to both axes. The stable equilibrium point is at the origin (disk) and the saddle points are on the x-axis at $x = \pm 1$ (circles). The thick curve is the characteristic that emanates from the point $x = -1/3$, $y = 0.3$ and converges to the saddle point from the interior of the largest domain.

therefore also similar to the calculations of Section 5.2. To see the analogy between the two trackers, we begin with a description of the noiseless dynamics (6.21)

$$\dot{x} = -\beta y - H(x), \quad \dot{y} = -a + H(x). \tag{6.37}$$

If $a < a_{\max}$ (recall (6.19)), the system (6.37) has a stable equilibrium at the point (x_0, y_0), where

$$H(x_0) = a, \quad H'(x_0) > 0, \quad y_0 = -\frac{H(x_0)}{\beta} = -\frac{a}{\beta}. \tag{6.38}$$

At the point (x_1, y_1), where

$$H(x_1) = a, \quad H'(x_1) < 0, \quad y_1 = y_0, \tag{6.39}$$

the system has an unstable equilibrium point (a saddle point). The domain of attraction D of the stable point (x_0, y_0) is shown in Figures 6.12–6.14 for various values of a and β. Figure 6.12 shows the domain D and its boundary ∂D for $a = 0$. The boundaries of the domains corresponding to $\beta = 0.5, 1, 3$, and 5 consist of the negative x-axis and the upper curves, in descending order, and their mirror images with respect to both axes. The stable equilibrium point is at the origin (disk) and

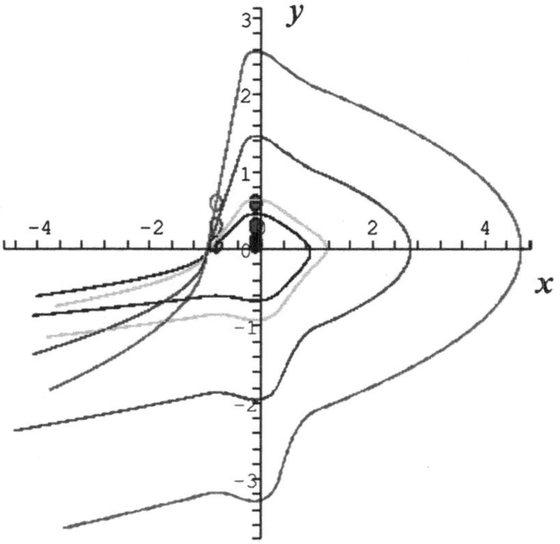

Fig. 6.13 The lock domain D and its boundary ∂D for $a = -0.3$ for $\beta = 0.5, 1, 3,$ and $\beta = 5$. The stable equilibrium points (disks) are $x_0 = a/3 = -0.1$ and $y_0 = -a/\beta$. The saddle points (circles) are on the boundary at $x_1 = -1 - 2a/3 = -0.8$ and $y_1 = y_0$. The dashed vertical line is at $x = -1/3$.

Fig. 6.14 The lock domain D and its boundary ∂D for $a = -0.6$ for $\beta = 0.5$ (the largest domain) and in descending order, for $\beta = 1, 3,$ and 5 (the innermost domain). The stable equilibrium points (disks) are $x_0 = a/3 = -0.2$ and $y_0 = -a/\beta$. The saddle points (circles) are on the boundary at $x_1 = -1 - 2a/3 = -0.6$ and $y_1 = y_0$.

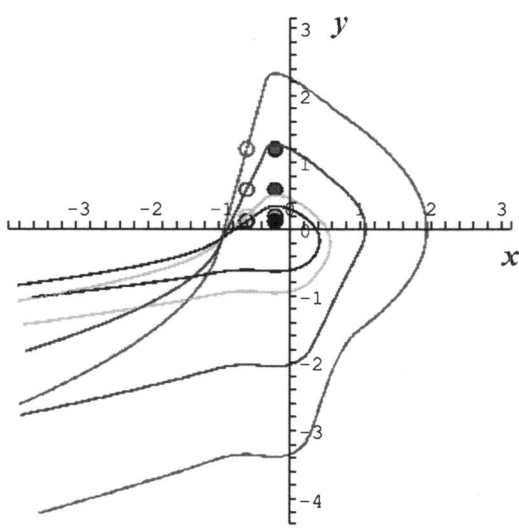

the saddle points are on the x-axis at $x = \pm 1$ (circles). The thick curve is the characteristic that emanates from the point $x = -1/3, y = 0.3$ and converges to the saddle point from the interior of the largest domain. Note that as $a \to 0$ the domain D assumes the familiar form given in [67]). Figure 6.13 shows the same for $a = -0.3$ with $\beta = 0.5$ (the largest domain), and in descending order $\beta = 1, 3,$ and $\beta = 5$ (the innermost domain). The stable equilibrium points (disks) are $x_0 = a/3 = -0.1$ and $y_0 = -a/\beta$. The saddle points (circles) are on the boundary at $x_1 = -1 - 2a/3 = -0.8$ and $y_1 = y_0$. The dashed vertical line is at $x = -1/3$. Finally, Figure 6.14 shows the same for $a = -0.6$ with $\beta = 0.5, 1, 3,$ and $\beta = 5$. Here $x_0 = a/3 = -0.2$ and $y_0 = -a/\beta$, the saddle points (circles) are on the boundary at $x_1 = -1 - 2a/3 = -0.6$ and $y_1 = y_0$. The distance between the stable equilibrium and the saddle point is $x_0 - x_1 = 1 + a$. At $a = -1$ they coalesce, and lock is lost instantaneously.

6.2.1 The Mean Time to Lose Lock

The asymptotic evaluation of the MTLL for small ε in the range tracker is essentially the same as in Theorem 5.3.1. In the case at hand, however, the solution of the characteristic equations for the system (6.21) is much simpler than in the case of the second-order PLL, as described in Chapter 5, and an explicit expression for the exponent $\hat{\Psi}$ in (5.53) can be found. Specifically, the stationary pdf of the system (6.21) has the WKB form (5.135), where the eikonal function $\psi(x, y)$ is the solution of the eikonal equation

$$(\psi_x - \psi_y)^2 - (H(x) + \beta y)\psi_x + (H(x) - a)\psi_y = 0. \qquad (6.40)$$

Inside D, in the strip $-1/3 < x < 1/3$, the drift is linear, because $H(x) = 3x$, so the stationary pdf is given by

$$\tilde{p} = c \exp\left\{-\frac{\varphi(x, y)}{\varepsilon}\right\}, \qquad (6.41)$$

where the solution of the eikonal equation is

$$\varphi(x, y)$$
$$= \frac{9}{2\beta}\left[\left(x - \frac{a}{3}\right)^2 + 2\left(x - \frac{a}{3}\right)\left(y + \frac{a}{\beta}\right) + \left(1 + \frac{\beta}{3}\right)\left(y + \frac{a}{\beta}\right)^2\right] \qquad (6.42)$$

and c is a normalization constant, which can be evaluated by integrating the density asymptotically by the Laplace method.

Exercise 6.5 (The normalization constant). Find an explicit asymptotic expression for c for small ε. □

The solution of the FPE outside the strip is still to be found by the WKB method, and the eikonal function $\Psi(x, y)$ has to be found outside the strip by the method of characteristics (see (5.79)). The characteristic equations in D outside the strip, on the side of the saddle point, are given by

$$\dot{x} = 2(p - q) + \frac{3}{2}(x + 1) - \beta y, \quad \dot{y} = -2(p - q) - a - \frac{3}{2}(x + 1),$$

$$\dot{p} = -\frac{3}{2}p + \frac{3}{2}q, \quad \dot{q} = \beta p, \quad \dot{\Psi} = (p - q)^2. \qquad (6.43)$$

The function $\Psi(x, y)$ and its first-order partial derivatives are identical to $\varphi(x, y)$ and its first-order derivatives on the lines $x = \pm 1/3$, so initial conditions for the characteristics outside the strip, on the side of the saddle point, are given on the line $x = -1/3$ and are parametrized by the initial choice of y. They are

$$x(0) = -\frac{1}{3}, \quad y(0) = y',$$

$$p(0) = \frac{9}{\beta}\left[-\frac{1}{3} - \frac{a}{3} + y' + \frac{a}{\beta}\right],$$

$$q(0) = \frac{9}{\beta}\left[-\frac{1}{3} - \frac{a}{3} + \left(1 + \frac{\beta}{3}\right)\left(y' + \frac{a}{3}\right)\right],$$

$$\Psi(0) = \varphi\left(-\frac{1}{3}, y'\right). \tag{6.44}$$

Note that the last three equations in (6.43) are decoupled from the first two, so they can be solved explicitly with the given initial conditions (6.44). The solutions $p(t), q(t)$ are linear combinations of the exponential functions $\exp\{\lambda_\pm t\}$, where the eigenvalues are given by

$$\lambda_\pm = -\frac{3}{4}\left(1 \pm \sqrt{1 + \frac{8\beta}{3}}\right).$$

The partial derivatives $\Psi_x(x(t), y(t)) = p(t)$, $\Psi_y(x(t), y(t)) = q(t)$ on the characteristic $(x(t), y(t))$ that converges to the saddle point are therefore given by

$$p(t) = p(0)e^{\lambda_- t}, \quad q(t) = q(0)e^{\lambda_- t}, \quad q(0) = \frac{\beta p(0)}{\lambda_-}. \tag{6.45}$$

The last equality in (6.45) is obtained by integrating the second equation on the second line of (6.43) from $t = 0$ to $t = \infty$. This equality and the initial conditions (6.44) determine y' as

$$y' = -\frac{a}{\beta} - \frac{(1 + a)\dfrac{\beta}{3} + \lambda_-}{\beta - \lambda_-\left(1 + \dfrac{\beta}{3}\right)}. \tag{6.46}$$

Now we integrate the last equation in (6.43) with the initial condition (6.44) and with the values (6.45) to obtain

$$\hat{\Psi} = \Psi(0) - \frac{p^2(0)}{2\lambda_-}\left(1 - \frac{\beta}{\lambda_-}\right). \tag{6.47}$$

The transport equation is solved as in Sections 5.3.5 and 5.3.7. Next we discuss the scaled MTLL $\bar{\tau}(x, y)$. This time is twice the mean time to hit ∂D for the first time, starting at $(x, y) \in D$, because a trajectory that hits ∂D goes across the boundary or returns to D with equal probabilities. It is given in dimensionless units in Theorem 5.3.1 as

$$\bar{\tau}(x, y) \sim K(\varepsilon)e^{\hat{\Psi}/\varepsilon} \quad \text{for } \varepsilon \ll 1, \tag{6.48}$$

and $K(\varepsilon)$ has an asymptotic expansion in powers of ε. In dimensional units the mean escape time is given by

$$\text{MTLL} = \frac{3}{K}\bar{\tau}(x, y) \tag{6.49}$$

(recall that $\bar{\tau}(x, y)$ is independent of (x, y), except for trajectories that start inside a boundary layer near ∂D; see the scaling of time in (6.19)).

6.2.2 Evading Radar by Jamming and Maneuvering

The above results can be applied to the quantification of the value of maneuvering in terms of equivalent jamming that leads to the same mean time to loss of lock on a maneuvering target [19], [131].

Exercise 6.6 (The MTLL is a decreasing function of β and $|a|$). Plot $\bar{\tau}$ vs ε for various values of β and a, as given by (6.48). Show that $\bar{\tau}$ is a monotone decreasing function of β and $|a|$. ☐

It turns out, however, that the dimensional MTLL at a constant $\ddot{r}$ and a constant $(J/S)_{IF}$ is not necessarily a monotone function of β. Next we turn to the discussion of the dimensional quantities, because they are the ones of engineering interest. A typical problem in second-order loop design is the choice of the damping factor. The usual choice in noiseless loop design is $\zeta = \sqrt{1/2}$. This is, however, not necessarily the optimal choice for the nonlinear noisy loop considered here (see below). In the investigation of the influence of jamming noise on the loop performance, the equivalent loop noise bandwidth $B_L \equiv (K_0 + K_1)/4$ Hz is kept constant at first [152]. In this case the linearized error variance $\{r(t) - \hat{r}(t)\}^2$ is constant for all ζ, as long as the loop is locked.

Exercise 6.7 (An "ideal" second-order tracking loop). Plot the dimensional MTLL (6.49) vs $(J/S)_{IF}$ for various values of β (i.e., various values of ζ) and various values of the normalized acceleration a in a loop without AGC, assuming an ideal detector. This provides a description of an "ideal" second-order tracking loop. Obtain the values of $\ddot{r}$ from (6.19). Choose the values of r_0 and PRF as in Example 6.1. ☐

Exercise 6.8 (Converting maneuvering to jamming.). Plot $(J/S)_{IF}$ vs $\ddot{r}$ (in m/sec^2) for a fixed MTLL$= 10$ sec. Use the fact that the graphs are steep to conclude that from an operations research point of view it hardly matters what value of the MTLL is chosen. $\square$

A practical application of these results is the evaluation of the influence of maneuvering, that is, of velocity or acceleration, in terms of equivalent jamming power. Thus, for a target trying to evade a range tracking loop, sharp maneuvering allows for a lower jamming level. Of course, this fact is well known in practice, though not quantitatively. It is interesting to note that from a loop designer point of view, for high acceleration $\ddot{r}$ it is preferable to choose large β, that is, small ζ.

Exercise 6.9 (Keeping constant bandwidth). Plot $(J/S)_{IF}$ vs $\ddot{r}$ under the assumption of constant B_L to show that the monotone dependence on β is reversed for small values of $\ddot{r}$. $\square$

6.2.3   The Influence of a Detector and AGC on the MTLL

The detector and AGC are often present in range trackers. While in a first-order loop (see Section 6.1.2) the loop gain decreases as the result of the AGC, in the second-order loop of Figure 6.7 both the open loop gain K and the damping factor ζ are affected by the AGC. This changes the values of $\hat{\Psi}$ and $K(\varepsilon)$ and thus also the value of $\bar{\tau}$. The combined effect of the AGC and the detector on ε is as described in Section 6.1.2. Thus, we set β_0 and a_0 to be the values of β and a in the absence of noise, respectively. Note that K_1 remains unchanged.

Exercise 6.10 (Keeping constant bandwidth). Compare the graphs of the jamming vs acceleration for a given MTLL for tracking loops with and without AGC. Show that, as in first-order loops, a drastic degradation in loop performance is observed even at relatively low values of $\ddot{r}$. Show that the values of $\ddot{r}$ at which track is lost in a loop containing a detector and AGC are about 10 times smaller then those in an "ideal" loop. It seems that this degradation in the practical loop performance is often underestimated by loop designers. $\square$

Exercise 6.11 (Keeping ω_n constant [10]).

(i) Show that if ω_n is kept constant rather than B_L, loop bandwidth is kept approximately constant in the absence of noise.
(ii) Show that in the absence of a detector and AGC this choice keeps $a_{\max}$ constant.
(iii) Show that in the plots of Exercise 6.10 in the absence of AGC and detector there is some dependence on β. However, in the presence of AGC and detector, where $K_0 K_1$ is kept constant, the dependence on β_0 practically disappears if $0.5 \leq \beta_0 \leq 5$. $\square$

Fig. 6.15 The acceleration $\ddot{r} = V^2h^2/r^3$ for constant target velocity V and constant altitude h.

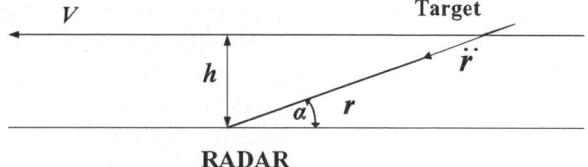

RADAR

The method presented in this section can be extended to other tracking loops as well. These may include angle tracking loops, noncoherent code tracking loops from spread spectrum communications, cellular telephony, GPS, and so on. The effect of the envelope detector and the AGC, which are essential for the practical realization of range trackers, are also taken into consideration in the analysis of the MTLL. These elements, while beneficial to the loop performance for nonmaneuvering targets, turn out to be detrimental to the loop performance if the target maneuvers. For a maneuvering target the level of jamming required to break lock within a prescribed period of time may be an order of magnitude lower in a practical loop containing these elements, as compared to an "ideal" loop without these elements. The practical significance of the jamming–maneuvering equivalence consists in the fact that under conditions of insufficient jamming power, a sufficiently maneuverable target may break tracking lock under realistic conditions.

Example 6.2 (Buzzing the radar). Consider a simplified scenario of a target moving on a rectilinear trajectory at a constant speed V as shown in Figure 6.15. The acceleration is $\ddot{r} = V^2h^2/r^3$ for a constant target velocity V and constant altitude h. Then

$$|\ddot{r}| = \frac{V^2h^2}{r^3} = V^2\frac{\sin^3\alpha}{h}, \tag{6.50}$$

where h is the distance between the radar and the rectilinear trajectory. Thus a low flying target may appear to the range tracker as a highly accelerating target. One possible conclusion of the present analysis is that in order to overcome the shortcomings of the AGC and the envelope detector, which affect K, one may try to dynamically adjust both K and K_1.   □

6.3   Spread Spectrum Transmission and Loss of Synchronization

Spread spectrum (SS) transmission is used in many applications, such as hiding the signal from detection by transmitting it over a wide spectrum at low power, or by frequency hopping (FH), that is, by changing the carrier frequency at random according to a sequence known only to the transmitter and receiver. Spread spectrum transmission is hard to jam, because of the large jamming power required to cover the entire spectrum. The direct sequence (DS) spreading of the signal spectrum is achieved by multiplying the low rate data signal by a high rate (i.e., short chip duration T_c) pseudonoise (PN) code. The direct sequence SS is used in iPhones and other cellular phones that are based on code division multiple access (CDMA) protocols, global positioning systems (GPS), and many more.

To recover the data at the receiver the spread signal is correlated with a synchronized replica of the PN code $c(t)$. Therefore successful recovery depends on a fine synchronization between the spreading and de-spreading PN codes. The PN code synchronization is achieved in two steps. The first is the code acquisition, which coarsely aligns the received and local PN codes to within T_c or $\frac{1}{2}T_c$. The second step is a fine alignment of the two by continuous code tracking. This process is quite similar to radar acquisition and tracking of a target, as described in the previous section (see calculations in [82], [154], [153], [13]).

6.3.1 The Delay-Locked Loop

The synchronizing tracker is the delay-locked loop (DLL), which is called coherent when the carrier frequency and phase are known, or noncoherent if they are not. The latter is used when carrier synchronization cannot precede de-spreading due to the low signal-to-noise ratio (SNR) of the spread signal. The DLL is used in global navigation satellite system receivers to estimate and track the delay between the received signal and the local replica of the transmitted signal. Specifically, the incoming spreading pseudonoise $c(t)$ is delayed by an unknown time $T(t)$ relative to the internal PN-code. To detect the delay and minimize it during tracking, the DLL creates an estimated delay $\hat{T}(t)$ and two versions of internal code, an early code $c(t - \hat{T}(t) + \Delta T_c)$ and late code $c(t - \hat{T}(t) - \Delta T_c)$, where $0 < \Delta \leq 1$.

The components of the DLL are the cross-correlators, marked $\otimes$ in the block diagram of Figure 6.16, the bandpass filters (BPF), the square-law devices, loop filter with transfer function $F(s)$, the voltage-controlled oscillator (VCO), and the

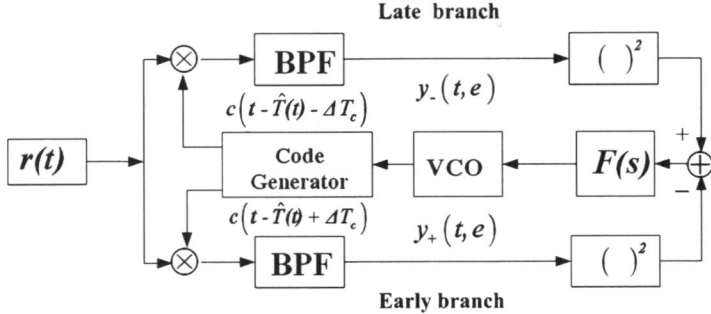

Fig. 6.16 Block diagram of the early–late DLL.

PN-code generator. The normalized tracking error is $e(t) = [T(t) - \hat{T}(t)]/T_c$. As in the PLL, the processed error adjusts the frequency of the VCO to decrease the local delay error until the DLL is locked on the incoming phase.

The random data stream modulated by binary phase shift keying (BPSK) $b(t) = \pm 1$ is a binary signal with bit duration T_b, the carrier frequency and phase are ω_0 and ϕ, respectively, and $n(t)$ is an additive noise (not necessarily white). Therefore the model of the received signal is

$$r(t) = \sqrt{2P}b(t)c(t - T(t))\cos(\omega_0 t + \phi) + n(t). \tag{6.51}$$

The incoming spreading PN-code $c(t)$ is assumed to have the symmetric triangle-shaped autocorrelation function

$$R_c(e) = \mathbb{E}[c(t)c(t + eT_c)] = \begin{cases} 1 - |e| & \text{for } |e| \leq 1, \\ 0 & \text{otherwise.} \end{cases} \tag{6.52}$$

Neglecting the effect of code self-noise $c(t)c(t + eT_c) - \mathbb{E}[c(t)c(t + eT_c)]$ on the loop, when the loop bandwidth is much smaller than the code chip rate $1/T_c$, the bandpass-filtered outputs of the correlators in the two branches are the convolutions with the impulse response $h_B(t)$ of the BPF,

$$y_-(t, e) = [\sqrt{2P}\, b(t)R_c(e - \Delta)\cos(\omega_0 t + \phi) + n(t)c(t - \hat{T}(t) - \Delta T_c)] * h_B(t), \tag{6.53}$$

$$y_+(t, e) = [\sqrt{2P}\, b(t)R_c(e + \delta)\cos(\omega_0 t + \phi) + n(t)c(t - \hat{T}(t) + \Delta T_c)] * h_B(t).$$

The effect of the BPF on the data-modulated carrier is approximated by the equivalent baseband response $h_L(t)$. The filtered data wave form is then given by $\bar{b}(t) = b * h_L(t)$ and (6.53) can be written as

$$y_-(t, e) = [\sqrt{2P}\, \bar{b}(t)R_c(e - \Delta)\cos(\omega_0 t + \phi) + n(t)c(t - \hat{T}(t) - \Delta T_c)] + n_L(t) \tag{6.54}$$

$$y_+(t, e) = [\sqrt{2P}\bar{b}(t)R_c(e + \Delta)\cos(\omega_0 t + \phi) + n(t)c(t - \hat{T}(t) + \Delta T_c)] + n_R(t),$$

where

$$n_L(t) = [n(t)c(t - \hat{T}(t) + \Delta T_c)] * h_B(t),$$

$$n_R(t) = [n(t)c(t - \hat{T}(t) - \Delta T_c)] * h_B(t).$$

It follows that the error going into the VCO is

$$y_-^2(t, e) - y_+^2(t, e) = S(e) + N(t), \tag{6.55}$$

Fig. 6.17 The S-curve $S(e)$ in (6.56) with $\Delta = 0.5$ and $PD = 1$ (dashed curve) and its piecewise linear approximation (solid curve).

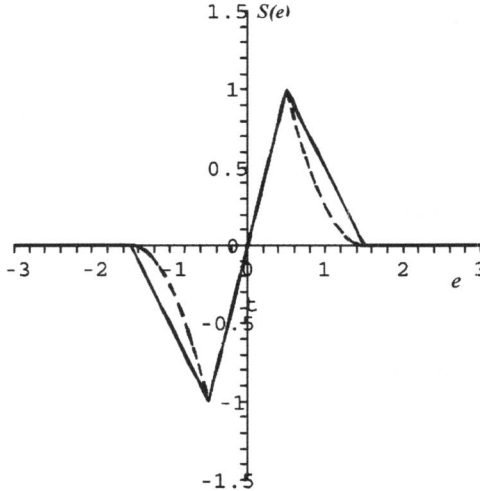

where

$$S(e) = PD[R_c^2(e - \Delta) - R_c^2(e + \Delta)], \tag{6.56}$$

D is the DC component of $\bar{b}^2(t)$, and the noise $N(t)$ consists of everything else. The loop low-pass filters the error signal that drives the VCO and corrects the the code delay of the local PN-code generator. The VCO model in operator form with $p = d/dt$ is given by

$$\hat{T} = KT_c F(p)[S(e) + N(t)]/p, \tag{6.57}$$

where K is the product of the VCO gain and multiplier gain and $F(s)$ is the transfer function of the loop filter. It follows that

$$\dot{e} = \frac{\dot{T}}{T_c} - KF(p)[S(e) + N(t)].$$

The graph of $S(e)$ and its piecewise linear approximation for $\Delta = 0.5$ is given in Figure 6.17, which is qualitatively the same as that of $H(x) = -g(e)$ in Figure 6.4. It follows that the MTLL due to Gaussian white noise can be calculated much the same way as in the case of range tracking.

6.3.2   Phase Noise in a Noncoherent DLL

The situation is different if on top of the additive Gaussian white noise there is also phase noise, whose spectral properties are quite different. Specifically, we

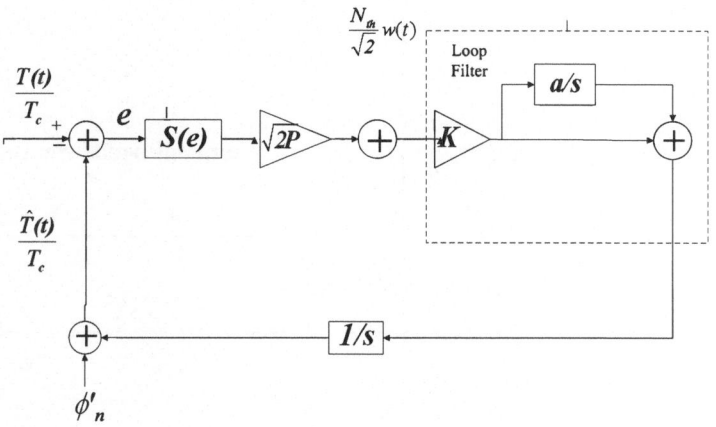

Fig. 6.18 Baseband equivalent model for nonlinear DLL.

consider the second-order nonlinear early–late DLL, shown in Figure 6.18 [169]. Its baseband equivalent model has normalized channel propagation delay $T(t)/T_c$, total power of received signal P, gains K and a that define the loop's filter, and phase noise $\tilde{\phi}_n$ with a given power spectrum (see (6.60) below). The loop filter used is of proportional integration type [36] with a zero at $s = -a$. The parameter $K[\sqrt{\text{Hz}}]$ determines the loop's gain, and the dimensionless parameter $a \neq 0$ stabilizes the loop. This type of loop can handle relative velocity $\dot{T}$ without a steady-state error, and relative acceleration $\ddot{T}$ with a steady-state error. The detailed derivation of the DLL S-curve and equations is given in [169], and a comprehensive study of the DLL is given in the dissertation [160]. The piecewise linear approximation to the S-curve is used for the early–late DLL used in spread spectrum synchronization of long PN sequences. Figure 6.19 shows the S-curve for an early–late DLL problem. The stable and unstable equilibrium points are the intersections of $S(e)$ with the line $S = T'' = \ddot{T}/a\sqrt{2PK_c}$. We assume the variant of the S-curve given in Figure 6.19,

$$
S(e) = \begin{cases}
2e, & |e| \leq \frac{1}{2}, \\
1.5 - e, & \frac{1}{2} < |e| < \frac{3}{2}, \\
-1.5 - e, & -\frac{3}{2} < |e| < -\frac{1}{2}, \\
0 & \text{otherwise.}
\end{cases}
$$

The resulting equations describing the system are

$$
e = \frac{T}{T_c} + \tilde{\phi}_n - \frac{\hat{T}}{T_c},
\tag{6.58}
$$

$$
dz = aK\sqrt{2P}\,S(e)\,dt + K\sqrt{\frac{N_{th}}{2}}\,dw(t),
$$

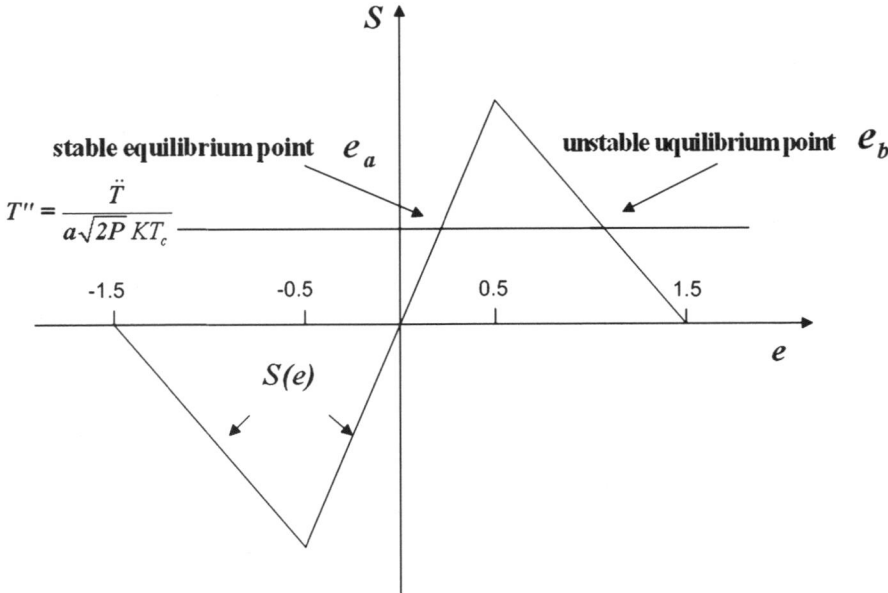

Fig. 6.19 S-curve for early–late DLL problem. The stable and unstable equilibrium points are the intersections of the $S(e)$ with the line $S = T'' = \ddot{T}/a\sqrt{2PK_c}$.

$$\frac{d\hat{T}}{T_c} = \left[z + K\sqrt{2P}S(e)\right]dt + K\sqrt{\frac{N_{th}}{2}}\,dw(t),$$

where e is the delay estimation error, dz/dt is the output of the loop filter, and the last equation is the output of the integrator. Here $w(t)$ is the standard Wiener process (Brownian motion), whose "derivative" $\dot{w}(t)$ is standard δ-correlated Gaussian white noise, independent of $v(t)$.

Differentiating e and setting $\tilde{z} = z - \dot{T}/T_c$, equations (6.58) become

$$de = -\left[\tilde{z} + K\sqrt{2P}S(e) + \dot{\phi}_n\right]dt - K\sqrt{\frac{N_{th}}{2}}\,dw(t), \qquad (6.59)$$

$$d\tilde{z} = aK\sqrt{2P}\left[S(e) + T''\right]dt + K\sqrt{\frac{N_{th}}{2}}\,dw(t),$$

where $T'' = \ddot{T}/a\sqrt{2P}KT_C$.

6.3.3 The Origin of $1/f$ Noise in Oscillators

The frequency generated by high-frequency oscillators contains a small but significant noise component, known as phase noise, also known as oscillator noise or phase jitter. The phase noise belongs to the family of stochastic processes with spectra $1/f^\alpha$, which exhibits scale invariance (or self-similarity) and a long-term correlation structure that decays polynomially in time. Both the phase and thermal noises cause errors in receivers that contain the oscillators. In particular, they cause losses of lock in phase-tracking systems such as the phase-locked loop in coherent systems, which include cellular phones, global positioning systems (GPS), and radar (e.g., synthetic aperture radar (SAR)), and in the DLL, which is an important component of code division multiple access (CDMA) receivers and interface to modern memory modules, such as double data rate synchronous dynamic random access memory (DDR-SDRAM).

Phase noise, also known as oscillator noise or phase jitter, is a well-known problem that does not yet have a full physical model (a recent example of a physical model is found in [102]) or extensive tools for mathematical manipulation. The phase drift and noise may be due to impurities, imperfections, thermal fluctuations, and other factors in the oscillator's crystal. The phase noise is usually described as having four parts [170]: the first is "frequency flicker" with power spectrum $1/f^3$; the second is "flat frequency" with power spectrum $1/f^2$; the third is "phase flicker" with power spectrum of the form $1/f$; and finally, the fourth is a "flat spectrum" phase. Often a white-noise term is added to represent thermal noise. This thermal noise should not be confused with the "flat spectrum" part of the phase noise. The family of stochastic processes with spectra $1/f^\alpha$ is of growing interest in many fields of research due to the wide variety of data for which they are inherently suited [108], [86]. This family of processes exhibits scale invariance (or self-similarity) and a long-term correlation structure that decays polynomially in time rather than exponentially, as is the case for the well-studied family of autoregressive moving average (ARMA) processes [109]. The long-term correlation structure of $1/f$ noise is due to the absence of a low-frequency cutoff in the spectrum, which results in nonstationarity. This means that an approximation that has a low-frequency cutoff (flattening of the spectrum below a certain frequency) has a finite correlation structure. Furthermore, "ideal" $1/f$ Gaussian noise cannot exist, because the single-time variance of the process is not finite. A model that has a low-frequency cutoff, as in the case at hand, can be Gaussian.

6.3.4   Diffusion Approximation to Phase Noise in a DLL

When the phase noise $\tilde{\phi}_n$ in (6.58) has a power spectrum, the stochastic system (6.59) is no longer of Itô type and is not amenable to the analysis developed in the previous sections. The system (6.59) can, however, be approximated by an Itô system by the approximation scheme used in Section 1.7. The resulting Itô system is now of higher order due to the auxiliary variables that define the approximating noise process. Thus the loss of lock problem for a DLL with $1/f^3$ phase noise is reduced to the classical exit problem, albeit at the expense of increased dimensionality.

The spectral power density

$$S_{\tilde{\phi}_n \tilde{\phi}_n}(\mathfrak{f}) = \frac{N_{ph}}{2 \, |2\pi \mathfrak{f}|^3} \tag{6.60}$$

indicates that the new noise process $\phi_n = \dot{\tilde{\phi}}_n$ is well defined and that its power spectral density function is given by

$$S_{\phi_n \phi_n}(\mathfrak{f}) = \frac{N_{ph}}{2 \, |2\pi \mathfrak{f}|}.$$

Thus, using (6.59), (1.177), we obtain the system

$$de(t) = -\left[\tilde{z}(t) + K\sqrt{2P}\,S(e(t)) - \phi_n(t)\right] dt - K\sqrt{\frac{N_{th}}{2}}\,dw(t), \tag{6.61}$$

$$d\tilde{z}(t) = \left[aK\sqrt{2P}\,S(e(t)) - \frac{\ddot{T}}{T_C}\right] dt + K\sqrt{\frac{N_{th}}{2}}\,dw(t),$$

$$dy_{2N}(t) = \left[y_{2N}(t) - 2\phi_n(t) - 4\sum_{k=1}^{N} y_{2k}(t)\right] dt + 2\sqrt{\frac{N_{ph}}{2}}\,dv(t),$$

$$dy_{2N-2}(t) = \left[y_{2N-2}(t) - 4y_{2N}(t) - 4\phi_n(t) - 8\sum_{k=1}^{N-1} y_{2k}(t)\right] dt$$
$$+ 4\sqrt{\frac{N_{ph}}{2}}\,dv(t),$$

$$dy_{2N-4}(t) = \left[y_{2N-4}(t) - 4y_{2N}(t) - 8y_{2N-2} - 6\phi_n(t) - 12\sum_{k=1}^{N-2} y_{2k}(t)\right] dt$$
$$+ 6\sqrt{\frac{N_{ph}}{2}}\,dv(t),$$

$$\vdots$$

$$dy_2(t) = \left[y_2(t) - 4\sum_{m=1}^{N-1} m y_{2(N-m+1)} - 2N\phi_n(t) - 4Ny_2(t)\right] dt$$
$$+ 2N\sqrt{\frac{N_{ph}}{2}}\,dv(t),$$

$$d\phi_n(t) = -\left[(2N+1)\phi_n(t) + 4\sum_{m=1}^{N} my_{2(N-m+1)}(t)\right] dt$$

$$+ 2(N+1)\sqrt{\frac{N_{ph}}{2}}\, dv(t).$$

Next, we normalize the equations so that the noise term converges to zero as the CNR term P/N_{ph} increases to infinity. The CNR, measured in Hz, is a well-accepted engineering quantity [78]. We introduce dimensionless time and define the auxiliary variables

$$\tau = \sqrt{2P}\,Kt, \quad \beta = \frac{\tilde{z}}{a}, \quad T'' = \frac{\ddot{T}}{a\sqrt{2P}\,KT_c} \tag{6.62}$$

to convert to the nondimensional system

$$de\,(\tilde{t}) = -\left[\frac{a}{\sqrt{2P}\,K}\beta\,(\tilde{t}) + S\,(e\,(\tilde{t})) - \frac{\phi_n\,(\tilde{t})}{\sqrt{2P}\,K}\right] d\tilde{t} - \sqrt{\frac{KN_{th}}{2\sqrt{2P}}}\, dw(\tilde{t}),$$

$$d\beta\,(\tilde{t}) = \left[S\,(e\,(\tilde{t})) - T''\right] d\tilde{t} + \sqrt{\frac{KN_{th}}{2\sqrt{2P}}}\, dw(\tilde{t}),$$

$$dy_{2N}\,(\tilde{t}) = \frac{1}{K\sqrt{2P}}\left[y_{2N}\,(\tilde{t}) - 2\phi_n\,(\tilde{t}) - 4\sum_{k=1}^{N} y_{2k}\,(\tilde{t})\right] d\tilde{t} + 2\sqrt{\rho}\, dv\,(\tilde{t}),$$

$$dy_{2N-2}\,(\tilde{t}) = \frac{1}{K\sqrt{2P}}\left[y_{2N-2}\,(\tilde{t}) - 4y_{2N}\,(\tilde{t}) - 4\phi_n\,(\tilde{t}) - 8\sum_{k=1}^{N-1} y_{2k}\,(\tilde{t})\right] d\tilde{t}$$

$$+ 4\sqrt{\rho}\, dv\,(\tilde{t}),$$

$$dy_{2N-4}\,(\tilde{t}) = \frac{1}{K\sqrt{2P}}\left[y_{2N-4}\,(\tilde{t}) - 4y_{2N}\,(\tilde{t}) - 8y_{2N-2}\,(\tilde{t}) - 6\phi_n\,(\tilde{t})\right.$$

$$\left. - 12\sum_{k=1}^{N-2} y_{2k}\,(\tilde{t})\right] d\tilde{t} + 6\sqrt{\rho}\, v\,(\tilde{t}),$$

$$\vdots \tag{6.63}$$

$$dy_2\,(\tilde{t}) = \frac{1}{K\sqrt{2P}}\left[y_2(\tilde{t}) - 4\sum_{m=1}^{N-1} my_{2(N-m+1)}\,(\tilde{t}) - 2N\phi_n(\tilde{t})\right.$$

$$\left. - 4Ny_2\,(\tilde{t})\right] d\tilde{t} + 2N\sqrt{\rho}\, dv\,(\tilde{t}),$$

$$d\phi_n\left(\tilde{t}\right) = \frac{1}{K\sqrt{2P}}\left[-(2N+1)\phi_n\left(\tilde{t}\right) - 4\sum_{m=1}^{N} my_{2(N-m+1)}\left(\tilde{t}\right)\right]d\tilde{t},$$
$$+ 2(N+1)\sqrt{\rho}\,dv\left(\tilde{t}\right),$$

where the dimensionless noise level is given by $\rho = N_{ph}/2K\sqrt{2P}$. For small values of ρ the system (6.63) can be viewed as a small stochastic perturbation of a nonlinear dynamical system that has a stable equilibrium at the point

$$e_a = \frac{T''}{2}, \quad \beta_a = -\frac{\ddot{T}}{T_c a^2}, \quad \phi_{n,a} = 0, \quad y_{2i,a} = 0, \quad 1 \le i \le N, \qquad (6.64)$$

and an unstable equilibrium point at

$$e_b = \frac{3}{2} - |T''|, \quad \beta_b = -\frac{\ddot{T}}{T_c a^2}, \quad \phi_{n,b} = 0, \quad y_{2i,b} = 0, \quad 1 \le i \le N, \quad (6.65)$$

which we refer to as the *saddle point*.

The stable equilibrium point (6.64) of the system (6.63) has a domain of attraction D. This means that any noiseless trajectory of (6.63) starting in D converges to the stable equilibrium point (6.64). The boundary of the region D is denoted by ∂D. As long as a trajectory of the stochastic system (6.63) remains in D, the DLL is said to be in a locked state. Upon exiting the region D through the boundary ∂D, the DLL is said to have lost lock. The exact description of the boundary ∂D is complex and is omitted here; however, in the limit of weak noise, the exit from D occurs in the immediate neighborhood of the saddle point. Thus the calculation of the MTLL is the classical exit problem of a dynamical system from the domain of attraction of a stable point under the influence of small noise [137, Chapter 10].

We denote a trajectory of (6.63) by

$$\boldsymbol{x}^T(t) = [e(t),\ \beta(t),\ y_{2N}(t),\ y_{2N-2}(t),\ \ldots,\ y_2(t),\ \phi_n(t)].$$

For each trajectory of (6.63) that starts in D, we denote by τ_D the first time it reaches the boundary ∂D (the first passage time to the boundary),

$$\tau_D = \inf\{t \ge 0 \,|\, \boldsymbol{x}(t) \in \partial D,\ \boldsymbol{x}(0) \in D\}, \qquad (6.66)$$

and its conditional expectation by

$$\bar{\tau}_D(\boldsymbol{x}) = \mathbb{E}[\tau_D \,|\, \boldsymbol{x}(0) = \boldsymbol{x}].$$

The MTLL is defined as

$$\bar{t}_L(x) = 2\bar{\tau}_D(x) \tag{6.67}$$

because once on ∂D, a trajectory is equally likely to return to D immediately or to leave D for a long time [138]. In the case of small ρ an analytic approximation to the MTLL can be obtained, as described below.

6.3.5 The Exit Problem for a DLL with $1/f^3$ Phase Noise

The eikonal equation (5.54) corresponding to the stochastic system (6.63) is given by

$$
H = \left(-\frac{a}{\sqrt{2PK}}\beta - S(e) + \frac{\phi_n}{\sqrt{2PK}} \right) \frac{\partial \Psi}{\partial e} + \left(S(e) - \frac{T''}{2} \right) \frac{\partial \Psi}{\partial \beta}
$$

$$
+ \frac{1}{\sqrt{2PK}} \sum_{i=1}^{N} \left[y_{2i} - 4 \sum_{l=1}^{N-i} l y_{2(N-l+1)} + 2(N-i+1) \right.
$$

$$
\times \left. \left(-\phi_n - 2\sum_{l=1}^{i} y_{2l} \right) \right] \frac{\partial \Psi}{\partial y_{2i}} + \frac{1}{\sqrt{2PK}} \left(-(2N+1)\phi_n \right.
$$

$$
\left. - 4\sum_{l=1}^{N} l y_{2(N-l+1)} \right) \frac{\partial \Psi}{\partial \phi_n} + \left(\sum_{l=1}^{N} 2(N-l+1)\frac{\partial \Psi}{\partial y_{2l}} + 2(N+1)\frac{\partial \Psi}{\partial \phi_n} \right)^2
$$

$$
+ \left(-K\sqrt{\frac{N_{th}}{N_{ph}}} \frac{\partial \Psi}{\partial e} + K\sqrt{\frac{N_{th}}{N_{ph}}} \frac{\partial \Psi}{\partial \beta} \right)^2 = 0.
$$

The solution in the slab $-1/2 \le e \le 1/2$, corresponding to the linear part of the S-curve (see Figure 6.19), is the quadratic form (5.83), determined by the solution to Lyapunov's (Riccati's) equation (5.85). The system in the linear region can be written as

$$d x(t) = A x(t)\, dt + B\, dv(t), \tag{6.68}$$

where $v(t)$ is a vector of standard Brownian motions independent of $w(t)$ and $v(t)$ in (6.63) and the matrices A and B are given by

$$A = \frac{1}{\sqrt{2P}\,K}$$

$$\times \begin{pmatrix}
-2\sqrt{2P}\,K & -a & 0 & 0 & 0 & \cdots & 0 & -1 \\
2\sqrt{2P}\,K & 0 & 0 & 0 & 0 & \cdots & 0 & 0 \\
0 & 0 & -3 & -4 & -4 & \cdots & -4 & -2 \\
0 & 0 & -4 & -7 & -8 & \cdots & -8 & -4 \\
0 & 0 & -4 & -8 & -11 & \cdots & -12 & -6 \\
\vdots & & & & & & & \\
0 & 0 & -4 & -8 & \cdots & -4(N-1) & -4N-1 & -2N \\
0 & 0 & -4 & -8 & -12 & \cdots & -4N & -2N-1
\end{pmatrix}$$

and

$$\boldsymbol{B}^T = \sqrt{\frac{N_{ph}}{2K\sqrt{2P}}} \begin{pmatrix} -K\sqrt{\dfrac{N_{th}}{N_{ph}}} & K\sqrt{\dfrac{N_{th}}{N_{ph}}} & 0\ 0 \cdots & 0 & 0 \\ 0 & 0 & 2\ 4 \cdots & 2N & 2(N+1) \end{pmatrix},$$

respectively. The explicit solution of (5.85) can be obtained using standard symbolic mathematical packages such as Maple or Mathematica.

To solve (6.68) outside the strip $|e| < 1/2$, we use the method of characteristics, as described in Section 5.3 above. We define the components of the vector $\boldsymbol{p}$ by the equations

$$p = \frac{\partial \Psi}{\partial e}, \quad q = \frac{\partial \Psi}{\partial \beta}, \quad \alpha_1 = \frac{\partial \Psi}{\partial y_2}, \quad \alpha_2 = \frac{\partial \Psi}{\partial y_4}, \quad \cdots$$

$$\alpha_N = \frac{\partial \Psi}{\partial y_{2N}}, \quad r = \frac{\partial \Psi}{\partial \phi_n}. \tag{6.69}$$

Now, taking the total derivative of H with respect to time, we get

$$0 = \frac{dH}{dt} = \frac{\partial H}{\partial e}\frac{de}{dt} + \frac{\partial H}{\partial \beta}\frac{d\beta}{dt} + \frac{\partial H}{\partial \phi_n}\frac{d\phi_n}{dt} \tag{6.70}$$

$$+ \sum_{i=1}^{N} \frac{\partial H}{\partial y_{2i}}\frac{dy_{2i}}{dt} + \frac{\partial H}{\partial p}\frac{dp}{dt} + \frac{\partial H}{\partial q}\frac{dq}{dt} + \frac{\partial H}{\partial r}\frac{dr}{dt} + \sum_{i=1}^{N} \frac{\partial H}{\partial \alpha_i}\frac{d\alpha_i}{dt}.$$

The characteristic equations (5.80)–(5.82) are given by

$$\frac{de}{dt} = \frac{\partial H}{\partial p}, \quad \frac{dp}{dt} = -\frac{\partial H}{\partial e}, \quad \frac{d\beta}{dt} = \frac{\partial H}{\partial q}, \quad \frac{dq}{dt} = -\frac{\partial H}{\partial \beta} \tag{6.71}$$

$$\frac{dy_2}{dt} = \frac{\partial H}{\partial \alpha_1}, \quad \frac{d\alpha_1}{dt} = -\frac{\partial H}{\partial y_2}, \quad \frac{dy_4}{dt} = \frac{\partial H}{\partial \alpha_2}, \quad \frac{d\alpha_2}{dt} = -\frac{\partial H}{\partial y_4},$$

$$\vdots$$

$$\frac{dy_{2N}}{dt} = \frac{\partial H}{\partial \alpha_N}, \quad \frac{d\alpha_N}{dt} = -\frac{\partial H}{\partial y_{2N}}, \quad \frac{d\phi_n}{dt} = \frac{\partial H}{\partial r}, \quad \frac{dr}{dt} = -\frac{\partial H}{\partial \phi_n}.$$

Inserting (6.68) and (6.69) into (6.71), we get

$$\frac{de}{dt} = -\frac{a}{K\sqrt{2P}}\beta - S(e) + \frac{\phi_n}{K\sqrt{2P}} - 2K^2\frac{N_{th}}{N_{ph}}(-p+q), \tag{6.72}$$

$$\frac{d\beta}{dt} = S(e) - T'' + 2K^2\frac{N_{th}}{N_{ph}}(-p+q),$$

$$\frac{dy_{2i}}{dt} = \frac{1}{K\sqrt{2P}}\left\{y_{2i} - 4\sum_{l=1}^{N-i}ly_{2(N-l+1)} - 2(N-i+1)\left(\phi_n + 2\sum_{l=1}^{i}y_{2l}\right)\right\}$$

$$+ 4(N-i+1)\left(\sum_{l=1}^{N}2(N-l+1)\alpha_l + 2(N+1)r\right) \quad \text{for all } 1 \le i \le N,$$

$$\frac{d\phi_n}{dt} = \frac{1}{K\sqrt{2P}}\left\{-(2N+1)\phi_n - \sum_{l=1}^{N}4ly_{2(N-l+1)}\right\}$$

$$+ 4(N+1)\left(\sum_{l=1}^{N}2(N-l+1)\alpha_l + 2(N+1)r\right),$$

$$\frac{dp}{dt} = (p-q)S'(e), \quad \frac{dq}{dt} = \frac{a}{K\sqrt{2P}}p,$$

and

$$\frac{d\alpha_i}{dt} = -\frac{1}{K\sqrt{2P}}\left\{\alpha_i - 4\sum_{l=i+1}^{N}(N-l+1)\alpha_l - 4(N-i+1)\sum_{l=1}^{i}\alpha_i\right\}$$

$$+ \frac{r}{K\sqrt{2P}}4(N-i+1) \quad \text{for all } 1 \le i \le N,$$

$$\frac{dr}{dt} = -\frac{p}{K\sqrt{2P}} + \frac{1}{K\sqrt{2P}}\sum_{i=1}^{N}2(N-i+1)\alpha_i + \frac{r}{K\sqrt{2P}}(2N+1).$$

To complete the solution of (6.68), we must show that $H = 0$ for at least one point. Taking the total derivative $d\Psi/dt$ along a characteristic, and using (6.68) and (6.72), we get

$$\frac{d\Psi}{dt} = \frac{\partial\Psi}{\partial e}\frac{de}{dt} + \frac{\partial\Psi}{\partial\beta}\frac{d\beta}{dt} + \frac{\partial\Psi}{\partial\phi_n}\frac{d\phi_n}{dt} + \sum_{i=1}^{N}\frac{\partial\Psi}{\partial y_{2i}}\frac{dy_{2i}}{dt} \tag{6.73}$$

$$= H + \sum_{i=1}^{N} 2\,(N-i+1)\,\alpha_i \left(\sum_{l=1}^{N} 2\,(N-l+1)\,\alpha_l + 2\,(N+1)\,r\right)$$

$$+ 2\,(N+1)\left(\sum_{l=1}^{N} 2\,(N-l+1)\,\alpha_l + 2\,(N+1)\,r\right) r$$

$$= H + \left(\sum_{l=1}^{N} 2\,(N-l+1)\,\alpha_l + 2\,(N+1)\,r\right)^2.$$

Thus, $H = 0$ if

$$\frac{d\Psi}{dt} = \left(\sum_{l=1}^{N} 2\,(N-l+1)\,\alpha_l + 2\,(N+1)\,r\right)^2 + K^2\frac{N_{th}}{N_{ph}}(-p+q)^2. \tag{6.74}$$

Equations (6.72) and (6.74) represent the solution of (6.68) on each characteristic curve. The boundary ∂D is spanned by characteristic curves that converge to the saddle point, and Ψ decreases on each characteristic to its value $\hat{\Psi}$ at the saddle point [137, Section 10.2].

6.3.6 MTLL in a Second-Order DLL with $1/f^3$ Noise

The S-curve $S(e)$ for a DLL is given in Figure 6.19. The stable equilibrium point of the system, in the absence of relative motion between the transmitter and the receiver, is the point where the S-curve vanishes with positive slope, and the two unstable equilibrium points (the saddle points) are the points where it vanishes with negative slopes. In case there is relative motion with constant acceleration $\ddot{T}$ (see (6.62)), the equilibrium points of the dynamics (6.63) are the points where the S-curve intersects the line $S = T''$ (see Figure 6.19). For $\ddot{T} \neq 0$ there are one stable equilibrium point and one unstable equilibrium point, given by (6.64) and (6.65), respectively.

The minimum value $\hat{\Psi}$ determines the leading-order term (or small ρ exponential growth rate) of the MTLL (see (5.53)), so we have to determine it by finding the characteristic that hits the saddle point and the limiting value of Ψ there. To this end, we start the characteristic on the hyperplane $e = 1/2$, where Ψ is given explicitly by (5.83). Because the characteristic equations are linear in the half-space to the right of the hyperplane $e = 1/2$, all characteristics diverge exponentially fast, except one that corresponds to the only negative eigenvalue of the system matrix (see below).

Thus, the starting point of the desired characteristic is the column corresponding to this eigenvalue in the matrix that reduces the system matrix to its Jordan canonical form.

Specifically, we observe in (6.72) and (6.74) that the quasipotential Ψ is dependent only on the variables $\{p, q, \alpha, r\}$. We define the state vector

$$v = \{p, q, \alpha, r\}^T. \tag{6.75}$$

In the strip $|e| < 1/2$ the S-curve is linear, so that the system (6.72) is linear, and with the notation (6.75) it can be written as

$$\dot{v} = Mv. \tag{6.76}$$

Because we are looking for the minimum $\hat{\Psi} = \Psi(x_b)$, and x_b is the saddle point of the system (6.63), we need only find the starting point of the characteristic that hits the saddle point (6.65). In the linear strip $|e| < 1/2$ we can use (5.83) and start shooting characteristic trajectories (6.71) from the hyperplane $e = 1/2$. For clarity, we explain the method for finding $\hat{\Psi}$ by considering the simplest case of a noise approximation of order $N = 1$, loop parameters $P = 1/2$, $K = a = 1$, and without thermal noise ($N_{th} = 0$). For these parameters the value of $\Psi(x)$ in the linear domain $|e| < 1/2$ is given by

$$\Psi(x) = \frac{1}{64} \left\{ -424\phi_n e + 1408 y_2 \beta + 832 y_2 e - 704 y_2 \phi_n + 900\beta \frac{\ddot{T}}{T_c} \right.$$

$$+ 388 e \frac{\ddot{T}}{T_c} + 784 e \beta - 436\phi_n \frac{\ddot{T}}{T_c} - 648\phi_n \beta + 992 y_2 \frac{\ddot{T}}{T_c} + 646\beta^2$$

$$\left. + 792 y_2^2 + 170\phi_n^2 + 396 e^2 + 353 \left(\frac{\ddot{T}}{T_c} \right)^2 \right\}. \tag{6.77}$$

The coordinate of the point on the hyperplane $e = 1/2$, where the shooting begins, is denoted by v_0. It is chosen as the coordinate $(\beta_0, y_0, \phi_{n,0})$ of a point on the unique stable characteristic trajectory of (6.76). We write (6.76) in the linear domain $e \geq 1/2$ as the linear system

$$\dot{u} = \Lambda u, \tag{6.78}$$

where Λ is diagonal with the eigenvalues of M on its diagonal. We can write

$$v = Pu, \quad \Lambda = P^{-1} M P. \tag{6.79}$$

The columns of P are the eigenvectors of the matrix M with respect to the eigenvalues on the diagonal Λ. The matrix M has only one negative eigenvalue,

$$\lambda_1 = -\frac{1}{2} - \frac{\sqrt{5}}{2}, \tag{6.80}$$

and thus only the eigenvector corresponding to that eigenvalue leads to a stable solution of (6.78). Assuming that the negative eigenvalue is the first element in Λ, we need only take the first element in u, replacing the others by zeros.

The initial values are defined for all $1 \leq i \leq N$ by

$$p_0 = \frac{\partial \Psi \left(\frac{1}{2}, \beta_0, y_0, \phi_{n,0} \right)}{\partial e}, \quad q_0 = \frac{\partial \Psi \left(\frac{1}{2}, \beta_0, y_0, \phi_{n,0} \right)}{\partial z},$$

$$r_0 = \frac{\partial \Psi \left(\frac{1}{2}, \beta_0, y_0, \phi_{n,0} \right)}{\partial r}, \quad \alpha_{i,0} = \frac{\partial \Psi \left(\frac{1}{2}, \beta_0, y_0, \phi_{n,0} \right)}{\partial y_{2i}}. \tag{6.81}$$

Using (6.77), (6.80), and (6.81), we get

$$p_0 = -\left(\frac{11}{8} + \frac{7}{8}\sqrt{5} \right) u_{1,0}, \quad q_0 = \left(\frac{3}{2} + \frac{\sqrt{5}}{4} \right) u_{1,0},$$

$$r_0 = -\left(\frac{7}{8} + \frac{\sqrt{5}}{8} \right) u_{1,0}, \quad \alpha_{1,0} = u_{1,0}. \tag{6.82}$$

It follows from (6.78) that $u_1 = u_{1,0}\exp\left\{ \lambda_1 \tilde{t} \right\}$. Now we can solve for the initial conditions $(\beta_0, y_0, \phi_{n,0})$ from equations (6.77), (6.81), and (6.82). Having found the initial conditions, we can proceed to integrate (6.74) to find the minimal value $\hat{\Psi}$ on ∂D,

$$\hat{\Psi} = \Psi \left(\frac{1}{2}, \beta_0, y_0, \phi_{n,0} \right)$$

$$+ 4 \int_0^\infty \left\{ \sum_{l=1}^N (N - l + 1)\alpha_l(t) + (N + 1)r(t) \right\}^2 dt, \tag{6.83}$$

where $\alpha_l(t)$ and $r(t)$ are calculated on the characteristic that starts at $(\frac{1}{2}, \beta_0, y_0, \phi_{n,0})$. For the case of zero acceleration between transmitter and receiver, i.e., $\ddot{T}/T_c = 0$, we get $\hat{\Psi} = 0.9120$. The analogous computation for Nth order approximations can be done by solving the Lyapunov equation numerically or symbolically (e.g., with Maple or Mathematica), finding the negative eigenvalue, and determining the matrix P. This was done with the results in Table 6.1. As can be seen in Table 6.1, the minimum value $\hat{\Psi}$ changes very slightly for $N \geq 5$. Thus, approximation of order $N = 5$ for the noise is sufficient for the calculation of the MTLL of

Table 6.1 $\hat{\Psi}$ and MTLL for different N.

N	$\hat{\Psi}$	MTLL	MTLL	MTLL
		CNR $= 1, \rho = 0.25$	CNR $= 1.5, \rho = 0.1667$	CNR $= 2.5, \rho = 0.1$
1	0.91204	77	476	18280
2	0.90916	76	468	17761
3	0.91027	76	471	17959
4	0.91077	76	472	18049
5	0.91094	76	473	18080
6	0.91102	76	473	18094
8	0.91108	77	473	18105
10	0.91110	77	473	18109
20	0.91111	77	473	18111
30	0.91111	77	473	18111

100–1000 seconds. This range of values of the MTLL is chosen because for MTLL less than 100 seconds the leading-order approximation is insufficient. The problem of maximizing the MTLL is most critical at low CNR, where the majority of losses of lock occur and the MTLL is still below 1000 seconds. The MTLL increases exponentially with the CNR, so higher accuracy of $\hat{\Psi}$ is needed. Long MTLLs are of less interest in the optimization process. The range of validity of the leading-order approximation is for values of the CNR that result in $\rho \ll \hat{\Psi}$. We have disregarded the pre-exponential factor in the asymptotic formula (5.53) for the MTLL, because the main contribution to the MTLL comes from the exponential term. Furthermore, because we assume a constant prefactor, the results of the simulations might be slightly displaced from the theoretical line. The prefactor can be resolved by simulations for small MTLL and then applied to large MTLL, where simulations are impractical.

The result can be understood as follows. The loop's noise equivalent bandwidth is $\frac{1}{2}$ Hz, which is much smaller than the region of validity of the truncated continued fraction approximation to $1/\mathfrak{f}$. Furthermore, for MTLL of order 100–1000 seconds, the corresponding frequency range is 10^{-3} Hz $\leq \mathfrak{f} \leq 10^{-2}$ Hz. Because the region of validity of the approximation for $N = 5$ is in the range 10^{-3} Hz $\leq \mathfrak{f} \leq 10$ Hz (Figure 1.5), it is understandable that using the approximation for the $1/\mathfrak{f}$ noise with $N = 5$ results in an accurate value of $\hat{\Psi}$, which in turn accurately approximates the exponential growth rate of the MTLL.

In general, one would expect that as more phase noise enters the DLL, the MTLL will become smaller. In Figure 1.5 we see that as the approximation order becomes larger, more energy enters at very low and very high frequencies. However, we see in Table 6.1 that $\hat{\Psi}$ actually increases monotonically for $N \geq 2$. The resolution of the apparent paradox consists in taking a closer look at the transfer function of our $1/\mathfrak{f}$ approximation. Before flattening out below a certain low frequency, the transfer function displays a "knee" that is above the $1/\mathfrak{f}$ curve. Further, for a specific MTLL, only frequencies larger than 1/MTLL need be considered and only frequencies smaller than the loop's bandwidth should be accounted for. In this frequency band

Table 6.2 Energy in $0.01\,\text{Hz} \leq$
$\mathfrak{f} \leq 0.5\,\text{Hz}$ frequency band for
approximation orders N.

N	Energy
1	4.7995
2	4.3968
3	4.1909
4	4.0139
5	3.9280
6	3.9050
7	3.9048
8	3.9086
9	3.9112
10	3.9121

the noise entering the loop actually decreases as the phase noise approximation order is increased, because the "knee" moves to lower frequencies that are irrelevant to the problem at hand. For example, let us consider an MTLL of 100 seconds. The frequency band in question is $0.01 \leq \mathfrak{f} \leq 0.5$. In Table 6.2 the energy in the $0.01 \leq \mathfrak{f} \leq 0.5$ frequency band is given for different approximation orders N. Because for $N < 6$ the "knee" is above $f = 0.01\,\text{Hz}$, the decrease in energy for increasing N explains the increasing $\hat{\Psi}$. From Table 6.1 we learn that for $N \geq 6$ the changes in $\hat{\Psi}$ result in an insignificant rise in MTLL. In fact, the difference between MTLL for CNR that gives MTLL of 100 seconds for approximation order $N = 6$ and MTLL given for approximation order $N = 30$ with the same CNR is less than 0.1%.

Monte–Carlo simulation results for the MTLL are shown in Figure 6.20 for a second-order DLL under the influence of $1/\mathfrak{f}^3$ noise approximation of order $N = 5$. The loop parameters were taken as $P = \frac{1}{2}, a = K = 1$. The solid line is the derived analytic leading-order of the MTLL, and the asterisks denote the Monte–Carlo simulation results (each asterisk represents the mean result of 50 trials). Similar results for simulations with the "exact" $1/\mathfrak{f}$ discrete noise, generated according to [83], are presented in Figure 6.21 along with the results of the MTLL under the influence of $1/\mathfrak{f}$ noise approximation with $N = 20$.

Figures 6.20 and 6.21 show that the analytic calculation of the leading-order term of the MTLL results in a model that fits well the Monte–Carlo simulation results. Furthermore, the similarity of the analytic calculation of the leading-order term of the MTLL to those calculated by Monte–Carlo simulations with the "exact" discrete $1/\mathfrak{f}$ shows that the truncation of our model at the appropriate N provides very accurate results for the calculation of the MTLL for the second-order DLL. The dependence of the loop's parameters on the power P can be eliminated by a proper AGC (automatic gain control) loop.

Example 6.3 (Optimization of loop parameters). Loop parameters can be optimized by finding the values of a and K that yield the maximum $\hat{\Psi}$. In the case

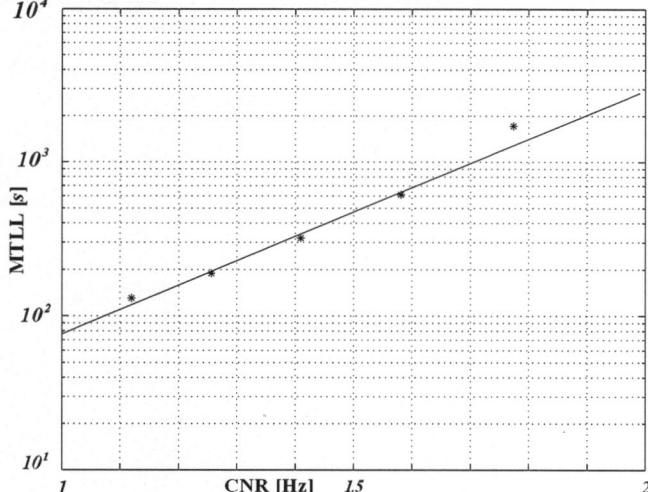

Fig. 6.20 The MTLL of a second-order DLL under the influence of $1/f^3$ noise approximation of order $N = 5$. The loop parameters were taken as $P = \frac{1}{2}, a = K = 1$. The solid line is the derived analytic leading-order of the MTLL, and the asterisks denote the Monte–Carlo simulation results (each asterisk represents the mean result of 50 trials).

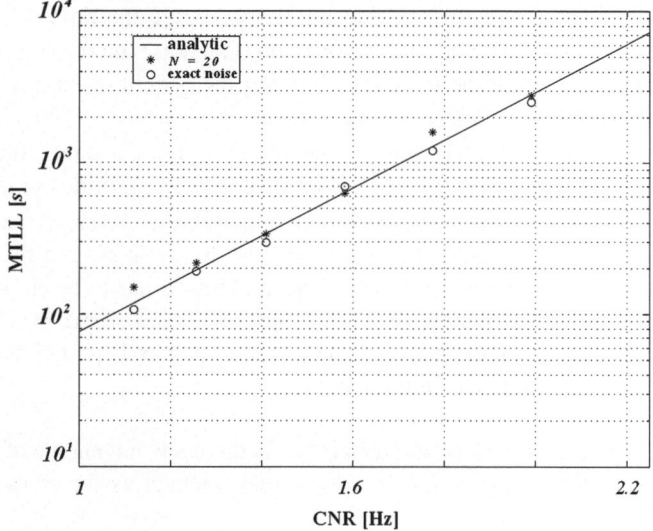

Fig. 6.21 The MTLL of a second-order DLL under the influence of $1/f^3$ noise approximation of order $N = 20$ (denoted by asterisks—mean of 50 trials each) along with the MTLL under the influence of exact discrete noise (denoted by pluses—mean of 50 trials each). The loop parameters were taken as $P = \frac{1}{2}, a = K = 1$. The solid line is the derived analytic leading-order of the MTLL.

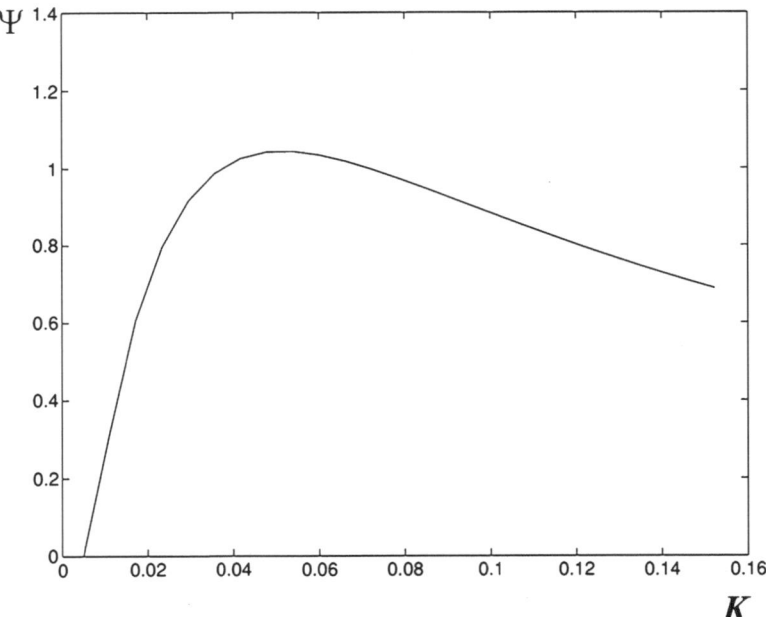

Fig. 6.22 Optimizing for loop filter parameter K for $a = 100$.

at hand it is easy to see that the best result is obtained in the limit $K \to \infty$ (see (6.63)). In a real system with additive channel thermal noise, increasing K increases the thermal noise entering the loop, thus limiting the benefit of increasing K. The analysis of the case with additive thermal noise by choosing $N_{th} = N_{ph}/10$, $P = \frac{1}{2}$, and $\ddot{T} = 0.5$ shows (see (6.63)) that for the system to remain stable, the condition $aK \geq \ddot{T}/\sqrt{2PT_c}$ has to be satisfied, which in our case simplifies to $aK \geq \frac{1}{2}$. It can be easily shown that $\hat{\Psi}$ increases monotonically as a increases, but only to a very slight extent, for example, beyond $a = 100$. In a real system the loop filter coefficients cannot be chosen arbitrarily large, and thus a has to be chosen as large as possible for any given realizable K. The plot of $\hat{\Psi}$ vs K for $a = 100$ is given in Figure 6.22. Additional elements of phase noise, such as $1/f^2$, $1/f$, flat segment, and so on can be handled in a similar manner.   □

Exercise 6.12 (Loss of lock in smoothers*). Is there any advantage of smoothing over filtering of FM transmission in a low-noise channel as far as the MTLL is concerned?   □

6.4 Annotations

Section 6.2 is based on [19]. The exposition in Section 6.3.1 is based on the standard texts [169], [141] and on the more recent dissertation [160]. The MTLL in DLLs has been calculated for Gaussian white noise in [82], [154], [153], [13]. The case of $1/f^{\alpha}$ phase noise considered in Sections 6.3.2–6.3.5 was considered in [98], which is the basis for this exposition. A considerable body of work has been devoted to $1/f^{\alpha}$ processes (see [86], [109], [35], [9], [168], [157], [158], [159], [83], [37], [46], and references therein). An approximation to $1/f$ noise by an output of a linear system of first-order stochastic differential equations, driven by a vector of white noises, is given in [115].

Chapter 7
Phase Tracking with Optimal Lock Time

The MTLL is the fundamental performance criterion in phase tracking and synchronization systems. Thus, for example, a phase-tracking system is considered locked as long as the estimation error $e(t) = x(t) - \hat{x}(t)$ is in $(-\pi, \pi)$. When the error exceeds these limits, the tracker is said to be unlocked, and it relocks on an erroneous equilibrium point, with a deviation of 2π. Another example is an automatic sight of a cannon. The sight is said to be locked on target if the positioning error is somewhere between certain limits. There are similar problems in which the maximization of exit time is an optimality criterion [114].

If maximizing the MTLL is chosen as the optimality criterion for phase estimation, then the PLL, which is the MMSEE estimator, may no longer be optimal. Several problems arise with optimal MTLL phase estimation: does the optimal estimator ever lose lock? If so, how does its MTLL compare with that of the PLL? Can the threshold (Exercise 5.5) be moved? By how much (in dB)?

The main result of this chapter is Theorem 7.4.1, which identifies the minimum MTLL filter at low noise as the MNE filter $x_{\mathrm{MNE}}(t)$. The sections leading to this theorem provide its derivation and proof. The remaining sections are devoted to applications of this theorem to benchmark phase trackers. They provide answers to the above questions.

7.1 State Equations

We consider the system (3.22), (3.23). For any adapted process $\hat{x}(t) \in \mathcal{C}(\mathbb{R}^+)$ (measurable with respect to the filtration generated by $y(t)$), we define the error process

$$e(t) = x(t) - \hat{x}(t) \tag{7.1}$$

and the first time to lose lock

$$\tau = \inf \{t \mid e(t) \in \partial L\} . \tag{7.2}$$

Z. Schuss, *Nonlinear Filtering and Optimal Phase Tracking*, Applied Mathematical Sciences 180, DOI 10.1007/978-1-4614-0487-3_7, © Springer Science+Business Media, LLC 2012

The optimal filtering problem is to maximize $\mathbb{E}[\tau \mid y_0^\tau]$ (see definition (7.14) below) with respect to all adapted continuous functions $\hat{x}(t)$. For example, if $h(x, t) = \sin x$ in a phase estimation problem, then $L = (-\pi, \pi)$ and lock is lost when $e(t) = \pm\pi$.

We can rewrite the model equations (3.22), (3.23) in terms of the error process $e(t)$ as

$$de(t) = M_{\hat{x}}(e(t), t) \, dt + \varepsilon\sigma \, dw(t), \tag{7.3}$$

$$dy(t) = H_{\hat{x}}(e(t), t) \, dt + \varepsilon\rho \, dv(t), \tag{7.4}$$

where

$$M_{\hat{x}}(e(t), t) = m(\hat{x}(t) + e(t)) - \dot{\hat{x}}(t),$$

$$H_{\hat{x}}(e(t), t) = h(\hat{x}(t) + e(t)),$$

and the filtering problem is to find $\hat{x}(t)$ such that $\mathbb{E}[\tau \mid y_0^\tau]$ is maximal. In higher-dimensional systems, e, m, M, y, h, H are replaced with the vectors e, m, M, y, h, and H, respectively.

The survival probability of a trajectory $(e(t), y(t))$ of (7.3) with absorption at ∂L and (7.4) can be expressed in terms of the pdf $p_\varepsilon(e, y, t \mid \xi, \eta, s)$ of the two-dimensional process with an absorbing boundary condition on ∂L. It is the solution of the FPE

$$\frac{\partial p_\varepsilon(e, y, t \mid \xi, \eta, s)}{\partial t}$$

$$= -\frac{\partial M_{\hat{x}}(e, t) p_\varepsilon(e, y, t \mid \xi, \eta, s)}{\partial e} - \frac{\partial H_{\hat{x}}(e, t) p_\varepsilon(e, y, t \mid \xi, \eta, s)}{\partial y}$$

$$+ \frac{\varepsilon^2\sigma^2}{2} \frac{\partial^2 p_\varepsilon(e, y, t \mid \xi, \eta, s)}{\partial e^2} + \frac{\varepsilon^2\rho^2}{2} \frac{\partial^2 p_\varepsilon(e, y, t \mid \xi, \eta, s)}{\partial y^2} \tag{7.5}$$

for $e, \xi \in L$, $y, \eta \in \mathbb{R}$, with the boundary and initial conditions

$$p_\varepsilon(e, y, t \mid \xi, \eta, s) = 0 \text{ for } e \in \partial L, \ y \in \mathbb{R}, \ \xi \in L, \ \eta \in \mathbb{R}, \tag{7.6}$$

$$p_\varepsilon(e, y, s \mid \xi, \eta, s) = \delta(e - \xi)\delta(y - \eta) \text{ for } e \in L, \ y \in \mathbb{R}, \ \xi \in L, \ \eta \in \mathbb{R}. \tag{7.7}$$

The pdf is actually the joint density and probability function

$$p_\varepsilon(e, y, t \mid \xi, \eta, s) = \Pr\{e(t) = e, y(t) = y, \tau > t \mid \xi, \eta, s\},$$

and thus the survival probability is

$$\Pr\{\tau > t \mid \xi, \eta, s\} = S_{e(\cdot), y(\cdot)}(t) = \int_{\mathbb{R}} \int_{\mathbb{R}} p_\varepsilon(e, y, t \mid \xi, \eta, s) \, de \, dy, \tag{7.8}$$

and it decays in time.

7.2 Simulation with Particles

To simulate the filtering problem on a finite interval $0 \le t \le T$, we discretize (3.22), (3.23) on a sequence of grids

$$\left\{ t_i = i\Delta t, \quad i = 0, 1, \ldots, N, \quad \Delta t = \frac{T}{N} \right\},$$

and define discrete trajectories by the Euler scheme

$$x_N(t_{i+1}) = x_N(t_i) + \Delta t \, m(x_N(t_i), t_i) + \varepsilon \sigma \, \Delta w(t_i), \tag{7.9}$$

$$y_N(t_{i+1}) = y_N(t_i) + \Delta t \, h\left(x_N(t_i), t_i\right) + \varepsilon \rho \, \Delta v(t_i), \tag{7.10}$$

for $i = 0, 1, \ldots, N - 1$, where $\Delta w(t_i)$ and $\Delta v(t_i)$ are independent zero-mean Gaussian random variables with variance Δt. The discretized version of (7.3), (7.4) is

$$e_N(t_{i+1}) = e_N(t_i) + \Delta t \, M_{\hat{x}}\left(e_N(t_i), t_i\right) + \varepsilon \sigma \, \Delta w(t_i), \tag{7.11}$$

$$y_N(t_{i+1}) = y_N(t_i) + \Delta t \, H_{\hat{x}}\left(e_N(t_i), t_i\right) + \varepsilon \rho \, \Delta v(t_i). \tag{7.12}$$

The particle filter imitates (3.52) and (3.63). Given an observed trajectory $\{y_N(t_i)\}_{i=0}^{N}$, we sample n trajectories $\{\{x_{j,N}(t_i)\}_{i=0}^{N}\}_{j=1}^{n}$, according to the scheme (7.9), which produce error trajectories $\{\{e_{j,N}(t_i)\}_{i=0}^{N}\}_{j=1}^{n}$, and determine their first exit times from L, denoted by $\{\tau_{j,N}\}_{j=1}^{n}$ (we set $\tau_{j,N} = T$ if $\{e_{j,N}(t_i)\}_{i=0}^{N}$ does not exit L by time T). Setting $\tilde{\tau}_{j,N} = [(\tau_{j,N} \wedge T)/\Delta t]$ and

$m_{j,N}$

$$= \exp\left\{ \frac{1}{\varepsilon^2 \rho^2} \sum_{k=0}^{\tilde{\tau}_{j,N}} \left[H(e_{j,N}(t_{k-1}), t_{k-1})\Delta y_{k,N} - \frac{1}{2}H^2(e_{j,N}(t_{k-1}), t_{k-1})\Delta t \right] \right\},$$

the conditional MTLL is defined on the ensemble by

$$\frac{\sum_{j=1}^{n}(\tau_{j,N} \wedge T)m_{j,N}}{\sum_{j=1}^{n} m_{j,N}}. \tag{7.13}$$

It follows that

$$\mathbb{E}[\tau \mid y_0^\tau] = \lim_{T \to \infty} \lim_{n \to \infty} \lim_{N \to \infty} \frac{\sum_{j=1}^{n} (\tau_{j,N} \wedge T) m_{j,N}}{\sum_{j=1}^{n} m_{j,N}}. \tag{7.14}$$

The conditional MTLL $\mathbb{E}[\tau \mid y_0^\tau]$ is a random variable on the σ-algebra of events generated by the measurements process $y(\cdot)$. Our purpose is to find $\hat{x}(t)$ that maximizes $\mathbb{E}[\tau \mid y_0^\tau]$ in the class of continuous adapted functions.

7.3 The Joint pdf of the Discrete Process

We proceed to derive a Zakai equation for the a posteriori pdf of the error, when trajectories are terminated at ∂L, much as in Section 3.5. The pdf of a trajectory of $(e_N(t), y_N(t))$ is the Gaussian

$$p_N(e_1, e_2, \dots, e_N; y_1, y_2, \dots, y_N; t_1, t_2, \dots, t_N)$$

$$= \prod_{k=1}^{N} \frac{\exp\left\{ -\dfrac{\mathcal{B}_k(x_k, x_{k-1})}{2\varepsilon^2 \Delta t} \right\}}{2\pi \varepsilon^2 \rho \sigma \Delta t}, \tag{7.15}$$

where the exponent is the quadratic form

$$\mathcal{B}_k(x_k, x_{k-1}) = [x_k - x_{k-1} - \Delta t a_{k-1}]^T \, B \, [x_k - x_{k-1} - \Delta t a_{k-1}],$$

such that

$$x_k = \begin{bmatrix} e_k \\ y_k \end{bmatrix}, \quad a_k = \begin{bmatrix} M_{\hat{x}}(e_k, t_k) \\ H_{\hat{x}}(e_k, t_k) \end{bmatrix}, \quad B = \begin{bmatrix} \sigma^{-2} & 0 \\ 0 & \rho^{-2} \end{bmatrix}.$$

The Wiener path integral [137, Chapter 3]

$$p_\varepsilon(e, y, t \mid \xi, \eta, s)$$

$$= \lim_{N \to \infty} \underbrace{\int_L de_1 \int_L de_2 \cdots \int_L de_{N-1}}_{N-1} \underbrace{\int_{\mathbb{R}} dy_1 \int_{\mathbb{R}} dy_2 \cdots \int_{\mathbb{R}} dy_{N-1}}_{N-1}$$

$$\times \prod_{k=1}^{N} \frac{\exp\left\{ -\dfrac{\mathcal{B}_k(x_k, x_{k-1})}{2\varepsilon^2 \Delta t} \right\}}{2\pi \varepsilon^2 \rho \sigma \Delta t}, \tag{7.16}$$

with $e_N = e$, $y_N = y$, $e_0 = \xi$, $y_0 = \eta$, is the solution of the FPE (7.5) with the boundary and initial conditions (7.6) and (7.7). Note the difference between (7.16) and (3.33).

As in Section 3.5, the pdf (7.15) is broken into

$$
p_N(e_1, e_2, \ldots, e_N; y_1, y_2, \ldots, y_N; t_1, t_2, \ldots, t_N)
$$

$$
= \prod_{k=1}^{N} \left[\frac{1}{\sqrt{2\pi \Delta t}\, \varepsilon \sigma} \exp \left\{ -\frac{[e_k - e_{k-1} - \Delta t M_{\hat{x}}(e_{k-1}, t_{k-1})]^2}{2\varepsilon^2 \sigma^2 \Delta t} \right\} \right.
$$

$$
\left. \times \exp \left\{ \frac{1}{\varepsilon^2 \rho^2} H_{\hat{x}}(e_{k-1}, t_{k-1})(y_k - y_{k-1}) - \frac{1}{2\varepsilon^2 \rho^2} H_{\hat{x}}^2(e_{k-1}, t_{k-1})\Delta t \right\} \right]
$$

$$
\times \prod_{k=1}^{N} \frac{\exp \left\{ -\frac{(y_k - y_{k-1})^2}{2\varepsilon^2 \rho^2 \Delta t} \right\}}{\sqrt{2\pi \Delta t}\, \varepsilon \rho}, \tag{7.17}
$$

where by the Feynman–Kac formula (1.142), the first product gives in the limit the function

$$
\varphi(e, t, \rho)
$$

$$
= \lim_{N \to \infty} \underbrace{\int_L de_1 \int_L de_2 \cdots \int_L de_{N-1}}_{N-1}
$$

$$
\times \prod_{k=1}^{N} \left[\frac{1}{\sqrt{2\pi \Delta t}\, \varepsilon \sigma} \exp \left\{ -\frac{[e_k - e_{k-1} - \Delta t M_{\hat{x}}(e_{k-1}, t_{k-1})]^2}{2\varepsilon^2 \sigma^2 \Delta t} \right\} \right.
$$

$$
\left. \times \exp \left\{ \frac{1}{\varepsilon^2 \rho^2} H_{\hat{x}}(e_{k-1}, t_{k-1})(y_k - y_{k-1}) - \frac{1}{2\varepsilon^2 \rho^2} H_{\hat{x}}^2(e_{k-1}, t_{k-1})\Delta t \right\} \right],
$$

which is the solution of the Zakai equation in Stratonovich form

$$
d_S \varphi(e, t, \rho)
$$

$$
= \left\{ -[M_{\hat{x}}(e, t)\varphi(e, t, \rho)]_e + \frac{1}{2}[\varepsilon^2 \sigma^2 \varphi(e, t, \rho)]_{ee} - \frac{\varphi(e, t, \rho)H_{\hat{x}}^2(e, t)}{2\varepsilon^2 \rho^2} \right\} dt
$$

$$
+ \frac{\varphi(e, t, \rho)H_{\hat{x}}(e, t)}{\varepsilon^2 \rho^2} d_S y(t), \tag{7.18}
$$

with the boundary conditions

$$
\varphi(e, t, \rho) = 0 \quad \text{for } e \in \partial L. \tag{7.19}
$$

Therefore the joint density

$$p_N(e_N, t_N; y_1, y_2, \ldots, y_N)$$
$$= \Pr\{e_N(t_N) = e_N, \tau > t; y_N(t_1) = y_1, y_N(t_2) = y_2, \ldots, y_N(t_N) = y_N\}$$

can be written at $t = t_N, e_N = e$ as

$$p_N(e, t; y_1, y_2, \ldots, y_N)$$
$$= [\varphi(e, t, \rho) + o(1)] \prod_{k=1}^{N} \frac{1}{\sqrt{2\pi \Delta t \varepsilon \rho}} \exp\left\{-\frac{(y_k - y_{k-1})^2}{2\varepsilon^2 \rho^2 \Delta t}\right\}, \tag{7.20}$$

where $o(1) \to 0$ as $N \to \infty$. Equivalently,

$$\varphi(e, t, \rho) = \frac{p_N(e, t; y_1, y_2, \ldots, y_N)}{\prod_{k=1}^{N} \frac{1}{\sqrt{2\pi \Delta t \varepsilon \rho}} \exp\left\{-\frac{(y_k - y_{k-1})^2}{2\varepsilon^2 \rho^2 \Delta t}\right\}} + o(1), \tag{7.21}$$

which can be interpreted as follows: $\varphi(e, t, \rho)$ is the joint conditional density of $e_N(t)$ and $\tau > t$, given the entire trajectory $\{y_N(t_i)\}_{i=0}^{N}$. However, the probability density of the trajectories $\{y_N(t_i)\}_{i=0}^{N}$,

$$p_N^B(y_0^t) = \prod_{k=1}^{N} \frac{1}{\sqrt{2\pi \Delta t \varepsilon \rho}} \exp\left\{-\frac{(y_k - y_{k-1})^2}{2\varepsilon^2 \rho^2 \Delta t}\right\},$$

is Brownian, rather than the a priori density imposed by (7.3), (7.4).

Now,

$$\Pr\{\tau > t_N, y_N(t_1) = y_1, y_N(t_2) = y_2, \ldots, y_N(t_N) = y_N\}$$
$$= \Pr\{\tau > t_N \mid y_N(t_1) = y_1, y_N(t_2) = y_2, \ldots, y_N(t_N)\}$$
$$\times \Pr\{y_N(t_1) = y_1, y_N(t_2) = y_2, \ldots, y_N(t_N) = y_N\},$$

which we abbreviate to

$$\Pr\{\tau > t, y_0^t\} = \Pr\{\tau > t \mid y_0^t\} p_N(y_0^t), \tag{7.22}$$

where the density $p_N(y_0^t) = \Pr\{y_N(t_1) = y_1, y_N(t_2) = y_2, \ldots, y_N(t_N) = y_N\}$ is defined by the system (3.30), (3.31), independently of $\hat{x}(t)$.

We now use the abbreviated notation (7.22) to write

$$\Pr\{\tau > t \mid y_0^t\} = \frac{\Pr\{\tau > t, y_N(t_1) = y_1, y_N(t_2) = y_2, \dots, y_N(t_N) = y_N\}}{p_N(y_0^t)}$$

$$= \int_L \frac{p_N(e, t; y_1, y_2, \dots, y_N)}{p_N(y_0^t)} \, de$$

$$= \frac{p_N^B(y_0^t)}{p_N(y_0^t)} \int_L \{\varphi(e, t, \rho) + o(1)\} \, de. \tag{7.23}$$

As $N \to \infty$, both sides of (7.23) converge to a finite limit, which we write as

$$\Pr\{\tau > t \mid y_0^t\} = \alpha(t) \int_{\mathbb{R}} \varphi(e, t, \rho) \, de, \tag{7.24}$$

where $\alpha(t)$ is given by (3.43), as above. Note that now

$$\alpha(t) \neq \frac{1}{\displaystyle\int_{\mathbb{R}} \varphi(e, t, \rho) \, de}, \tag{7.25}$$

because $\varphi(e, t, \rho)$ satisfies the boundary condition (7.19), while the solution $\varphi(x, t)$ of (3.25) is defined on the entire line. More specifically, the a posteriori density $p_\varepsilon(e, t \mid y_0^t) = \alpha(t)\varphi(e, t, \rho)$ is defective, because its integral

$$P(\tau > t \mid y_0^t) = \int_{\mathbb{R}} p_\varepsilon(e, t \mid y_0^t) \, de \tag{7.26}$$

decays in time, unlike the a posteriori density (3.42), which integrates to 1. The conditional expectation of a function $H(e(t), t)$, given $\tau > t$, is

$$\hat{H}(t) = \int_L H(e, t) p_\varepsilon(e, t, \rho \mid y_0^t, \tau > t) \, de$$

$$= \frac{\displaystyle\int_L H(e, t) p_\varepsilon(e, t, \rho \mid y_0^t) \, de}{\displaystyle\int_L p_\varepsilon(e, t, \rho \mid y_0^t) \, de} = \frac{\displaystyle\int_L H(e, t) \varphi(e, t, \rho) \, de}{\displaystyle\int_L \varphi(e, t, \rho) \, de}. \tag{7.27}$$

Next, we show that $\mathbb{E}[\tau \mid y_0^\tau]$, as defined in (7.14), is given by

$$\mathbb{E}[\tau \mid y_0^\tau] = \int_0^\infty \Pr\{\tau > t \mid y_0^t\} \, dt. \tag{7.28}$$

Indeed, because $\Pr\{\tau > t \mid y_0^t\} \to 0$ exponentially fast as $t \to \infty$, we can write

$$\int_0^\infty \Pr\{\tau > t \mid y_0^t\}\,dt = \lim_{T \to \infty} \int_0^T t\,d\Pr\{\tau < t \mid y_0^t\}$$

and

$$\int_0^T t\,d\Pr\{\tau < t \mid y_0^t\} = \lim_{N \to \infty} \sum_{i=1}^N i\,\Delta t\,\Delta\Pr\{\tau < i\,\Delta t \mid y_0^{i\,\Delta t}\},$$

where

$$\Delta\Pr\{\tau < i\,\Delta t \mid y_0^{i\,\Delta t}\} = \Pr\{\tau < i\,\Delta t \mid y_0^{i\,\Delta t}\} - \Pr\{\tau < (i-1)\,\Delta t \mid y_0^{(i-1)\,\Delta t}\}.$$

Now we renumber the sampled trajectories $e_{j,N}(t_i)$ in the numerator in (7.14) according to increasing $\tau_{i,N}$, so that in the new enumeration $\tau_{i,N} = i\,\Delta t$. Then we group together the terms in the sum that have the same $\tau_{i,N}$ and denote their sums by $\tilde{m}_{i,N}$, so that (7.13) becomes

$$\frac{\displaystyle\sum_{j=1}^n (\tau_{j,N} \wedge T) m_{j,N}}{\displaystyle\sum_{j=1}^n m_{j,N}} = \frac{\displaystyle\sum_{i=1}^N i\,\Delta t\,\tilde{m}_{i,N}}{\displaystyle\sum_{i=1}^N \tilde{m}_{i,N}}. \tag{7.29}$$

Finally, we identify

$$\Delta\Pr\{\tau < i\,\Delta t \mid y_0^{i\,\Delta t}\} = \frac{\tilde{m}_{i,N}}{\displaystyle\sum_{i=1}^N \tilde{m}_{i,N}}\,(1 + o(1)),$$

where $o(1) \to 0$ as $N \to \infty$. Hence (7.28) follows.

7.4 Asymptotic Solution of Zakai's Equation

For small ε, the solution of (7.18) with the boundary conditions (7.19) is constructed by the method of matched asymptotics (see [137, Section 11.1]). The outer solution is given by the WKB method or large deviations theory (see [137, Section 11.2]) as

$$\varphi_{\text{outer}}(e, t, \rho) = \exp\left\{-\frac{\psi(e, t, \rho)}{\varepsilon^2}\right\}, \tag{7.30}$$

where

$$\psi(e, t, \rho)$$
$$= \inf_{\mathcal{C}_e^1([0,t])} \int_0^t \left\{ \left[\frac{\dot{e}(s) - M_{\hat{x}}(e(s), s)}{\sigma} \right]^2 + \left[\frac{\dot{y}(s) - H_{\hat{x}}(e(s), s)}{\rho} \right]^2 \right\} ds, \tag{7.31}$$

and

$$\mathcal{C}_e^1([0, t]) = \left\{ e(\cdot) \in \mathcal{C}^1([0, t]) : e(0) = e \right\}.$$

We denote by $\tilde{e}(t)$ the minimizer of the integral on the right-hand side of (7.31). The outer solution $\varphi_{\text{outer}}(e, t)$ does not satisfy the boundary conditions (7.19), so a boundary layer correction $k(e, t, \varepsilon)$ is needed to obtain a uniform asymptotic approximation,

$$\varphi(e, t, \rho) \sim \varphi_{\text{uniform}}(e, t, \rho) = \varphi_{\text{outer}}(e, t, \rho) k(e, t, \varepsilon)$$
$$= \exp\left\{ -\frac{\psi(e, t, \rho)}{\varepsilon^2} \right\} k(e, t, \varepsilon). \tag{7.32}$$

The boundary layer function has to satisfy the boundary and matching conditions

$$k(e, t, \varepsilon) = 0 \quad \text{for} \quad e \in \partial L, \quad \lim_{\varepsilon \to 0} k(e, t, \varepsilon) = 1 \quad \text{for} \quad e \in L, \tag{7.33}$$

uniformly on compact subsets of the interior of L.

Because the survival probability is

$$\Pr\left\{ \tau > t \mid y_0^t \right\} = \int_L \alpha(t) \exp\left\{ -\frac{\psi(e, t, \rho)}{\varepsilon^2} \right\} k(e, t, \varepsilon) \, de,$$

the MTLL, according to (7.28), is given by

$$\mathbb{E}[\tau \mid y_0^\tau] = \int_0^\infty \int_L \alpha(t) \exp\left\{ -\frac{\psi(e, t, \rho)}{\varepsilon^2} \right\} k(e, t, \varepsilon) \, de \, dt. \tag{7.34}$$

7.4.1 The Asymptotically Optimal Filter

In view of (7.1), the minimizer $\tilde{e}(t)$ of the integral on the right-hand side of (7.31) can be represented as $\tilde{e}(t) = \tilde{x}(t) - \hat{x}(t)$, where $\tilde{x}(t)$ is the minimizer of the integral

$\Psi(x,t,\rho)$

$$= \inf_{\mathcal{C}_x^1([0,t])} \int_0^t \left\{ \left[\frac{\dot{x}(s) - m(x(s),s)}{\sigma} \right]^2 + \left[\frac{\dot{y}(s) - h(x(s),s)}{\rho} \right]^2 \right\} ds, \quad (7.35)$$

where

$$\mathcal{C}_x^1([0,t]) = \left\{ x(\cdot) \in \mathcal{C}^1([0,t]) : x(0) = x \right\}.$$

Writing $\psi(e,t,\rho) = \Psi(x,t,\rho)$ and $k(e,t,\varepsilon) = K(x,t,\varepsilon)$, we rewrite (7.34) as

$$\mathbb{E}[\tau \mid y_0^\tau] = \int_0^\infty \int_{L+\hat{x}(t)} \alpha(t) \exp\left\{ -\frac{\Psi(x,t,\rho)}{\varepsilon^2} \right\} K(x,t,\varepsilon) \, dx \, dt. \quad (7.36)$$

The integral in (7.36) is evaluated for small ε by the Laplace method, in which the integrand is approximated by a Gaussian density with mean $\tilde{x}(t)$ and variance proportional to ε^2. It is obviously maximized over the functions $\hat{x}(t)$ by choosing $\hat{x}(t)$ so that the domain of integration covers as much as possible of the area under the Gaussian bell. If L is an interval, then the choice $\hat{x}(t) = \tilde{x}(t)$ is optimal. Thus the main result of this chapter can be formulated as the following theorem.

Theorem 7.4.1 (The asymptotic maximum MTLL filter). *For small noise, the the minimum noise energy filter $x_{\mathrm{MNE}}(t)$ is the maximal MTLL filter, which is the minimizer $\tilde{x}(t)$ in (7.35).*

7.5   The MTLL of the MNE Phase Estimator

To evaluate the MTLL, we recall that the instantaneous rate of escape of error trajectories from L is defined as the ratio of the boundary flux to the population of trajectories in L [137, Section 11.2],

$$\kappa(t) = \frac{\dfrac{\partial}{\partial t} \displaystyle\int_L p_\varepsilon(e,t,\rho \mid y_0^t) \, de}{\displaystyle\int_L p_\varepsilon(e,t,\rho \mid y_0^t) \, de} = \frac{\dot{\alpha}(t) \displaystyle\int_L \varphi(e,t,\rho) \, de + \alpha(t) \displaystyle\int_L \dot{\varphi}(e,t,\rho) \, de}{\alpha(t) \displaystyle\int_L \varphi(e,t,\rho) \, de},$$

where $p_\varepsilon(e,t,\rho \mid y_0^t)$ is the a posteriori density of error trajectories in L and $\varphi(e,t,\rho)$ is the solution of the Zakai equation in Stratonovich form (7.18) with the boundary condition (7.19). Using (3.49) and (7.27), we obtain

$$\kappa(t) = -\frac{J \cdot v|_{\partial L}}{\displaystyle\int_L \varphi(e,t,\rho) \, de} - \frac{\widehat{H^2}(t) - \hat{h}^2(t)}{2\varepsilon^2 \rho^2} + \frac{\hat{H}(t) - \hat{h}(t)}{\varepsilon^2 \rho^2} \dot{y}_S(t), \quad (7.37)$$

where the derivatives are in the sense of Stratonovich and the flux density vector J is defined by the Fokker–Planck operator, (3.26)

$$\mathcal{L}\varphi = -\nabla \cdot J.$$

Next, we estimate the difference $\hat{H}(t) - \hat{h}(t)$. First, we estimate the difference

$$\int_{\mathbb{R}} h(x,t)\varphi(x,t)\,dx - \int_L H(e,t)\varphi(e,t,\rho)\,de$$

$$= \int_{\mathbb{R}} h(x,t)e^{-\Psi(x,t,\rho)/\varepsilon^2}\,dx - \int_{x-\hat{x}(t)\in L} h(x,t)e^{-\Psi(x,t,\rho)/\varepsilon^2} K(x,t,\varepsilon)\,dx$$

$$- \int_{x-\hat{x}(t)\notin L} h(x,t)e^{-\Psi(x,t,\rho)/\varepsilon^2}\,dx$$

$$+ \int_{x-\hat{x}(t)\in L} h(x,t)e^{-\Psi(x,t,\rho)/\varepsilon^2}[1 - K(x,t,\varepsilon)]\,dx.$$

The Laplace expansion of the integral gives

$$\lim_{\varepsilon \to 0} \varepsilon^2 \log \int_{x-\hat{x}(t)\notin L} h(x,t)e^{-\Psi(x,t,\rho)/\varepsilon^2}\,dx = -\inf_{x-\hat{x}(t)\in \partial L} \Psi(x,t,\rho)$$

$$= -\inf_{e\in\partial L} \psi(e,t,\rho) \qquad (7.38)$$

and

$$\lim_{\varepsilon \to 0} \varepsilon^2 \log \int_{x-\hat{x}(t)\in L} h(x,t)e^{-\Psi(x,t,\rho)/\varepsilon^2}[1 - K(x,t,\varepsilon)]\,dx$$

$$= \lim_{\varepsilon \to 0} \varepsilon^2 \log \int_{e\in L} H(e,t)e^{-\psi(e,t,\rho)/\varepsilon^2}[1 - k(e,t,\varepsilon)]\,de. \qquad (7.39)$$

The boundary ∂L is not characteristic for a generic trajectory $\hat{x}(t)$, so that $\psi'_{\partial L} = \partial\psi(e,t,\rho)/\partial e \neq 0$ for $e \in \partial L$ and

$$1 - k(e,t,\varepsilon) = \exp\left\{-\frac{|\psi'_{\partial L}|\,\mathrm{dist}(e,\partial L)}{\varepsilon^2}\right\}. \qquad (7.40)$$

Substituting (7.40) in (7.39) and evaluating the integral by the Laplace method, we obtain

$$\lim_{\varepsilon \to 0} \varepsilon^2 \log \int_{e\in L} H(e,t)e^{-\psi(e,t,\rho)/\varepsilon^2}[1 - k(e,t,\varepsilon)]\,de = -\inf_{e\in\partial L} \psi(e,t,\rho). \qquad (7.41)$$

On the other hand, (4.5), (7.30), and (7.31) indicate that

$$\lim_{\varepsilon \to 0} \varepsilon^2 \log \int_{\mathbb{R}} \varphi(x,t)\,dx = \lim_{\varepsilon \to 0} \varepsilon^2 \log \int_L \varphi(e,t,\rho)\,de$$

$$= -\inf_{x \in \mathbb{R}} \Psi(x,t,\rho) = -\Psi(\tilde{x}(t),t,\rho)$$

$$= -\psi(\tilde{e}(t),t,\rho), \tag{7.42}$$

where $\tilde{x}(t)$ is the minimizer in (7.35). Obviously,

$$\lim_{\varepsilon \to 0} \varepsilon^2 \log \left\{ -\frac{\boldsymbol{J} \cdot \boldsymbol{v}|_{\partial L}}{\displaystyle\int_L \varphi(e,t,\rho)\,de} \right\} = -\left[\inf_{e \in \partial L} \psi(e,t,\rho) - \psi(\tilde{e}(t),t,\rho) \right]. \tag{7.43}$$

Therefore the instantaneous escape rate of error trajectories from the lock region L is

$$\lim_{\varepsilon \to 0} \varepsilon^2 \log \kappa(t) = -[\inf_{e \in \partial L} \psi(e,t,\rho) - \psi(\tilde{e}(t),t,\rho)]. \tag{7.44}$$

It follows from renewal theory (see [137, Section 7,2]) that

$$\lim_{\varepsilon \to 0} \varepsilon^2 \log \mathbb{E}[\tau \mid y_0^\tau] = \inf_{e \in \partial L, t > 0} [\Psi(\tilde{x} + e, t) - \Psi(\tilde{x}, t)]. \tag{7.45}$$

7.6   Optimal MTLL Tracking of Phase

The benchmark model (3.10) (or its scaled version (3.11)) of a first-order phase tracking system consists of a linear model (3.11) of the phase $\boldsymbol{x}(t) = [x(t), x_2(t), \ldots, x_N(t)]^T$,

$$\dot{\boldsymbol{x}} = \boldsymbol{A}\boldsymbol{x} + \varepsilon \boldsymbol{B}\,\dot{\boldsymbol{w}}, \tag{7.46}$$

and the simplest nonlinear model of the noisy measurements

$$\boldsymbol{y}(t) = \begin{pmatrix} y_s(t) \\ y_c(t) \end{pmatrix}, \quad \dot{\boldsymbol{y}} = \boldsymbol{h}(\boldsymbol{x}) + \varepsilon \dot{\boldsymbol{v}}, \tag{7.47}$$

with

$$\boldsymbol{h}(\boldsymbol{x}) = \begin{pmatrix} \sin x \\ \cos x \end{pmatrix}.$$

To find $\Psi(x, t, \rho)$, we have to minimize the functional

$$I[z(\cdot)] \equiv \int_0^t \left[|\dot{y} - h(z)|^2 + |u|^2 \right] dt, \tag{7.48}$$

with the equality constraint

$$\dot{z} = Az + Bu, \quad u \in L^2[0, T], \tag{7.49}$$

Note that the integral $I[x(\cdot)]$ contains the white noises $\dot{w}(t)$ and $\dot{v}(t)$, which are not square integrable. To remedy this problem, we begin with a model in which the white noises $\dot{w}(t)$, $\dot{v}(t)$ are replaced with square integrable wideband noises, and at the appropriate stage of the analysis, we take the white-noise limit (see below). We have to minimize

$$I(x(\cdot), e(\cdot)) = \frac{1}{2} \int_0^t \left\{ \left| \begin{bmatrix} \sin x \\ \cos x \end{bmatrix} - \begin{bmatrix} \sin(x + e) \\ \cos(x + e) \end{bmatrix} + \varepsilon \begin{bmatrix} \dot{v}_1 \\ \dot{v}_2 \end{bmatrix} \right|^2 + |u|^2 \right\} ds,$$

or, in view of (7.45), we need to minimize the difference

$$\Delta I(x(\cdot), e(\cdot)) = I(x(\cdot), e(\cdot)) - I(x(\cdot), 0)$$

$$= \frac{1}{2} \int_0^t \left[4 \sin^2\left(\frac{e}{2}\right) + |u|^2 \right] ds + \varepsilon \int_0^t [\sin x - \sin(x + e)] \dot{v}_1 \, ds$$

$$+ \varepsilon \int_0^t [\cos x - \cos(x + e)] \dot{v}_2 \, ds + \varepsilon \int_0^t u^T \dot{w} \, ds. \tag{7.50}$$

Now we let the bandwidth of the noise terms in (7.50) become infinite, so that the integrals containing the noise variables become Itô integrals, and we also use the fact that the noise variables are independent, to write them as a single Itô integral

$$\Delta I(x(\cdot), e(\cdot)) = \frac{1}{2} \int_0^t \left[4 \sin^2\left(\frac{e}{2}\right) + |u|^2 \right] ds$$

$$+ \varepsilon \int_0^t \sqrt{4 \sin^2\left(\frac{e}{2}\right) + |u|^2} \, d\xi(s), \tag{7.51}$$

where $\xi(t)$ is a standard Brownian motion that depends on $w(t)$ and $v(t)$. In the limit $\varepsilon \to 0$, we are left with a simple problem in the calculus of variations. Defining the value function

$$V(t_0, e(t_0)) = \frac{1}{2} \int_0^t \left[4 \sin^2\left(\frac{e}{2}\right) + u^2 \right] ds,$$

the Hamilton–Jacobi–Bellman equation is given by

$$-\frac{\partial V}{\partial t_0} = \min_{u(t_0)} \left\{ \frac{1}{2} \left[4 \sin^2\left(\frac{e}{2}\right) + u^2 \right] + \frac{\partial V}{\partial e} u \right\}$$

with boundary conditions

$$e(0) = 0, \quad e(t) = 2\pi.$$

For the infinite horizon (i.e., no limitation on slip duration), we obtain the solution

$$V^*(e) = 4\cos\left(\frac{e}{2}\right) + 4,$$

$$u^* = -\frac{\partial V^*}{\partial e} = 2\sin\left(\frac{e}{2}\right),$$

because $V^*(e) \geq 0$. Thus the exponential rate of the MTLL in the MNE filter for a first-order phase model is asymptotically given by

$$\lim_{\varepsilon \to 0} \varepsilon^2 \log \mathbb{E}\tau = V^*(0) = 8.$$

Exercise 7.1 (Improvement upon the PLL). Recall that the rate in the first-order PLL in Exercise 4.23 is

$$\lim_{\varepsilon \to 0} \varepsilon^2 \log \mathbb{E}\tau = 2.$$

(i) Convert this improvement into dB by requiring identical MTLLs in the PLL and the MNE filter. (HINT: Set ε_c, ε_{nc} to be the values of ε and get $\tau_c = \tau_{nc} \Rightarrow 2/\varepsilon_c = 8/\varepsilon_{nc}$).

(ii) Show that there exists a 12 dB performance gap in the MTLL between the estimators in terms of CNR. (HINT: Denote by $CNR_c[dB]$, $CNR_{nc}[dB]$ the carrier-to-noise ratio (CNR) in the PLL and MNE filter, respectively and get $CNR_c[dB] - CNR_{nc}[dB] = 10 \cdot 2\log_{10}(8/2) \approx 12\,dB$). ☐

Exercise 7.2 (The most likely escape trajectory). Show that the Euler–Lagrange equation for the minimizer is

$$\dot{e}^* = 2\sin\left(\frac{e^*}{2}\right), \tag{7.52}$$

whose solution that satisfies the initial condition $e^*(0) = 0$ is $e^*(t) = 0$. Investigate the solution of (7.52) with the initial condition $e^*(0) = \delta$ for small $\delta > 0$. ☐

7.7 Numerical Analysis of the MNE Estimator on a Grid

One possible way of implementing an approximation to the MNE filter (7.35) is to define a grid on which the time axis and trajectory-value axis are quantized. Denote the time quantization unit by Δt and that of the signal value by Δx. Then there are

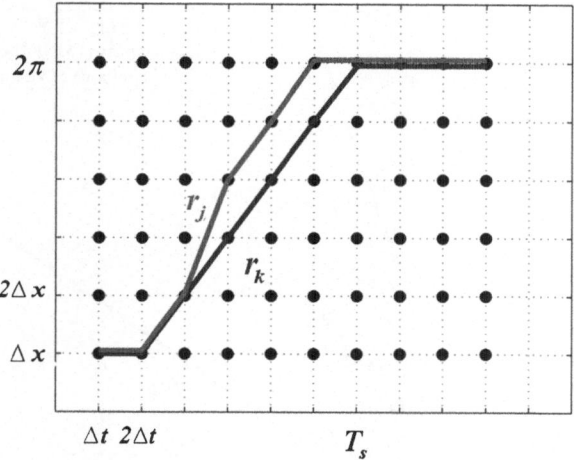

Fig. 7.1 Possible trajectories $r_j = e_j$ and $r_k = e^*$ on a grid.

only finitely many error trajectories e_j that differ from the minimizer e^*. One such possibility is depicted in Figure 7.1. To differ from $e_j(\cdot)$, the slip trajectory $e^*(\cdot)$ must deviate at least at one point from $e_j(\cdot)$. Setting $\Delta e_{j*}(\cdot) = e_j(\cdot) - e^*(\cdot)$ and expanding (7.51) about $e^*(\cdot)$ to leading-order, we get

$$\int_0^t \left[(\Delta e_{j*}) \sin^2 e^* + \dot{e}^* \Delta \dot{e}_{j*} \right] ds = - \int_0^t \left[(\Delta e_{j*})^2 + (\Delta \dot{e}_{j*})^2 \right] ds$$

$$\times \left[1 + O \left(\int_0^t (\Delta e_{j*})^2 \, ds \right) \right]$$

and define the ratio

$$\alpha_{j*} = - \frac{\int_0^t \left[(\Delta e_{j*})^2 + (\Delta \dot{e}_{j*})^2 \right] ds}{\int_0^t \left[(\Delta e_{j*}) \sin^2 e^* + \dot{e}^* \Delta \dot{e}_{j*} \right] ds} \left[1 + O \left(\int_0^t (\Delta e_{j*})^2 \, ds \right) \right].$$

$$(7.53)$$

For trajectories with small $\max |\Delta e_{j*}(\cdot)|$, but not identically zero, the differential terms are dominant relative to $\Delta e_{j*}(\cdot)$, so that we can bound α_{j*} from below by

$$\alpha_{j*} \geq \frac{\Delta x}{\max_t |\dot{e}_j(t)| \Delta t}.$$

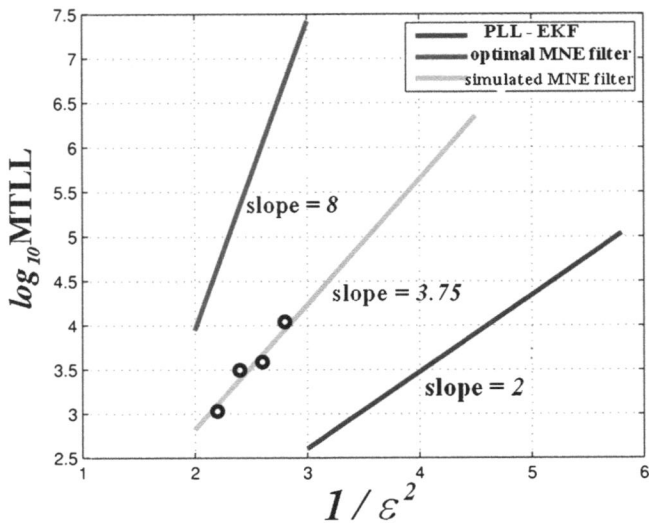

Fig. 7.2 Plot of $\log_{10}$ MTLL vs ε^{-2} of MNE filters and a PLL for a first-order model. The line with slope 2 is for PLL-EKF, slope 3.75 is for a simulated NME filter, and slope 8 is for the theoretical NME filter.

Specializing the trajectories $e_j(\cdot)$ to slip trajectories, we have from (7.52)

$$\alpha_{j*} \geq \frac{1}{2}\left(\frac{\Delta x}{\Delta t}\right) . \tag{7.54}$$

Thus the rate of the MTLL is given by

$$\lim_{\varepsilon \to 0} \varepsilon^2 \log \bar{\tau} = 8\left(1 - \frac{\alpha}{2}\right)^2, \quad 0 \leq \alpha \leq 1. \tag{7.55}$$

Figure 7.2 shows theoretical and simulation results for the first-order model [47]. Three graphs show $\log_{10}$ MTLL against ε^{-2}. The PLL has theoretical slope of 2, whereas the slope of the MNE filter is 8. The potential gain of 12 dB is not achieved, due to the discretization. For $\alpha = \frac{1}{2}$ the rate is only 4.5. The rate in the simulation is 3.75, which reflects 5.46 dB performance gain over the PLL. An important implementation issue is apparent from (7.54). In order to achieve improved performance, the simulation quantization steps have to be chosen such that the condition (7.54) is satisfied as well as the constraint $0 \leq \alpha \leq 1$. An inappropriate selection of the ratio $\Delta x / \Delta t$ may violate the above constraint, which consequently results in the saturated value $\alpha = 1$. Formula (7.55) means that this corresponds to the rate 2, which is exactly that of the PLL. This "brick-wall" effect means that unless the x-axis quantization is sufficiently fine relative to the maximum differential of a slip trajectory, we get $\alpha \equiv 1$, for which "conventional" performance is achieved. Once the quantization gets finer, the wall is crossed, and performance gain develops. In the simulation above we set $\Delta x = \Delta t = 0.01$, so that $\alpha = \frac{1}{2}$. However, setting $\Delta x = 0.01$ the grid must have 628 by 628 transitions, which means an enormous processing load. Thus there is yet room for efficient implementation algorithms to be developed. Note that the results in the above graphs present only the exponential dependence of the MTLL on the noise parameter. They do not show the pre-exponential values.

7.8   Second Order Phase Model

In the second-order phase model (see Exercise 5.5), the matrices A and B of (7.46) are

$$A = \begin{pmatrix} 0 & 1 \\ 0 & 0 \end{pmatrix}, \quad B = \begin{pmatrix} 0 \\ 1 \end{pmatrix}.$$

The optimization problem (7.48), (7.49) is to optimize the Hamiltonian

$$\mathscr{H} = \frac{1}{2}\left[\sin^2\left(\frac{e_1}{2}\right) + |u|^2\right] + \lambda^T(Ae + Bu)$$

and obtain the Hamiltonian system with the state, costate, and stationarity condition equations

$$\dot{e}_1 = e_2,$$
$$\dot{e}_2 = u_2,$$
$$\dot{\lambda}_1 = -\sin e_1,$$
$$\dot{\lambda}_2 = -\lambda_1,$$
$$u_1 = 0,$$
$$u_2 = -\lambda_2,$$

and the boundary conditions

$$e(0) = 0,$$
$$e(t) = \begin{bmatrix} 2\pi \\ 0 \end{bmatrix}.$$

This set of nonlinear first-order differential equations can be solved numerically (such as by shooting; see, e.g., [18]). The identity $\dot{V} = \dot{e} \cdot \nabla V$, together with the relation $\lambda = \partial V/\partial e$, gives

$$\dot{V} = \lambda_1 \dot{e}_1 - \lambda_2^2. \tag{7.56}$$

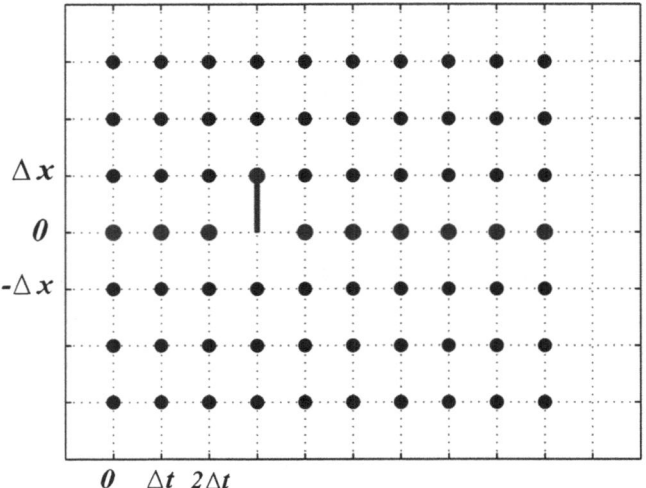

Fig. 7.3 A plot of quantized $\Delta \dot{e}_{j*}$ for the example in Figure 7.1.

Thus, to evaluate $V(t_0, e)$, (7.56) has to be integrated numerically (with a negative time step) along the extremals $e(\cdot)$ and $\lambda(\cdot)$, with the terminal condition $V(t, e(t)) = 0$. The extremal satisfies $\ddot{e}_1^* = u_2^* = -\lambda_2^* = \partial V/\partial e_2$, and therefore, as can be seen numerically, $\max_t \ddot{e}^*(t)$ is about 2.

We turn now to the numerical analysis of the optimization procedure. Optimizing on a grid, we find, as above, that

$$\alpha_{j*} \approx -\frac{\int_0^t \left[(\Delta e_{j*})^2 + (\Delta \ddot{e}_{j*})^2 \right] ds}{\int_0^t \left[(\Delta e_{j*}) \sin^2 e^* + \ddot{e}^* \Delta \ddot{e}_{j*} \right] ds}.$$

As above, we bound α_{j*} from below. In order to keep the numerator in the above expression minimal, while maintaining $\max |\Delta e_{j*}(\cdot)|$ small, we must consider trajectories for which $\Delta \dot{e}_{j*}(\cdot)$ is as shown in Figure 7.3. It shows the quantized deviation from a slip-trajectory slope $\Delta \dot{e}_{j*}$ of the trajectories shown in Figure 7.1. Consequently, the trajectories of $\Delta \ddot{e}_{j*}(\cdot)$ become as shown in Figure 7.4, which shows the quantized $\Delta \ddot{e}_{jk}$ of trajectories shown in Figure 7.3. Then we can write, in the sense of (7.53),

$$\alpha_{j*} \approx -\frac{\int_0^t (\Delta \ddot{e}_{j*})^2 \, ds}{\int_0^t \ddot{e}_* \Delta \ddot{e}_{j*} \, ds} \geq \frac{2\Delta x}{\max_t |\ddot{e}_*(t)| \Delta t}.$$

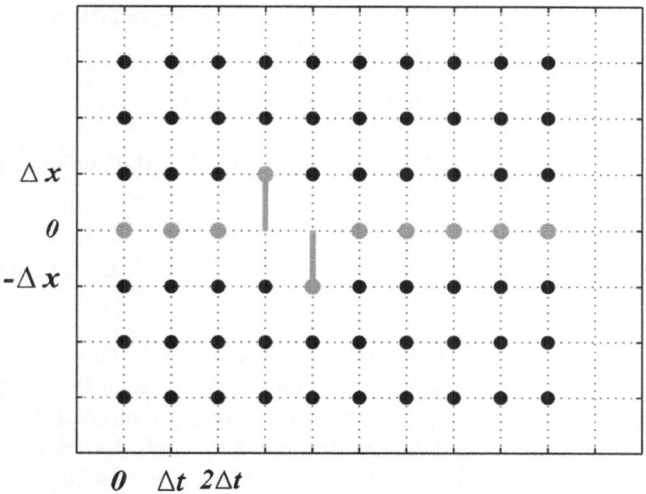

Fig. 7.4 A plot of quantized $\Delta \ddot{e}_{jk}$ for the example in Figure 7.3.

Therefore

$$\lim_{\varepsilon \to 0} \varepsilon^2 \log \bar{\tau} = 5 \left(1 - \frac{\alpha}{2}\right)^2, \quad 0 \le \alpha \le 1. \tag{7.57}$$

We conclude that

$$\alpha_{jk} \ge \frac{\Delta x}{\Delta t}.$$

With this α, we can bound the achievable MTLL performance of a discrete, Viterbi-type implementation of the MNE filter for a second-order phase model. Because the computing resources for the second-order phase model are significantly more stringent than those needed for first-order, we do not pursue simulation results any further. The formula for the MTLL is similar to (7.55), but with $V(0) = 5$ [43].

Exercise 7.3 (Improvement upon the PLL). Recall that in Exercise 4.25 it was shown that for the model (7.46), (7.47) of the second-order PLL (for FM transmission),

$$\lim_{\varepsilon \to 0} \varepsilon^2 \log \tau = 0.85. \tag{7.58}$$

Show that the CNR gap in this case is

$$\text{CNR}_c[\text{dB}] - \text{CNR}_{nc}[\text{dB}] = 40/3 \log_{10}(5/0.85) \approx 10.25 \, \text{dB}. \tag{7.59}$$

□

Exercise 7.4 (Maximizing the MTLL in a fixed-delay smoother).

 (i) Show that the MNE estimator maximizes the MTLL of fixed-delay smoothers
      of PM and FM transmission in a low-noise channel.
 (ii) Calculate that MTLL.
(iii) Find the CNR gap between the PLL and the MNE estimator [43], [44].   □

7.9   Annotations

The suboptimal phase trackers are known to lose lock (or slip cycles) [147]. The
MTLL in these filters is simply the mean first passage time of the estimation error
to the boundary of the lock region. The MFPT from an attractor of a dynamical
system driven by small noise has been calculated by large deviations and singular
perturbation methods [53], [112], [137], and in particular, for the PLL [18]. The
MTLL in particle filters for phase estimation was found in [48]. It has been found
in [43], [47] that minimizing the MNE leads to a finite, yet much longer, MTLL
than in the above-mentioned phase estimators.

   Problems in which the maximization of exit time is an optimality criterion, were
considered by several authors [167], [4], [114]. Optimal MTLL phase estimation
was studied in [49]. Particle filtering is described in [30], [2], [48], [3], [41].

References

1. Aggoun, L., R.J. Elliott. *Measure Theory and Filtering. Introduction and Applications.* Cambridge Series in Statistical and Probabilistic Mathematics, volume 15, x+258 pp. ISBN: 0-521-83803-7. Cambridge University Press, Cambridge, 2004.
2. Amblard, P.O., J.M. Brossier and E. Moisan. Phase tracking: what do we gain from optimality? particle filtering versus phase-locked loops. *Signal Process.*, 83:151–167, 2003.
3. Arulampalam, M.S., S. Maskell, N. Gordon and T. Clapp. A tutorial on particle filters for on-line nonlinear/non-Gaussian Bayesian tracking. *IEEE Trans. Signal Processing*, 50:174–188, 2002.
4. Atar, R, P. Dupuis, and A. Shwartz. An escape time criterion for queueing networks: Asymptotic risk-sensitive control via differential games. *Math. Op. Res.*, 28:801–835, 2003.
5. Athans, M., S.K. Mitter and L. Valavani. An interim report of research on stochastic and adaptive systems. Technical Report LIDS-IR-1081, Laboratory for Information and Decision Systems MIT, Cambridge, Massachusetts, 1981.
6. Bain, A., D. Crisan. *Fundamentals of Stochastic Filtering.* Stochastic Modelling and Applied Probability, volume 60, xiv+723 pp. ISBN: 978-0-8176-4301-0; 0-8176-4301-X. Springer, NY, 2009.
7. Balaji, B. Universal nonlinear filtering using Feynman path integrals ii: the continuous-continuous model with additive noise. *PMC Physics A*, 3:1–28, 2009.
8. Bar-Shalom, Y., X.R. Li, and T. Kurbarajan. *Estimation with Applications to Tracking and Navigation.* Wiley Inter-Science, NY, 2001.
9. Barnes, J.A. and D.W. Allan. A statistical model of flicker noise. *Proc. IEEE*, 54:176–178, 1966.
10. Barton, D.K. *Radar System Analysis and Modeling.* Artech House, Norwood, MA, 2005.
11. Basin, M. *New Trends in Optimal Filtering and Control for Polynomial and Time-Delay Systems*, volume 380, xxiv+206 pp. ISBN: 978-3-540-70802-5. Springer-Verlag Lecture Notes in Control and Information Sciences, Berlin, 2008.
12. Bender, C.M. and S.A. Orszag. *Advanced Mathematical Methods for Scientists and Engineers.* McGraw-Hill, NY, 1978.
13. Bernhard, U.P. and B.Z. Bobrovsky. Influence of Doppler effects and a random channel delay on the optimal dessign of a noncoherent DLL. *Proceedings of Symposium on Communication Theory and Applications*, Lake District, UK, 1993.
14. Bertsekas, D.P. *Dynamic Programming and Optimal Control.* Athena Scientific, Nashua, NH, 3rd edition, 2007.
15. Blackman, S. and R. Popoli. *Design and Analysis of Modern Tracking Systems.* Artech House, Boston, 1999.

Z. Schuss, *Nonlinear Filtering and Optimal Phase Tracking*, Applied Mathematical Sciences 180, DOI 10.1007/978-1-4614-0487-3, © Springer Science+Business Media, LLC 2012

16. Blankenship, G.L and J.S. Baras. Accurate evaluation of stochastic Wiener integrals with applications to scattering in random media and to nonlinear filtering. *SIAM J. Appl. Math.*, 41:518–552, 1981.

17. Bobrovsky, B.Z. and M. Zakai. Asymptotic a priori estimates for the error in the nonlinear filtering problem. *IEEE Trans. Inform. Theory*, IT-27:371–376, 1982.

18. Bobrovsky, B.Z. and Z. Schuss. A singular perturbation method for the computation of the mean first passage time in a nonlinear filter. *SIAM J. Appl. Math.*, 42(1):174–187, 1982.

19. Bobrovsky, B.Z. and Z. Schuss. Jamming and maneuvering-induced loss of lock in a range tracking loop. *Faculty of Engineering Internal Report, Tel-Aviv University*, 1984.

20. Brigo, D., B. Hanzon and F. LeGland. A differential geometric approach to nonlinear filtering: The projection filter. *IEEE Trans. Automat. Contr.*, 43(2):247–252, 1998.

21. Brigo, D., B. Hanzon and F. LeGland. Approximate nonlinear filtering by projection on exponential manifolds of densities. *Bernoulli*, 5(3):495–534, 1999.

22. Brockett, R.W. Nonlinear systems and nonlinear estimation theory, in *The Mathematics of Filtering and Identification and Applications*, M. Hazewinkel and J.C. Willems, eds., pp. 479–504, 1981.

23. Brockett, R.W. and J.M.C. Clark. The geometry of the conditional density functions, in, *Analysis and Optimization of Stochastic Systems*, O.L.R. Jacobs et al. (eds.), pp. 299–309, Academic Press, NY, 1980.

24. Bucy, R.S. Bayes' theorem and digital realizations for nonlinear filters. *J. Astronaut. Sci.*, 17:80–94, 1969.

25. Bucy, R.S. and J. Pages. A priori error bounds for the cubic sensor problem. *IEEE Trans. Autom. Control*, 23(1):88–91, 1978.

26. Bucy, R.S., P.D. Joseph. *Filtering for Stochastic Processes with Applications to Guidance*, xviii+217 pp. ISBN: 0-8284-0326-0, 93-02. Chelsea Publishing Co., NY, 2-nd edition, 1987.

27. Carravetta, F. and A. Germani. Suboptimal solutions for the cubic sensor problem. *Proc. of 36th IEEE Conf. Decision Contr., San Diego, CA*, pp. 4460–4461, 1997.

28. Chapman, S. On the Brownian displacements and thermal diffusion of grains suspended in a non-uniform fluid. *Proc. R. Soc. London, Ser. A*, 119:34-54, 1928.

29. Courant, R. and D. Hilbert. *Methods of Mathematical Physics*. Wiley-Interscience, NY, 1989.

30. Crisan, D. Exact rates of convergence for a branching particle approximation to the solution of the Zakai equation. *Annals of Probability*, 31:693–718, 2003.

31. Crisan, D., B. Rozovskii. *The Oxford Handbook of Nonlinear Filtering*. Oxford Handbooks in Mathematics, OUP. Oxford University Press, Oxford, 2011.

32. Davenport, W.B. and W.L. Root. *Random Signals and Noise*. McGraw-Hill, NY, 1958.

33. De Finetti, B. *Theory of Probability*. Wiley, NY, 3rd edition, 1993.

34. Dembo, A. and O. Zeitouni. *Large Deviations Techniques and Applications*. Jones and Bartlett, 1993.

35. Demir, A., A. Mehrotra, and J. Roychowdhury. Phase noise in oscillators: A unifying theory and numerical methods for characterization. *IEEE Trans. Circuits Systems I: Fund. Theory Appl.*, 47:655–674, 2000.

36. Dorf, R. and R.H. Bishop. *Modern Control Systems*. Prentice–Hall, Englewood Cliffs, NJ, 10th edition, 2004.

37. Duncan, T.E. and B. Pasik-Duncan. Fractional Brownian motion and stochastic equations in Hilbert spaces. *Stoch. Dyn.*, 2:225–250, 2002.

38. Dygas, M.M., B.J. Matkowsky, and Z. Schuss. A singular perturbation approach to non-Markovian escape rate problems. *SIAM J. Appl. Math.*, 46(2):265–298, 1986.

39. Einstein, A. *Investigations on the Theory of the Brownian Movement*. Translated and reprinted by Dover Publications, N.Y., 1956.

40. Elliott, R.J., L. Aggoun, and J.B. Moore. *Hidden Markov Models: Estimation and Control*. Springer, NY, 1995.

41. elMoral, P. and A. Guionnet. Large deviations for interacting particle systems. Applications to nonlinear filtering problems. *Stochast. Process. Applicat.*, 78:69–95, 1998.

42. Erban, R. and J. Chapman. Reactive boundary conditions for stochastic simulations of reaction–diffusion processes. *Phys. Biol.*, 4:16–28, 2007.
43. Ezri, D. *Loss of Lock and Steady-State Errors in a Non-causal Phase Estimator.* PhD dissertation, Tel Aviv University, 2006.
44. Ezri, D. *Loss of Lock and Steady-State Errors in a Non-causal Phase Estimator.* VDM Verlag, 2009.
45. Feller, W. Diffusion processes in one dimension. *Trans. AMS*, 77(1):1–31, 1954.
46. Feyel, D. and A.D.L. Pradelle. On fractional Brownian processes. *Potential Anal.*, 10:273–288, 1999.
47. Fischler, E. *Loss of Lock in Optimal and Suboptimal Phase Estimators.* PhD dissertation, Tel Aviv University, 2007.
48. Fischler, E. and B.Z. Bobrovsky. Mean time to lose lock of phase tracking by particle filtering. *Signal Process.*, 86:3481–3485, 2006.
49. Fischler, E. and Z. Schuss. Nonlinear filtering with optimal MTLL. *arXiv:math/0703524v1 [math.OC]*, 2007.
50. Fleming, W. H. Deterministic nonlinear filtering. *Ann. Scuola Normale Superiore Pisa Cl. Sci. Fis. Mat.*, 25:435–454, 1997.
51. Fleming, W.H. *Deterministic and Stochastic Optimal Control.* Springer Verlag, NY, 2005.
52. Freidlin, M.A. *Functional Integration and Partial Differential Equations.* Princeton University Press, Princeton, N.J., 1985.
53. Freidlin, M.A. and A.D. Wentzell. *Random Perturbations of Dynamical Systems.* Springer-Verlag, New York, 1984.
54. Friedman, A. *Partial Differential Equations of Parabolic Type.* Prentice-Hall, Englewood Cliffs, NJ, 1964.
55. Fristedt, B., N. Jain, N. Krylov. *Filtering and Prediction: A Primer. Student Mathematical Library,* volume 38, xii+252 pp. ISBN: 978-0-8218-4333-8. AMS, Providence, RI, 2007.
56. Fujisaki, M., G. Kallianpur and H. Kunita. Stochastic differential equations for the nonlinear filtering problem. *Osaka J. of Math*, 9:19–40, 1972.
57. Gantmacher, F.R. *The Theory of Matrices.* AMS, 1998.
58. Gardiner, C.W. *Handbook of Stochastic Methods.* Springer Verlag, NY, 2nd edition, 1985.
59. Gelb, A. *Applied Optimal Estimation.* MIT Press, Cambridge, MA, 1974.
60. Gihman, I.I. and A.V. Skorohod. *Stochastic Differential Equations.* Springer-Verlag, Berlin, 1972.
61. Hänggi, P., P. Talkner and M. Borkovec. 50 years after Kramers. *Rev. Mod. Phys.*, 62:251–341, 1990.
62. Haynes, L.W., A.R. Kay, K.W. Yau. Single cyclic GMP-activated channel activity in excised patches of rod outer segment membrane. *Nature*, 321(6065):66–70, 1986.
63. Hazewinkel, M., S.I. Marcus and H.J. Sussmann. Nonexistence of finite dimensional filters for conditional statistics of the cubic sensor problem. *Syst. Contr. Lett.*, 3(6):331–340, 1983.
64. Hazewinkel, S.I. Marcus and H.J. Sussmann. Nonexistence of finite dimensional filters for conditional statistics of the cubic sensor problem, in *Filtering and Control of Random Processes*, pp. 76–103. Springer Berlin, Heidelberg Book Series Lecture Notes in Control and Information Sciences 61, 1984.
65. Hida, T. *Brownian Motion.* Springer, NY, 1980.
66. Hijab, O. Asymptotic Bayesian estimation of a first-order equation with small diffusion. *The Annals of Probability*, 12:890–902, 1984.
67. Holmes, J.K. *Coherent Spread Spectrum Systems.* Wiley, NY, 1982.
68. Itô, K. and H.P. McKean, Jr. *Diffusion Processes and Their Sample Paths* (Classics in Mathematics). Springer-Verlag, 1996.
69. Jackson, J.D. *Classical Electrodymnics.* 2nd ed., Wiley, NY, 1975.
70. James, M.R. and J.S. Baras. Nonlinear filtering and large deviations: a PDE-control theoretic approach. *Stochastics: An International Journal of Probability and Stochastic Processes,* 23:391–412, 1988.
71. Jazwinski, A.H. *Stochastic Processes and Filtering Theory.* Dover, NY, 2007.

72. Johnson, J.B. Thermal agitation of electricity in conductors. *Phys. Rev.*, 32:97–109, 1928.
73. Jondral, F. Generalized functions in signal theory. *Acta Applicandae Mathematicae*, 63:1–3, 2000.
74. Kallianpur, G. *Stochastic Filtering Theory*. Applications of Mathematics, volume 13, xvi+316 pp. ISBN: 0-387-90445-X. Springer-Verlag, New York-Berlin, 1980.
75. Kalman, R.E. A new approach to linear filtering and prediction problems. *Trans. AMSE, Ser. D., J. Basic Engng*, 82:35–45, 1960.
76. Kalman, R.E. A new method and results in linear prediction and filtering theory. Technical Report 61, RIAS, Martin Co., Baltimore, MD, 1961.
77. Kalman, R.E. and R.S. Bucy. New results in linear filtering and prediction theory. *Trans. AMSE, Ser. D., J. Basic Engng*, 83:95–108, 1961.
78. Kaplan, E.D. ed. *Understanding GPS: Principles and Applications*. Artech House, Norwood, MA, 1996.
79. Karatzas, I. and S.E. Shreve. *Brownian Motions and Stochastic Calculus*. Springer Graduate Texts in Mathematics 113, NY, 2nd edition, 1991.
80. Karlin, S. and H.M. Taylor. *A First Course in Stochastic Processes*, volume 1. Academic Press, NY, 2nd edition, 1975.
81. Karlin, S. and H.M. Taylor. *A Second Course in Stochastic Processes*, volume 2. Academic Press, NY, 2nd edition, 1981.
82. Karmy, R., B.Z. Bobrovsky, and Z. Schuss. Loss of lock induced by Doppler or code rate mismatch in code tracking loops, in *Military Communications Conference - Crisis Communications: The Promise and Reality,* Proceedings of an International Conference, MILCOM 1987, volume 1. IEEE, 1987.
83. Kasdin, N.J. Discrete simulation of colored noise and stochastic processes and a $1/f$ power law noise generation. *Proc. IEEE*, 83:802–827, 1995.
84. Katzur, R. *Asymptotic Analysis of the Optimal Filtering Problem in One dimension,* PhD Dissertation. Tel-Aviv University, Israel, 1984.
85. Katzur, R., B.Z. Bobrovsky and Z. Schuss. Asymptotic analysis of the optimal filtering problem for one-dimensional diffusions measured in a low-noise channel, part i. *SIAM J. Appl. Math.*, 44(3):591–604, 1984.
86. Keshner, M.S. $1/f$ noise. *Proc. IEEE*, 70:212–218, 1982.
87. Kevorkian, J. and J.D. Cole. *Perturbation Methods in Applied Mathematics (Applied Mathematical Sciences)*. Springer-Verlag, Berlin and Heidelberg, 1985.
88. Kloeden, P.E. The systematic derivation of higher order numerical schemes for stochastic differential equations. *Milan J. Math.*, 70:187–207, 2002.
89. Kloeden, P.E. and E. Platen. *Numerical Solution of Stochastic Differential Equations*. Springer-Verlag, NY, 1992.
90. Knessl, C., B.J. Matkowsky, Z. Schuss and C. Tier. An asymptotic theory of large deviations for Markov jump processes. *SIAM J. Appl. Math.*, 45:1006–1028, 1985.
91. Knessl, C., B.J. Matkowsky, Z. Schuss and C. Tier. Boundary behavior of diffusion approximations to Markov jump processes. *J. Stat. Phys.*, 45:245–266, 1986.
92. Knessl, C., B.J. Matkowsky, Z. Schuss and C. Tier. A singular perturbations approach to first passage times for Markov jump processes. *J. Stat. Phys.*, 42:169–184, 1986.
93. Kolmogorov, A.N. Über die analytischen M ethoden in der Wahrscheinlichkeitsrechnung. *Math. Annal.*, 104:415, 1931.
94. Kramers, H.A. Brownian motion in field of force and diffusion model of chemical reaction. *Physica*, 7:284–304, 1940.
95. Kushner, H.J. Dynamical equations for optimal nonlinear filtering. *J. Diff. Equations*, 2:179–190, 1967.
96. Kwakernaak, H. and R. Sivan. *Linear Optimal Control Systems*. Wiley, NY, 1972.
97. Landau, L.D. and E.M. Lifshitz. *Statistical Physics*, Course of Theoretical Physics, volume 5. Pergamon Press, Elmsford, NY, 3rd edition, 1980.
98. Landis, S., B.Z. Bobrovsky, and Z. Schuss. The influence of $1/f^{\alpha}$ phase noise on a second-order delay lock loop: calculation of the mean time to lose lock. *SIAM J. Appl. Math.*, 66.

99. Langevin, P. Sur la théorie du mouvement Brownien. *C.R. Paris*, 146:530–533, 1908.

100. Lawson, J.L. and G.E. Uhlenbeck (eds). *Threshold Signals*. Dover, NY, 1950.

101. Lee, R.C.K. *Optimal Estimation Identification and Control*. MIT Press, Cambridge, MA, 1964.

102. Lee, T.E. and A. Hajimiri. Oscillator phase noise: A tutorial. *IEEE J. Solid-State Circuits*, 35:326–336, 2000.

103. Lighthill, M.J. *Introduction to Fourier Analysis and Generalized Functions*. Cambridge University Press, NY, 1958.

104. Liptser, R., B.Z. Bobrovsky, Z. Schuss and Y. Steinberg. Fixed lag smoothing of scalar diffusions–part I: The filtering-smoothing equation. *Stochastic Processes and their Applications*, 64:237–255, 1996.

105. Liptser, R.S. and A.N. Shiryayev. *Statistics of Random Processes*, volumes I, II. Springer-Verlag, New York, 2001.

106. Ludwig, D. Persistence of dynamical systems under random perturbations. *SIAM Rev.*, 17(4):605–640, 1975.

107. Mahler, R.P.S. *Statistical Multisource-Multitarget Information Fusion*. Artech House, Boston, 2007.

108. Mandelbrot, B.B. *The Fractal Geometry of Nature*. Freeman, SF, 1982.

109. Mandelbrot, B.B. and J.W. Van Ness. Fractional Brownian motions, fractional noises and applications. *SIAM Rev.*, 10:422–437, 1968.

110. Mandl, P. *Analytical Treatment of One-Dimensional Markov Processes*. Springer Verlag, NY, 1968.

111. Marcus, S.I. Algebraic and geometric methods in nonlinear filtering. *SIAM J. Control and Optimization*, 22(6):817–844, 1984.

112. Matkowsky, B.J. and Z. Schuss. The exit problem for randomly perturbed dynamical systems. *SIAM J. Appl. Math.*, 33:365–382, 1977.

113. McKean, H.P., Jr. *Stochastic Integrals*. Academic Press, NY, 1969.

114. Meerkov, S.M. and T. Runolfsson. Residence time control. *IEEE Trans. Automatic Control*, 33:323–332, 1988.

115. Milotti, E. Linear processes that produce $1/f$ or flicker noise. *Phys. Rev. E*, 51(3):3087–3103, 1995.

116. Mitter, S.K. On the analogy between mathematical problems of nonlinear filtering and quantum physics. *Ricerche Automat.*, 10:163–216, 1979.

117. Mortensen, R.E. Maximum likelihood recursive nonlinear filtering. *J. Opt. Theory Appl.*, 2:386–394, 1968.

118. Naeh, T., M.M. Kłosek, B. J. Matkowsky and Z. Schuss. Direct approach to the exit problem. *SIAM J. Appl. Math.*, 50:595–627, 1990.

119. Najim, M. *Modeling, Estimation and Optimal Filtering in Signal Processing*. English edition of the 2006 French original. ISTE, London, xvi+392 pp. ISBN: 978-1- 84821-022-6. John Wiley & Sons, Inc., Hoboken, NJ, 2008.

120. Noble, B. *Methods Based on the Wiener–Hopf Technique for the Solution of Partial Differential Equations*. AMS/Chelsea Publication, 1988.

121. Nyquist, H. Thermal agitation of electric charge in conductors. *Phys. Rev.*, 32:110-Ű113, 1928.

122. Ocone, D. Probability densities for conditional statistics in the cubic sensor problem. *Math. Control Signals Systems*, 1:183–202, 1988.

123. Paley, R.N., N. Wiener and A. Zygmund. Note on random functions. *Math. Z.*, 37:647–668, 1933.

124. Papoulis, A. *Probability, Random Variables, and Stochastic Processes*. McGraw Hill, NY, 3rd edition, 1991.

125. Picard, J. Efficiency of the extended Kalman filter for nonlinear systems with small noise. *SIAM. J. Appl. Math.*, 51:843–885, 1991.

126. Pontryagin, L., A.A. Andronov and A.A. Vitt. On the statistical treatment of dynamical systems. *Noise in Nonlinear Dynamics*, 1:329–340, 1989.
127. Pontryagin, L.S., A.A. Andronov and A.A. Vitt. On the statistical treatment of dynamical systems. *J. Theor. Exper. Phys. (Russian)*, 3:165–180, 1933.
128. Proakis, J.G. Probability and stochastic processes, in *Digital Communications*, pp. 68–72. McGraw–Hill, NY, 3rd edition, 1995.
129. Protter, M.H. and H.F. Weinberger. *Maximum Principles in Differential Equations*. Prentice-Hall, Englewood Cliffs, NJ, 1967.
130. Ramo, S. Currents induced by electron motion. *Proc. IRE.*, 9:584–585, 1939.
131. Rennert, M., B.Z. Bobrovsky, and Z. Schuss. Taking loss of lock into account when designing radar tracking loops against jamming and maneuvering. *Proccedings of 18-th Convention of Electrical and Electronics Engineers in Israel*, Cat. No.95TH8044(3.3.6):1–5, 1995.
132. Rogers, L.C.G. and D. Williams. *Diffusions, Markov Processes, and Martingales*, volumes I, II. Cambridge University Press, paperback edition, 2000.
133. Rozovskii, B.L. *Stochastic Evolution Systems*. Kluwer, Dordrecht, 1990.
134. Ryter, D. and H. Meyr. Theory of phase tracking systems of arbitrary order. *IEEE Trans. Info. Theory*, IT-24:1–7, 1978.
135. Saberi, A., A.A. Stoorvogel, P. Sannuti. *Filtering Theory: With Applications to Fault Detection, Isolation, and Estimation*. Systems & Control: Foundations & Applications, xiv+723 pp. ISBN: 978-0-8176-4301-0; 0-8176-4301-X. Birkhäuser Boston, Inc., Boston, MA, 2007.
136. Schuss, Z. *Theory and Applications of Stochastic Differential Equations*. John Wiley, New York, 1980.
137. Schuss, Z. *Diffusion and Stochastic Processes: An Analytical Approach*. Springer NY, 2010.
138. Schuss, Z. and B.J. Matkowsky. The exit problem: a new approach to diffusion across potential barriers. *SIAM J. Appl. Math.*, 35(3):604–623, 1979.
139. Shaked, U. and B.Z. Bobrovsky. The asymptotic minimum variance of linear stationary single output processes. *IEEE Trans. Automat. Control*, 21:498–504, 1981.
140. Shockley, W. Currents to conductors induced by moving point charge. *J. Appl. Phys.*, 9:635–636, 1938.
141. Simon, M.K., J. Omura, R.A. Scholtz, and B. Levitt. *Spread Spectrum Communications*, volumes I–III. Computer Science Press, Rockville, MD, 1985.
142. Skorokhod, A.V. Stochastic equations for diffusion processes in a bounded region. *Theory of Probability and Applications*, 6(3):264–274, 1961.
143. Sneddon, I. *Elements of Partial Differential Equations*. McGraw-Hill International Editions (Mathematics Series), 1985.
144. Snyder, D.L. *The State Variable Approach to Continuous Estimation*. The M.I.T Press, 1969.
145. Spitzer, F. Some theorems concerning two-dimensional Brownian motion. *Trans. AMS*, 87:187–197, 1958.
146. Steinberg, Y., B.Z. Bobrovsky and Z. Schuss. On the optimal filtering problem for the cubic sensor. *Circuits, Systems, and Signal Processing*, 7(3):381–408, 1988.
147. Stensby, J. *Phase-Locked Loops, Theory and Applications*. CRC Press, NY, 1997.
148. Sussmann, H.J. Rigorous results on the cubic sensor problem, in *Stochastic Systems: The Mathematics of Filtering and Identification and Applications*, M. Hazewinkel and J.C. Willems, eds., pp. 637–648, *IEEE* 1981.
149. Üstünel, A.S. and M. Zakai. *Transformation of Measure on Wiener Space*. Springer, Berlin, 2000.
150. Van Trees, H.L. *Detection, Estimation and Modulation Theory*, vols. I–III. John Wiley, NY, 1970.
151. Verhaegen, M., V. Verdult. *Filtering and System Identification. A Least Squares Approach*, xvi+405 pp. ISBN: 978-0-521-87512-7. Cambridge University Press, Cambridge, 2007.
152. Viterbi, A.J. *Principles of Coherent Communication*. McGraw-Hill, New York, 1966.
153. Welti, A.L. and B.Z. Bobrovsky. Mean time to lose lock for a coherent second-order PN-code tracking loop – the singular perturbation approach. *IEEE J. Select. Areas Commun.*, 8:809–818, 1990.

154. Welti, A.L., U.P. Bernhard and B.Z. Bobrovsky. Third-order delay-locked loop: Mean time to lose lock and optimal parameters. *IEEE Trans. on Comm.*, 43:2540–50, 1995.
155. Wiener, N. Differential space. *J. Math. Phys.*, 2:131–174, 1923.
156. Wong, E. and M. Zakai. On the convergence of ordinary integrals to stochastic integrals. *Ann. Math. Stat.*, 36:1560–1564, 1965.
157. Wornell, G.W. A Karhunen-Loève-like expansion for $1/f$ processes via wavelets. *IEEE Trans. Inform. Theory*, 36:859–861, 1990.
158. Wornell, G.W. *Synthesis, analysis, and processing of fractal signals*. Technical Report RLE 566, MIT, Cambridge, MA, 1991.
159. Wornell, G.W. and A.V. Oppenheim. Estimation of fractal signals from noisy measurements using wavelets. *IEEE Trans. Signal Process.*, 40:611–623, 1992.
160. Wu, Y. *Spread-spectrum systems under multiuser environment and multipath fading channels*, PhD Dissertation. City University of Hong Kong, 2009.
161. Xiong, J. *An Introduction to Stochastic Filtering Theory*. Oxford Graduate Texts in Mathematics, volume 18, xiv+270 pp. ISBN: 978-0-19-921970-4. Oxford University Press, Oxford, 2008.
162. Yaesh, Y., B.Z. Bobrovsky and Z. Schuss. Asymptotic analysis of the optimal filtering problem for two-dimensional diffusions measured in a low-noise channel. *SIAM J. Appl. Math.*, 50(4):1134–1155, 1990.
163. Yau, S.T., R. Du, and L.X. Jia. Special solutions to some Kolmogorov equations arising from cubic sensor problems. *CI Information and Systems*, 7(2):195–206, 2007.
164. Zakai, M. On the optimal filtering of diffusion processes. *Z. Wahrsch. Verw. Geb.*, 11:230–243, 1969.
165. Zeitouni, O. An extension of the Beneš filter and some identification problems solved by nonlinear filtering methods. *Systems and Control Letters*, 5:9–17, 1984.
166. Zeitouni, O. and B.Z. Bobrovsky. On the joint nonlinear filtering-smoothing of diffusion processes. *Systems and Control Letters*, 7:317–321, 1986.
167. Zeitouni, O. and M. Zakai. On the optimal tracking problem. *SIAM J. Control Optim.*, 30:426–439, 1992.
168. Ziel, van der, A. On the noise spectra of semiconductor noise and of flicker effect. *Physica*, 16:359–372, 1950.
169. Ziemer, R. and R. Peterson. *Digital Communications and Spread Spectrum Systems*. Macmillan, NY, 1985.
170. Ziemer, R. and R. Peterson. *Introduction to Digital Communication*. Prentice–Hall, Englewood Cliffs, NJ, 10th edition, 2001.

Index

Z. Schuss, *Nonlinear Filtering and Optimal Phase Tracking*, Applied
Mathematical Sciences 180, DOI 10.1007/978-1-4614-0487-3,
© Springer Science+Business Media, LLC 2012